Relational Cohesion in Palaeolithic Europe: Hominin-Cave Bear Interactions in Moravia and Silesia, Czech Republic, During OIS3

Patrick J. Skinner

BAR International Series 2379
2012

Published in 2016 by
BAR Publishing, Oxford

BAR International Series 2379

Relational Cohesion in Palaeolithic Europe: Hominin-Cave Bear Interactions in Moravia and Silesia, Czech Republic, During OIS3

ISBN 978 1 4073 0969 9

BAR Publishing is the trading name of British Archaeological Reports (Oxford) Ltd. British Archaeological Reports was first incorporated in 1974 to publish the BAR Series, International and British. In 1992 Hadrian Books Ltd became part of the BAR group. This volume was originally published by Archaeopress in conjunction with British Archaeological Reports (Oxford) Ltd / Hadrian Books Ltd, the Series principal publisher, in 2012. This present volume is published by BAR Publishing, 2016.

Printed in England

BAR
PUBLISHING

BAR titles are available from:

BAR Publishing
122 Banbury Rd, Oxford, OX2 7BP, UK
EMAIL info@barpublishing.com
PHONE +44 (0)1865 310431
FAX +44 (0)1865 316916
www.barpublishing.com

Preliminaries and acknoweldgements

This book is largely the product of my PhD thesis that was carried out at the University of Cambridge, Department of Archaeology, and supervised by Dr Preston Miracle and advised by Prof. Graeme Barker. Any errors are of course of my own making.

There are a number of people who have contributed to this book. First I must thank my PhD supervisor Dr Preston Miracle and advisor Prof. Graeme Barker for their support throughout. Prof. Martin Jones was a key figure in making connections with important Czech researchers such as Dr. Lenka Lisa who introduced me to the landscape of the Moravian Karst. Prof. Karel Valoch provided essential help and direction with regards choosing cave-bear case-study sites and gaining access to cave-bear remains, and I am grateful to Prof. Rudolf Musil who provided me with key cave-bear literature. Thanks go to Dr Petr Neruda for his help in locating Pod hradem Cave and for providing important details regarding Kůlna Cave, to Dr. Martina Ábelová who was kind enough to introduce me to Brno on my first visit, and all member of the Anthropos Institute in Brno and the Budišov Museum, especially Gabriela Dreslerová, who provided me with a comfortable and welcoming place to work during my time in Brno. I must thank residents of Sloup and the owner of the Relaxa camp site, Karel, for making me feel more than welcome during my two-month stay, and in particular Milan and his wife for looking after my car during my time back in the UK. Dr Tamsin O'Connel, Dr Rhiannon Stevens, and Catherine Kneale provided guidance in the taking and processing of stable isotope samples. My gratitude goes to Wolfson College for awarding me the O'May studentship, and to AHRC for providing subsequent funding to undertake this research. Finally, I must thank my mum who has always provided encouragement for my studies, and to Yoshiko who has offered endless moral and intellectual support and who kindly employed her artistic skills to design and illustrate the front-cover of this book.

Abstract

Although archaeologists have considered the social significance of animals, this has mostly been from a dualistic perspective and based on metaphoric or symbolic reference to the human world, with animals largely seen as culturally neutral in their own right. Recent anthropological and ethnographic studies, however, have shown animal appearance and behaviour, and differences in the frequency of interactions with animals, communicate changing meanings that affect how humans understand and treat animals, and thus how they understand themselves and the world around them. Building on this, my book examines how interactions with now extinct cave bears (*Ursus spelaeus* Rosenmüller 1794) impacted on hominins (Neanderthals and Anatomically Modern Humans – AMH) living in Central Europe (Moravia and Silesia – Eastern Czech Republic) during OIS3 (*c.* 60,000 – 24,000 Cal. BP). Considering hominins and cave bears in Central Europe during this period is ideal for the proposed project: bears are universally revered within ethnographic and ethnohistoric sources and their remains are common among relevant archaeological sites; and these prehistoric hunter-gatherers commonly interacted with animals, yet demonstrated significant cultural differences. This book addresses three key questions: how can we trace interactions? What was the nature of those interactions? And what was the significance of those interactions for past humans? I use published literature together with odontometric (tooth measurement) and tooth-wear analyses to identify hominins and cave bears, their chronology, site locations, hominin hunting and lithic raw material procurement practices and use of animal remains and depictions of animals, and cave-bear demography, diet, appearance, perception, social affiliation, and predator/anti-predator behaviour. I use GIS (Geographic Information Systems) computer mapping techniques to create topographic, palaeohydrological, palaeovegetation, and friction maps, and to map hominin and cave-bear site locations and home ranges, hominin pathways, lithic raw material outcrops, cave-bear and prey species preferred habitats, prey species diversity, and cave-bear population densities during contrasting climatic and seasonal periods. Potential interactions are identified by comparing contemporaneous hominin and cave-bear distribution patterns, and associated frequencies of interactions are used together with relevant archaeological data and new theoretical concepts to assess the significance of interactions for past humans. I address questions such as what were the distribution patterns of hominins and cave bears and how frequently did they interact, and what evidence is there for human use of cave-bear remains and depictions of cave bears. The results show that Neanderthals interacted with cave bears more often than AMH; AMH used cave-bear remains more frequently than Neanderthals; and AMH depicted bears and other animals whereas Neanderthals did not. Interactions were important for both Neanderthals and AMH, but Neanderthals were governed more by the immediacy of the dwelt-in world, rather than memories of the perceived world as was the case for AMH, and depictions of animals were more important than the use of animal remains for AMH.

Contents

Figures

Tables

Chapter 1: Human-animal interactions and relational cohesion in Palaeolithic Europe

Introduction

This book is about human interactions. More specifically, it is concerned with how human interactions with animals, in particular now extinct cave bears (*Ursus spelaeus* Rosenmüller 1794), affected the social lives of prehistoric hunter-gatherers (hominins – Neanderthals and AMH) living in Central Europe (Moravia and Silesia – Eastern Czech Republic) during OIS3 (*c*. 60,000 – 24,000 Cal. BP). I use a multidisciplinary approach, using published literature, animal remains, digital data, and GIS (Geographic Information Systems), together with odontometric (measuring of teeth) and tooth-wear analyses, and spatial reconstruction techniques to identify potential interactions between hominins and cave bears. New theoretical concepts are used to interpret the results and as a means for making statements about the role that cave bears, and potential interactions with cave bears, played in the social lives of hominins.

The objective of this chapter is to derive a working hypothesis and a case study context for the book. I briefly discuss what interactions are, their significance for people's everyday lives, their place within archaeological discourse, and how I became interested in the subject. I discuss human-animal relations in the context of modern-day Western and non-Western societies, prehistoric societies, and with particular reference to hunter-gatherer societies in Palaeolithic Europe. I examine how archaeologists have approached human-animal relations, paying particular attention to apparent 'Western' and non-'Western' perspectives, and discuss the consequences of these for understanding the significance of human-animal interactions for past people. I propose a working hypothesis for this book, at the heart of which is the argument that the manner in which people interacted with animals was intrinsically related to how they understood and behaved towards them. I discuss the reasons for choosing cave bears, Moravia and Silesia, and OIS3 as a study context, and finally provide a chapter overview.

Interactions can be thought of as the human perceptual recognition of the other (animal, mineral or vegetable), and may be mediated through one, all, or any combination of the five available human perceptual sensory mechanisms (seeing, hearing, smelling, touching, and tasting); although within this book I am primarily interested in sensory mechanisms that do not require direct physical contact, i.e. seeing, hearing, and smelling. Interactions play an inevitable and important part in people's lives: simply being aware of the surrounding world necessitates that people employ their perceptual mechanisms and therefore interact with the world in which they live. The ability to see, hear, and smell allows for navigation between places, recognition of objects, people and animals, and provides the means by which humans can subsequently respond and behave towards things in an appropriate and meaningful manner. Simply, interactions are defined by the employment of human perceptual mechanisms and the recognition of the other, and are at the very heart of what allows us to live, to experience, to understand, and behave in the world in the way that we do.

The significance of interactions within archaeology was recognised as far back as the early part of the last century. With reference to prehistoric hunter-gatherers of Northern Europe, Childe (1931, 345) made the point that because of their mobility these people were brought into contact, not only with one another, but also with their food-producing neighbours, and as a result subsequently changed their behaviour and began tilling the land and acquiring domestic animals. Renfrew (1986, 1) commented that peer-polity interaction (imitation, emulation, competition, warfare, and the exchange of material goods and information between socio-political units) provided an explanation for the growth of socio-political systems and the emergence of cultural complexity: Minoan state formation and the physical structures of buildings in Crete during the 2nd millennium BC, for example, were to some extent governed by peer-polity interaction (Cherry 1986). Parker-Pearson and Richards (1994, 19-20) made the point that human interactions with built structures, such as houses, created and maintained the social structure of people living in them, and Tilley (1994, 11), in his promotion of a Phenomenological approach to Neolithic archaeology, commented that 'the key issue in any phenomenological approach is the manner in which people experience...the world'. More recently, Gibson (2007) looked at issues regarding human movement and subsequent interactions within Mediterranean landscapes, and in particular how movement and interactions with roads and paths embodied the experiences and social lives of the people who built and used them. Gibson (2007, 82) made the point that interactions with the very physicality of the roads and paths allowed people to define and redefine their social lives. Studies within archaeology that have looked at human interactions with animals *per se* are rare, but archaeologists such as Jones (2009) have considered the significance of human perception and experience of animals during the Early Aceramic Neolithic period in Cyprus. According to Jones (2009, 93), not only were interactions between human and non-human living beings, i.e. animals, central to people's lives, but it is through interactions with animals that the very concept of humanness was achieved.

My interest in interactions was sparked during my undergraduate studies at Leicester University when studying modules that dealt with space and concepts of space, and how people perceived space within the landscape. I was particularly interested in how people's perception of places and their interactions with things came together to affect how people thought about and acted in the world. This was combined with an interest in theoretical concepts such as Structuralism, Structuration, Phenomenology, more formal methods of spatial investigation such as Spatial Syntax and Access Analysis, and computer digital mapping techniques using GIS. Also, particularly poignant for me were ideas regarding non-Western approaches to the environment. In particular, articles written by Ingold (2000e; 1986b), Strathern (1992) and Howell (1996), within which the complexity and contrasting nature of non-Western, in comparison with Western, views of the world, landscape, and animals, were particularly significant. As a result, I became interested in trying to devise an approach that would enable me to investigate human interactions with animals from a non-Western perspective, using and building on established theoretical and methodological frameworks, and I was particularly interested in undertaking this within a prehistoric hunter-gatherer context. Thus, within this book I hope to have gone some way to achieving this, and provided some new understandings, not only of how this might be achieved, but also how such an approach can reveal new and exciting insights into past people, and more specifically, how such investigations can reveal an understanding of the social significance of human interactions with animals in the past.

Human-animal relations

In contemporary Western societies animals are a central part of people's lives. Manning and Serpell (1996, ix) commented that animals do and always will form a central feature of the human world. One of the most common ways in which animals are used in modern-day Western societies is for the production of food: animals such as cows, sheep, and poultry are commonly used for the production of beef, milk, lamb, poultry meat (e.g. chicken), and eggs, and in 2008 within the UK alone 5664 million litres of milk were produced, there were 5.5 million cattle, 15.5 million sheep, and 128 million poultry (www.statistics.defra.gov.uk). Animals are also regularly used for other purposes. Cats can be used to eradicate vermin, and dogs are used for defence, life saving, guiding the blind, sport, and herding sheep (Cansdale 1952, 114). Animals also commonly provide a means of companionship: recent surveys demonstrated that there were 36 million pet dogs and 35 million pet cats within the European Union in 1994 (Serpell 1996, 13). Animal remains are often used for clothing or for more industrial purposes. Leather is commonly used to make shoes, coats and belts, wool can be used to make jumpers, gloves, and socks, duck feathers can be used to insulate sleeping bags and duvets, and animal fat is often used for the production of cosmetics, soap, pet food, and even biodiesel (www.telegraph.co.uk). It is not just in the physical sense that animals are part of people's lives in present-day Western societies: animals are often referenced in spoken language and depicted in art. Somebody who is particularly unhappy may comment that 'life's a bitch'! In Orwell's (1946) '*Animal Farm*', animals are anthropomorphised and used as a means of portraying the realities of communism, and within the galleries of Tate Britain an exhibition currently explores Britain's apparent obsession with animals (www.tate.org.uk).

Animals are also important in contemporary non-Western societies. For the Kalahari hunter-gatherers of Africa, animals such as wildebeest, giraffe, springbok, ostrich, warthog, jackal, fox, bustard, hare, porcupine, genet, and guinea fowl are commonly consumed (Lee and DeVore 1976, 102 and 110). The Achuar of the Amazonian rainforest regularly consume animals such as birds, woolly monkey, squirrel, armadillo, and deer (Descola 1994, 246). In some present-day hunter-gatherer societies animals are kept as pets and sometimes breast fed and treated like children (Ritvo 1994, Plate 72), and the Inuit of north-western Arctic Alaska rely on animals such as dogs to pull sledges and to assist in hunting of animals such as Arctic fox, harbour seal, and walrus, and rely on caribou skin and bearded sealskin to make objects such as boots and coats, the tails of red or Arctic foxes to make chin protectors, and seal blubber to make oil for lanterns (Nelson 1969). Many contemporary hunter-gatherer societies also commonly refer to animals in language or depict them in art. The Chewong of the Malay tropical forest have a number of songs and stories that include animals, one of which includes the story of a dog who pretends to be a man married to a woman (Howell 1996, 140). Within Australian aboriginal art, along with humans, weapons, and vegetation, depictions of animals are common, and within Inuit societies animals are often depicted together with people (Ingold 2000e; Ucko and Rosenfeld 1967, 161).

Further back in history, animals were also important to people. During Medieval and Roman times, animals provided important sustenance for humans, with beef and mutton being the most common source of meat (O'Conner 1989, 15). In some cases, animals were treated as companions, sometimes sexually, and often treated in a similar way to humans: during medieval periods, although bestiality, along with incest, paedophilia, promiscuity and group sex was considered morally wrong, animals were sometimes used for sex (Lee and George 2008, 209). Also, during the 13th and 14th century in France it was common to bring judgment against an animal for killing a human (Cohen 1994, 74-75). Animal remains were commonly used for industry, craft, and clothing or for other purposes such as burials. During the 8th-11th centuries in England antler and animal bone were used as raw material for manufacturing; bone and the horns of sheep, goat and cattle were used to make combs, playing pieces, dice and handles (MacGregor 1989, 107-116). In medieval Europe, animal furs were often worn, and animal leather was used by tanners from at least the 15th century until the 1940s in England (Serjeantson 1989, 131 and 139). At the medieval site of Spong Hill (Norfolk, England) more than 100 human cremations were found with the bones of cattle, sheep, dogs, pigs, horses, beaver, fox, red deer, roe deer, bear, hare, fowl, domestic goose and fish (Bond 1996, 78). Animals were also often used in language or depicted. Medieval teachers often referred to pupils collectively as animals, and disorderly ones were identified as being 'cuckoo' (Ziolkowski 1993, 146). Anglo-Saxon shields were decorated with animal motifs (Dickinson 2005, 161), and animals such as horses were painted on codices and are commonly depicted on the 11th century Bayeux Tapestry (Gathercole 1995, 53).

In prehistoric times, farming communities kept animals such as sheep, goats, cattle and pigs for the production of food (Barker 2006, 139, 156, 346), as is the case with the prehistoric Neolithic B farming communities of South-West Asia (Barker 2006, 139). Animal skins were used to make leather (Barker 2006, 158), and bones were processed and used to make tools such as hooks at places such as Çatal Hüyük, in Turkey, Soufli in Magoula, and Nea Nikomedeia in Greece (Barker 2006, 347). Animals were also commonly depicted: at Çatal Hüyük paintings of deer, wild boar, felines, and vultures have been found, together with sculpted animals such as clay figurines of bovines (Mellaart 1967, 134-169). Also, rock-art engravings of domestic cattle and sheep associated with prehistoric farming communities have been discovered in Messak Settafet, south-west Libya (Barker 2006, 100), and in association with prehistoric farming communities of the Altai mountains in the former Soviet Union, where buried human remains have also been found with tattooed animal motifs (Parker-Pearson 1999, 64-65).

Animals were also a central part of prehistoric hunter-gatherer societies. Human-made butchery marks have been found on animal bones from African deposits dating back to as far as 2 million years ago (Bunn 1981), and remains of animals such as pig hunted by prehistoric hunter-gatherers have been found at sites such as Niah Cave in Sarawak, Borneo (Rabbett and Barker 2007). Remains of hunted animals such as reindeer, beaver, pig, red and roe deer, wild cattle, and wild horse have been found at numerous Mesolithic (*c.* 10,000 – 3500 Cal. BP) hunter-gatherer sites in the Baltic region of Europe (Zvelebil 2008, 19), and at the Mesolithic hunter-gatherer site of Star Carr in Yorkshire, England, the remains of hunted and butchered animals such as roe and red deer are commonly found (Clark 1954). Prehistoric hunter-gatherers also used animal remains to make objects such as the bone harpoon found at the site of Katanda in West Africa, dating to *c.* 90,000 years ago (Henshilwood and d'Errico 2005, 252). In Northern Bohemia, Czech Republic, at Pod Zubem rock shelter and the cave of Martina, Mesolithic hunter-gatherers used animal bones to make tools such as pierced axes of deer antler, and bone awls (Svoboda 2008, 234-235). Also, items such as red deer 'masks' have been found at the Mesolithic site of Bedburg-königshoven, Germany (Verhart 2008), and depictions of animals have been found associated with hunter-gatherer societies of the same period; at the site of Addaura Cave, near Palermo in northern Sicily, engravings of animals have been found (Pluciennik 2008, 349).

During the Palaeolithic period in Europe (*c.* 900,000 – 10,000 years ago), animals again were central to the lives of people. At the lower Palaeolithic (*c.* 900,000 – 300,000 years ago) site of Boxgrove in the UK, remains of animals such as horse have been found with cut marks (Pope and Roberts 2005, 93). In south-western France, remains of hunted animals such as reindeer, red deer, horse and bovids have been found in association with hunter-gatherer remains (Mellars 1996, 165), and at the Middle Palaeolithic (*c.* 300,000 – 40,000 years ago) site of Salzgitter-Lebenstedt, near Hannover, Germany, remains of human hunting activities include more than 86 individual reindeer (Gamble and Gaudzinski 2005, 174). Also, at the Upper Palaeolithic (*c.* 40,000 – 10,000 years ago) site of Abri Pataud (Dordogne, France), remains of animals such as reindeer that were hunted by humans are commonly found (Gamble 1999, 343). Animal remains were also often used for other purposes during this period in Europe. In Lower Palaeolithic deposits at Bilzingsleben in Germany, bone tools that were re-touched and turned into a number of other different tools have been found (Mania and Mania 2005, 106); perforated bones and teeth have been recovered from Middle Palaeolithic levels at Repolsthöhle in Austria and Bocksteinschmiede in Germany (Mellars 1996, 373). In Upper Palaeolithic deposits at the cave of Arcy-sur-Cure, France, wolf and fox canines made into pendants have been discovered (Bahn and Vertut 1988, 72). Similar finds have appeared at other Upper Palaeolithic sites such as Mamutowa Cave in Poland (Wojtal 2007, 85), and a boomerang made of a mammoth tusk has been found at Oblazowa Cave in Poland associated with Upper Palaeolithic human remains (Wojtal 2007, 150). A 'flute' made from a swan's radius has been recovered from the Upper Palaeolithic levels at Geissenklösterle, southern Germany (Gamble 1999, 338), and animal figurines have been found in association with Upper Palaeolithic deposits at Vogelherd, Geissenklösterle, and Hohenstein in Germany (Svoboda *et al.* 1996, 128). Parietal art is also found in certain regions of Europe in association with Upper Palaeolithic hunter-gatherers (Bahn and Vertut 1999, 75), and animals such as bison, rhino, horse, mammoth, oxen, deer, and wild boar are commonly portrayed (Ucko and Rosenfeld 1967, 39); two of the most important sites are Chauvet Cave and Lascaux Cave in France, which contain hundreds of animal depictions (Ucko and Rosenfeld 1967; Chauvet *et al.* 1996).

Clearly, animals are and have always been central to the lives of humans regardless of the character of their society (i.e. contemporary Western or non-Western, historic, or prehistoric farming/hunter-gatherer societies), providing a source of companionship, food, raw materials, and often used and referred to in spoken, written, and depicted cognitive expression.

Approaches to animals

Traditionally, within Western societies animals have mostly been seen from a dualistic perspective. Dualistic worldviews probably originated during Classical times when Western modern philosophy set humans apart from and above all animals (Lippit 1994, 809). From this perspective, humans stand to animals as do heaven to earth, soul to body, and culture to nature (Pálsson 1996, 65), and humans are considered to be radically different and superior to all other creatures: animals are automata, not acting from knowledge, but only from the disposition of their organs (Ingold 1986c, 18). From this perspective, animals are not considered to be culturally significant in their own right in the same way that a human being might be: animals are culturally neutral, and have no autonomy in the way they are thought about by humans. That is not to say, however, that from this perspective animals are culturally inactive, simply that any cultural significance stems from and indeed necessitates the cultural generosity of people, and symbolic or metaphoric reference to the human world. As Ingold (2000c, 44) commented with reference to Bird-David's assessment of the Batek, Mbuti, and Nayaka peoples, 'hunter gatherers are supposed to call upon their experience of relations in the human world in order to model their relations with the non-human one'. This point is similarly made by Pepper (1942, 91) who commented that in an attempt to understand the world, a person will use a list of structural characteristics that are analogies or metaphors to their original world, and Gudeman (1986, 37-38) made the point that humans are modellers, and in order to understand a world, humans will project a model of their own world on to the world of an object.

The large majority of archaeological work that has been undertaken within the context of human-animal relations, in particular within the context of Palaeolithic Europe, has been undertaken from a dualistic perspective. The result of this has been that for the most part animals have been considered as largely culturally neutral beings, seen either as economically, or metaphorically/symbolically significant. Mellars (1996, 193-244) has dedicated a whole chapter to the economic significance of animals for hunter-gatherers in Europe during the Middle Palaeolithic with little or no mention of their cultural significance. This is similarly the case with Musil (2003), whose overall objective was to understand the economic significance of animals for hunter-gatherers in Europe during the Middle and Upper Palaeolithic. Also, Adams' (1998, 95) analysis of the Middle and Upper Palaeolithic transition in Central Europe discusses animals mainly in terms of their economic significance for people, and Chase's (1986) study of hunter-gatherers at the site of Combe Grenal in France looked explicitly at attempting to understand the economic significance of animals for Middle Palaeolithic people. Bailey *et al.* (1983) have investigated Upper Palaeolithic human relations with animals in Epirus (Greece), looking particularly at their economic significance, and Davidson (1983) has considered animals in terms of their economic significance for hunter-gatherers occupying areas of the Levante region of Eastern Spain during the late Upper Palaeolithic. Also, d'Errico and Vanhaeren (2007) have discussed the use of animal remains in Europe by Middle and Upper Palaeolithic hunter-gatherers as a sign and indication of apparent symbolic behaviour by the people who made them. A similar approach is taken by White (2007), who has examined personal ornamentation made with animal remains by Upper Palaeolithic hunter-gatherers in Europe in an attempt to understand the symbolic significance of the objects for the people who wore them. My point here is not to demonstrate that animals were *not* of economic, metaphoric or symbolic significance (a dualistic approach is just as valid an ethnographic approach as any other); I simply wish to highlight the fact that the vast majority of research interested in human-animal relations in the context of Palaeolithic Europe has been undertaken from a dualistic perspective.

Recently there has been a tendency to be critical of such approaches. Ingold (1986c, 18) made the point that academics have still not progressed beyond a Cartesian view of animal existence, and Strathern (1992, 76) commented, with reference to modern-day Melanesian societies, 'instead of dismantling holistic systems through inappropriate analytical categories, perhaps we should strive for a holistic apprehension of the manner in which our subjects dismantle their own constructs'. Tilley (1994, 23) also made the point that most indigenous world views are unlike those in Western societies, and it is unhelpful to think of these in terms of a binary culture/nature distinction. In a similar vein, Jones (2009, 77) commented that non-human animals require not only reconsideration and acknowledgement, but also freedom from anthropocentric claims and narratives. Also, Hallowell (1926, 10) commented, 'it becomes apparent that the categories of rational thought, by which we are accustomed to separate human life from animal life and the supernatural from the natural, are drawn upon lines which the facts of primitive cultures do not fit. If we are to understand or interpret peoples who entertain such notions we must rebuild the specific content of these categories upon the foundation of their belief not ours. The truth or falsity of the categories is not an issue but simply the inapplicability of our concepts of them as a point of departure for a comprehension of primitive thought'.In contrast to a Western dualistic mode of thought, most anthropological and ethnographic data suggest that many non-Western societies see the world from a different, singular perspective. Barker (2006, 58) commented that it is clear from modern-day ethnographic research that most foragers conceptualise relations between humans and their world in a very different way from our own Cartesian approach. The world is not divided in terms of culture and nature, and the cultural significance of something does not make reference to the human world through symbols and metaphor, relying on the cultural generosity of humans: animals, by default, are culturally alive and active in their own right in a similar way that another person might be. It would be wrong to think of this simply in terms of assigning cultural significance to elements of the environment: the distinction should be seen in terms of moving away from the anthropocentric focus of relational attribution; all things may exist as part of a singular world within which they are empowered and saturated with personal powers, regardless of their biological or intellectual ability (Ingold 2000e). From this perspective, animals can be regarded as persons, with the only difference being their outward appearance: the environment, and individual objects in the environment, embody 'personages', each with language, reason, intellect, moral consciousness, and knowledge regardless of the outer shape (Barker 2006, 59). Viveiros de Castro (2004, 465-466) commented that for Amazonian peoples, categories of human-human relations are also evident in most contexts in which humans and non-humans confront each other. Similarly, Coward and Gamble (2010, 50) commented, 'Mobile hunter-gatherers engage in a thoroughly relational epistemology of relations with animals and even inanimate entities in their environments...in the course of day-to-day activities'. For the Navajo of Africa, animals are seen as partners: the wolf, coyote, wild cat, screech owl, and crow are seen as fellow hunters, and the Zuni people (Africa) regard and treat game animals as living relatives (Hill 1938, 104). Within Amazonian culture, cultivated plants may be understood as blood relatives of the women who tend them (Viveiros de Castro 2004, 465). Kohn (2007, 4) stated that for the Upper Amazonian Runa, all beings engage with the world and with each other as selves, that is, as beings with a point of view; Hornborg (2006, 22) made the point that entities such as plants, rocks and non-human animals may be understood as communicative beings rather than inert objects; and within Puebloan society animals such as birds are viewed as warriors and hunters (Potter 2004, 328).

One important aspect of a singular perspective is that the physical is seen as permeable. Wagner (1991, 165) pointed out that people and their identities may be understood as fractal rather than singular and the human-ness of persons is extended to objects. As such, human identity and personhood may be embodied in non-human things. Subsequently, the person becomes a 'dividual', as opposed to an individual, and potentially divisible and composed of elements that exist elsewhere in the world (Fowler 2004, 122), transcending the limits of human biological existence (Humphrey 1996, 99-100). The mind seeps out into the world and becomes coextensive with the world (Knappett 2006, 239), and other people, places and things become part and parcel of human identity (Ingold 1986d, 239). This is the case for Duar Mongols, whose personhood transcends the distinction between human and animal (Humphrey 1996, 105). Also, Harris (1998, 79) demonstrated that the seasonal changes in the river levels of the Amazonian Floodplain were not simply ecological happenings for the Parúaros people who experienced them, but embodied the very essence of their identities, and Ingold (2000f, 204) while examining *The Harvester* (painted by Pieter Brugel the Elder 1565), commented that the pear tree within the picture embodied the entire history of its development, including the people who nurtured it, picked its fruit and leant against it: 'the tree is embodied with the personhood of the individuals that have engaged with it'.

At the same time that the human physical is seen as being culturally permeable, allowing personhood and identity, in a very real way, to be embodied in other elements of the world, this is also the case for other physical entities: humans do not have a monopoly on the dividualisation of identity. Rocks, animals, and plants have the potential to deposit or absorb identity from other things in the world; identity is mutually inter changeable, and can flow to and from all objects and things, from human to animal and equally from animal to human. Jones (2009, 93) made the point that humans are not the only consciousness constituting the world, and there are other contributors constituting their own worlds and influencing ours; some of them are human and some of them are not. Physicality, then, simply provides the container for identity, and the conceptualisation of 'persons' is not associated with permanent human form any more than it is associated with any other physical form (Ingold 1986a, 248-249). As Viveiros de Castro (1998, 482) commented, 'there is no doubt that bodies are discardable and exchangeable'.

Thus, traditional approaches to animals have largely viewed them as culturally neutral, reliant on the cultural generosity of humans, and metaphoric or symbolic reference to the human world for social recognition. From a non-Western singular perspective, however, animals are not waiting to be culturally empowered by humans: animals *are* culturally significant in their own right. Moreover, the identities of humans and animals, and all other things, are not confined to the physical, but transcend physical form.

Understanding human-animal interactions

The contrasting nature of Western dualistic and non-Western singular approaches to animals does not have mundane consequences for the significance of human interactions with animals. From a dualistic perspective, human interactions with animals can be no more than an opportunity for economic exploitation, or a chance to re-affirm or deny cultural models based on metaphoric and symbolic reference to the human world. From a singular perspective, however, interactions provide the means by which identity can be mediated: by perceiving the world, identities are exchanged, deposited, and absorbed, resulting in physical bodies that always embody an ever-changing, fluid, and dynamic kaleidoscope of identities.

However, it is not simply that, from a singular perspective, interactions provide the opportunity for the mediation of identities: interactions, and the subsequent mediation and perpetuation of identity between physical things, allow the world to be maintained in a state of relational cohesion. Hannerz (1992, 37) refers to the world as a global ecumene, and Descola (1992, 116) commented that 'basic to many Amazonian animistic systems is a view of the Universe as a gigantic closed circuit within which there is a constant circulation of the substances, souls and identities held to be necessary for the conservation of the world and the perpetuation of social order'. This is similar to what Giddens (1984, 375) refers to as ontological security; the confidence or trust that the world, self and identity are as they appear to be. Although Giddens is ultimately coming from a theoretically dualistic perspective, he remarks that it is interactions with things that provide the reassurance for people that the world is as they believe it to be. For Giddens, ontological security can be perpetuated and maintained through the enactment of daily routines and interaction with important elements of the world such as other people and material objects (Giddens 1984, 87). However, while for Giddens interactions provide the means by which people can establish a sense of ontological security through 'mental' cohesion with the world, i.e. physical interactions allow people to *think* about the world in a particular way, from what I now term a singular-relational perspective, interactions provide, in a very practical sense, the means by which people can maintain and create the world through the mediation of elements of that world, i.e. via the mediation of identities. Thus, rather like a rotating bicycle wheel that may be balanced horizontally on its central ax, where it is the movement and rotation of the wheel that maintains its balance and stopping the wheel will result in the wheel becoming unbalanced, from a singular-relational perspective, it is the rotation of identities through physical things that maintains the world in a state of relational cohesion and stability. Without the perpetuation and circulation of identities the world becomes unbalanced and in a state of disequilibrium. From a singular-relational perspective it is not simply that interactions provide the means by which identities can be mediated: interactions are at the very heart of creating and maintaining the world, and without the ability to interact there would be no world to interact with; interactions are the life-force of the world.

Interactions, though, do not have universal significance in maintaining relational cohesion: the significance of human interactions with things depends on their perceived affordance. Affordance does not equate to meaning and is not intrinsic to the object of interest: affordance is both reliant on the physical nature of the object and how the object is perceived and engaged with. According to Gibson (1982, 397-398) it is the looking, listening, touching and sniffing that goes on when the perceptual systems are at work' that is important, and Tilley (2004, 24) commented that affordance is neither imposed on things or pre-given, but is discovered in the course of practical activity. Also, 'meanings are not attached by the mind to objects…rather these objects take on their significance…they afford what they do by virtue of their incorporation into a characteristic pattern of day-to-day activities…[and] meaning is imminent in the relational contexts of people's practical engagement with their lived-in environments' (Ingold 2000b, 168).

With respect to animals, two of the most important factors governing affordances are physical appearance and behaviour. For the Duar Mongols, important qualities of horses are their colour, temperament and stamina (Humphrey 1996, 98; 1974, 478), foxes are considered as 'living devils' because of their unpredictable and mysterious, sly, and deceitful nature, tigers are considered important because of their strength, the hare for its winding tracks that are difficult to follow, and the elk and deer for their fighting antlers (Humphrey 1996, 102). The Huaorani of the Amazonian Ecuador eat monkey in a very different manner to the way in which they eat other animals, as the behaviour of the monkey is very different to other animals: monkeys are particularly unpredictable in their behaviour (Rival 1996, 146-151). Also, for the Cree people of north-eastern Canada, it is the particular behaviour of deer when hunted that provides a particular affordance allowing them to be killed: the action of a deer that stops and stares directly at a human or non-human predator upon realising it is being hunted, is seen by the Cree people as an intentional communicative act, an offering of itself to the hunter, therefore giving permission and allowing the hunter to kill the deer (Ingold 2000a, 13). Similarly, Thackeray (1983, 41) commented that for the /Xam Bushmen of South Africa, it is the particular behaviours of the ostrich that appears to wait for the hunter that allow the animal to be killed. Also, according to Ellen (1972, 231), for the Nuaulu of central Seram, Indonesia, the cuscus is embodied with particular significance because of the sum of its attributes, which make this animal appropriate and meaningful, and the Barasana Indians of Colombia apparently become anxious about killing and eating game as the size of the animal increases (Serpell 1996, 180).

It is not simply the appearance or behaviour of animals that are important, but rather how animals look and behave in relation to human appearance and behaviour. According to Tilley (1996, 63), animals are distinctive from humans in their anatomies, but animals are similar to humans in their basic anatomical plan and behavioural characteristics, and as such are simultaneously akin to humanity; 'they are part of our biological ancestry, yet by definition different'; it is the fundamental ambiguity of animals, when considered in relation to humans, 'on which the human mind sets to work'. Indeed, with reference to the Mesolithic Ertebølle societies in Zealand and Skåne (Denmark), Tilley (1996, 63) remarked that red deer were extremely important animals as particular human characteristics (e.g. coughing, yawning, sleeping, eating, affection for their young, fear, pain, pleasure) were evident within their behaviour, and Jones (2009, 91) commented that dogs were important for the Aceramic Neolithic people in Cyprus, because of the apparent likeness between humans and canines. Whittle (2000, 252) commented that animals behave like people in some respects and often live in social groups, mate and are sexually differentiated.

Moreover, it is an animal's ability to be similar yet to surpass human qualities that provides the scope for animals to be afforded particularly important significance. According to Viveiros de Castro (1998, 470), while Ameridian cosmologies do not attribute the same degree of personhood to each type of entity in the world, the emphasis seems to be on species with particualr characteristics such as great predators. Jones (2009, 89) made the point that it was the super-human qualities of cats (speed, balance and agility) that made them particularly important for Neolithic people in Cyprus. Whittle (2000, 252) also made this point, commenting that animals can have greater strength and speed, they can run and swim in ways that people cannot, and can appear and disappear in the landscape in ways that people cannot match. Thus, although all interactions can be understood as being involved in the mediation of identity and subsequent maintenance of relational cohesion, in the context of human interactions with animals it is those animals that resemble humans in appearance and behaviour yet possess qualities that somehow go beyond those of humans that afford the most significance for perpetuating identities and maintaining a sense of relational cohesion in the world.

However, the spatial and temporal context in which humans and animals inhabit the world is, inevitably, heterogeneous, and as a result human interactions with 'important' animals will not be uniform. As a result, in the same way that relational cohesion might be achieved through regular interactions with important animals, irregular and unpredictable interactions with these creatures are likely to result in a sense of relational instability or insecurity. Giddens (1984, 62) made the point that ontological insecurity might arise through the stretching of social life and the knowing of the unknown. For instance, in modern-day Western societies such a distinction seems to be manifested in the way people understand wild and domestic animals: wild animals are commonly understood as species that are spatially remote and rarely encountered. As Whatmore (2002, 33) commented, 'The enduring coincidence between the species and spaces of wildlife as the antipodes of human society means that to ask what is wild is always simultaneously a question of its whereabouts'. Thus, from a singular-relational perspective, to be aware of important elements of the world such as super-human, anthropomorphic animals, but not to have the ability to regularly interact with these animals, would result in a state of relational instability.

The role that important animals play in maintaining a sense of relational cohesion, however, is not restricted to the immediacy of the 'source' object (e.g. animal): identities may be enchained via memory and imagination through spoken language or behaviour, and subsequently deposited in new and existing objects such as pictures, figurines and sculptures, and animal and human remains. The extensions of self and identity in time and space, what has been referred to as 'the release from proximity' (Gamble 2010, 31), i.e. going beyond the information available, is a hallmark of humans (Gamble 1998, 431; Rodseth *et al.* 1991). For instance, Donald (2010, 77) commented that much of the work of modern society centres on the maintenance of external memories, and on the operation of the external symbolic storage system that human beings have built. Within the context of non-Western societies, knowledge transmitted through communication (e.g. oral histories, stories, ceremonies and customs, everyday discourse and oratory, experiential, teaching, and learning) allowed aboriginal people of British Columbia to pass on their relationship with the environment from generation to generation (Turner *et al.* 2000, 1277). Melanesian people are able to externalise relationships that constitute a part of themselves in gift objects that are taken and absorbed by others (Fowler 2004, 67). Whittle (2003, 93) commented that among pastoralists in western Mongolia, sheep tibiae represent patrilineal descent or genealogical descent, and are particularly important in communications between people and ancestors.

Thus, identities embodied in an animal and mediated to a person via interaction, may subsequently be embodied in another object (e.g. animal remains, figurines or paintings). As a result, from a singular-relational perspective, identity that has been mediated from a living animal to an object via the human physical form can be used to maintain a sense of relational cohesion: through regular interaction with such objects relational cohesion can be maintained and this does not require interaction with the 'source' object. Identity deposited in the 'new' object *is* the identity that was bound up in the original object, and so regular interactions with that new object suffice to maintain the known world in its state of equilibrium and relational cohesion.

In summary, dualistic approaches maintain that animals are culturally neutral, and in order for them to obtain social significance, rely on the cultural generosity of humans and metaphoric or symbolic reference to the human world. A singular-relational approach, on the other hand, sees the world from a different perspective where animals are just as involved in the social creation of the world as humans. Furthermore, from this perspective identities are not bound by the physical, but can be mediated between objects. Thus, human interactions with animals provide the means by which identities can be mediated, thereby providing the life-force of the world, and allowing for a sense of relational cohesion to be maintained. Moreover, interactions with animals that demonstrate particular anthropomorphic as well as 'super-human' qualities are likely to be particularly significant both for the way people understand the world and their ability to maintain a sense of relational cohesion. However, an inability to interact with such animals is likely to manifest a sense of relational instability or insecurity, but this may be compensated for by interacting with other associated objects (e.g. animal remains and/or animal depictions) that are empowered with important enchained identities, thereby allowing a sense of relational cohesion with the world to be maintained.

This represents the theoretical point of departure, and it is from this perspective that I wish to set up the working hypothesis for the remaining part of the book. Animals that are particularly anthropomorphic in nature and have qualities that go beyond those of people are considered of prime importance for past hunter-gatherers in their attempt to maintain a sense of relational cohesion, and as a result require regular interactions with such animals. In cases where such animals are known but not often interacted with, identity that would otherwise be mediated through direct interaction must be deposited in other third party objects (e.g. animal remains, animal figurines, animal depictions), thus providing the means by which regular interactions with associated identities can be achieved. If this was the case in the past, then important animals that are interacted with less often should be apparent in objects such depictions and sculptures and/or their remains. On the other hand, if this is not the case then there should be little difference in the degree to which identities are deposited and enchained in third-party objects (i.e. the frequency at which animals are depicted and their remains used), regardless of how often a particular animal is encountered.

Deriving a case study context

Most archaeological investigations interested in human-animal relations in Palaeolithic Europe have focused largely on hunted animals. For instance, Boyle (1998) in her overview of Middle Palaeolithic southern France has focused on human relations with commonly hunted animals such as horse, red deer, ibex, chamoix, and roe deer. This is similarly the case with Chase (1986) who, within the context of Middle Palaeolithic humans at the site of Combe Grenal in France, focused on human relations with commonly hunted animals such as horse, red deer, reindeer, and *Bos*/bison. Also, Soffer (1985, 262), in her overview of the Upper Palaeolithic of the Central Russian Plain, focused on commonly hunted fauna such as mammoth, horse, rhino, *Bos*/bison, and reindeer. However, although it is understandable and clearly reasonable for archaeologists interested in human-animal relations in Palaeolithic Europe to focus on commonly hunted animals (remains of hunted animals are common, and provide irrefutable evidence of human-animal relations), commonly hunted animals were not the only animals to have inhabited Europe during the Palaeolithic period. One of the most common, but rarely hunted animals found in association Palaeolithic Europe is the cave bear. Large numbers of cave-bear remains have been recovered from various archaeological and palaeontological sites throughout Europe (Musil 1980a; 1980b; 1980c), and in some cases cave-bear remains have been found with human-made cut marks (Münzel 2001; Münzel and Conrad 2004; Valensi and Psathi 2004, 268-269; Withalm 2004, 221; Wojtal 2007, 44-52, 142) (Fig. 1.1). Moreover, anthropological and ethnographic data indicate that in modern-day and historic societies, the bear is one of the most commonly revered animals. Among the modern-day Gilyak (Russia) and Ainu (Japan) peoples there is no animal that is more highly revered (Hallowell 1926, 151), and according to Hallowell (1926, 148), with reference to global ethnographic accounts of human relations with bears, 'no other animal was found to attain such universal prominence as the bear, nor to have associated with it, over such a wide geographical area, such a large series of customs'.

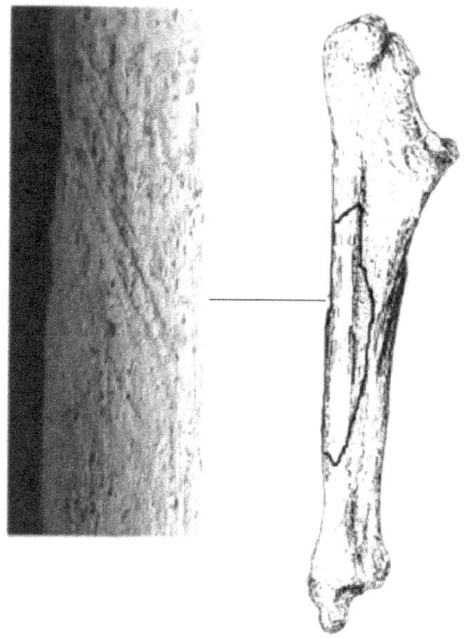

Fig. 1.1. Filleting marks on a cave-bear cub radius, found at Caverna Delle Fate, in Italy. (After Valensi and Psathi 2004, 269.)

Most archaeological investigations concerned with human social relations with animals in Europe during the Palaeolithic have focused on Western Europe after about *c.* 30,000 years ago, largely governed by the geographical and chronological distribution of archaeological data apparently indicative of such relations (e.g. animal depictions and sculptures). However, an alternative and potentially promising spatial and temporal arena within which to study human social relations with animals is the Eastern Czech Republic (Moravia and Silesia), during OIS3. Evidence of hominins (Neanderthals and AMH) in Moravia and Silesia is particularly prevalent during OIS3 (Svoboda *et al.* 1996), and associated archaeological data demonstrate that the people inhabiting the region during this period commonly hunted animals, used their remains for various purposes (e.g. to build structures, to make personal ornaments, to engrave depictions upon, and to make utensils and tools), and often depicted their form (Svoboda *et al.* 1996, 1). Furthermore, remains of cave bears are commonly found in this area, and it is within the context of OIS3 that their remains are most commonly recovered (Musil 1980b). In this book then, I focus on hominins (Neanderthals and AMH) and cave bears in Eastern Czech Republic (Moravia and Silesia) during OIS3.

Chapter outline

In the next chapter, I explore what interactions are, discuss and highlight the main issues associated with human-animal interactions, and outline how I intend to approach those issues most relevant and important for this book. Chapter 3 presents an overview of cave bears, discussing issues such as the history of cave-bear research, cave-bear phylogeny, evidence for their presence, their geographical and chronological distribution, important elements of their ecology, biology, physiology, and ethology, and existing evidence for human use of their remains and cave-bear depictions. Chapter 4 presents a thorough overview of the case study, looking in particular at issues such as climate, geology, topography, flora, and fauna, hominins and cave bears in the study region, and the specific case study sites chosen for this book. Chapter 5 creates a digital framework, mapping site locations, lithic raw material outcrops, topography, palaeohydrology, palaeovegetation, friction maps, prey species distribution and diversity maps. Chapters 6 and 7 map cave-bear and hominin distribution patterns, and Chapter 8 brings the results of Chapters 6 and 7 together, identifying potential interactions between the two. In the final chapter, I discuss the significance of the results of the book both in terms of hominins and cave bears within the study region during OIS3, and within a wider zoological, spatial and temporal context.

Summary and conclusion

In this chapter, I have attempted to introduce the topic of interactions, to demonstrate the significance of human-animal relations, different ways (dualistic/singular-relational) in which animals, and human relations with animals, may be considered, and the significance of different approaches for our understanding of the importance of interactions with animals for past people. I subsequently established a theoretical point of departure, derived a working hypothesis, and provided a chapter outline. In the next eight chapters, I attempt to construct a logical and progressive discussion that hopefully provides the reader with some understanding, not only of the extent to which hominins in Moravia and Silesia interacted with cave bears, but also of the significance of those interactions for people in the study region during OIS3, and for other prehistoric, historic, and contemporary non-Western and Western humans.

Chapter 2: A methodology for understanding interactions

Introduction

The objective of this chapter is to discuss the main issues and approaches for the book. The chapter revolves around three main questions and associated subsidiary questions.

1. How can we identify human-animal interactions/what factors affected human-animal interactions?

2. What was the nature of those interactions/what factors affected the nature of those interactions?

3. And what was the significance of interactions for past human sense of identity/what factors affected the significance of interactions for past human sense of identity?

Most of this chapter is focused on the first two key questions, looking to reveal those issues most important in governing human-animal interactions, the nature of those interactions, and how such issues can be approached. However, I also explore approaches to factors such as animal appearance and behaviour, which are important not only for governing whether or not interactions take place, but also for how people interpret interactions with animals and the significance of those interactions for people (Chapter 1). I also look specifically at how insights into the significance of interactions with animals for past humans can be approached. I attempt to highlight important influential factors associated with the key questions, and then discuss how these can be approached, indicating where in the remaining part of the book these approaches take place. Before addressing the main elements of this chapter however, I briefly explore what interactions are.

What are interactions?

In the last chapter, I proposed that interactions could be understood as the human perceptual recognition of the other. However, although the employment of perceptual mechanisms is clearly necessary for interactions to take place, it is how those perceptual mechanisms are employed that governs the nature of interactions. For instance, interactions might be described as focused or unfocused. Focused interactions can be thought of as situations where two or more actants are involved in some form of mutual and intentional communication. Goffman (1961, 7, 17-18) suggested that these types of interactions occur when actants effectively agree to sustain for a period of time cognitive and sensory attention, and Giddens (1984, 72) suggested that they occur where two or more individuals co-ordinate their activity through a continued encounter. Humans can be described as engaging in focused interactions when talking to each other during face-to-face encounters. Also, a predator (human or non-human) might be thought of as having a focused interaction with their prey at the point when predator and prey both become acutely aware of each other. However, the employment of perceptual mechanisms in a focused manner by one actant does not necessitate that this is also the case for other actants. A person or animal may be intently focused on another person or animal; however, the observed actant may be only vaguely or indeed completely unaware of the presence of the observer. This is the case in human activities such as ornithology, where the intention of the bird watcher is to observe intently the behaviour of birds, while at the same time trying to ensure that the birds remain unaware of being watched, or at least in a state of unfocused interaction. Unfocused interactions have been described by Goffman (1967, 133) as situations where an actant is aware of the presence of another actant, but their behavioural patterns are largely unaffected. For instance, most people when walking through their local town centre or nearby woodland are probably generally aware of their surroundings in an unfocused manner, only engaging in focused interactions when particular situations arise, such as meeting a friend, or seeing an unusual animal. However, focused and unfocused interactions are not mutually exclusive: an individual actant may engage in both focused and unfocused interactions at the same time. Going back to the previous example, although a person may be engaged in a focused interaction while talking to a friend or observing an animal, they are inevitably also aware of the rest of their surroundings, unconsciously observing the other people that pass by or the animals and plants surrounding them.

It is not just the manner in which perceptual mechanisms are employed that governs the nature of interactions, but also the number of actants involved. For instance, interactions may be described as one-to-one, one-to-many, many-to-one, or many-to-many (Eerkens and Lipo 2007, 250). Face-to-face interactions are the most obvious type of human one-to-one interaction, but these types of interactions can also occur between humans and other non-human actants such as animals. For instance, a person may be involved in a one-to-one interaction when petting a dog, or when hunting a specific individual animal. One-to-many interactions are also common in modern Western societies, occurring in places such as academic conferences where speakers present their ideas to audiences. They also occur within the context of predator-prey relations, for instance when a predator (human or non-human) is stalking a group of animals. Conversely, the same contexts provide the conditions for many-to-one interactions: an audience observing a speaker for instance (Eerkens and Lipo 2007, 251), or a group of animals observing a human or non-human predator. Many-to-many type interactions also occur within the context of modern-day Western societies, in places such as football stadiums where fans associated with each team sing in unison at each other, or where a panel of experts pass on information to a classroom of students (Eerkens and Lipo 2007, 251). Within a hunter-gather context, these types of interactions might occur in situations where groups of people hunt groups of animals. However, as is the case with focused and unfocused interactions, one-to-one, one-to-many, many-to-one, and many-to-many interactions are not mutually exclusive. For instance, two people engaged in a one-to-one, face-to-face interaction will inevitably be aware of the other things that surround them, such as people and animals. An individual can still be part of an audience, engaging in a many-to-one type of interaction with the speaker, but at the same time be engaged in a conversation with the person sitting next to them. The same can be said for actants involved in many-to-many type interactions, where being part of a group interacting with another group does not necessitate that one-to-one type interactions cannot take place between adjacent individuals.

Whatever the particular nature of an interaction, it cannot be isolated from the temporal, spatial, and situational context within which it exists. For instance, interactions are affected by prior interactions or knowledge of interactions. Goffman (2006, 129) made the point that prior interactions or knowledge of interactions can provide insights into the expected behaviour of an actant, thus affecting the manner in which that individual behaves during the interaction. Also, the interaction event itself is temporally dynamic and fluid. Giddens (1984, 71) suggested that interactions may be very loose and transitory in form. In a gathering of people, an individual may engage in a one-to-one interaction, which is constantly interrupted by people or things they see, hear or smell outside of their immediate focused-interaction. Furthermore, interactions perpetuate into the future via the memory and imagination of actants, subsequently affecting the way they think and behave in the future. As previously alluded to, it is the memory of interactions that provides an actant with the framework for knowing how to behave during future interactions. Spatially, apart from the last 100 years or so when electronic communication has allowed interactions to take place over long distances (Meyrowitz 1990, 86-87), interactions have largely been governed by the spatial proximity of actants, and the associated perceptual capabilities of actants. Also, the particular situational context within which interactions take place is important in determining their nature. According to Goffman (1961, 118), context includes the physical environment of an interaction but is not something merely 'in which' interaction occurs. For instance, within the context of a church ceremony, the particular arrangement of the furniture means that the spatial alignment and proximity of actants is mostly fixed, promoting particular types of interactions (i.e. many-to-one and one-to-many) between the congregation and the priest. In a similar vein, the type of interactions that a human or non-human predator might have with their prey is very much governed by the characteristics of the landscape. One-to-many type interactions, for example, might be much easier in a flat, open landscape, where prey can be observed from a distance, in comparison with hilly, wooded landscapes where perceptual capabilities are likely to be more restricted. Furthermore, different situational contexts also offer different opportunities for temporal fluidity during interactions. The particular structured timeframe within which some events, such as church ceremonies, take place means that there is little opportunity to engage in certain types of interactions (e.g. one-to-one interactions). On the other hand, a group of hunter-gatherers hunting a prey animal are likely to experience a much more fluid and dynamic type of interaction. The unpredictability of the behaviour of the prey might afford human actants the opportunity to engage with one another on an ad-hoc basis, depending on the behaviour of the animal and the associated circumstances that arise.

Although interactions may be described as focused, unfocused, one-to-one, one-to-many, many-to-one, and/or many-to-many, and temporally, spatially, and context dependent, ultimately the nature of interactions depends on how they are understood and experienced by the actants involved in them. For humans, the significance of interactions depends, at least to some extent, on the cultural norms to which they are accustomed. Ultimately, however, interactions are experienced through the body, and those experiences invoke a whole range of emotional responses such as composure, intimacy, happiness, pride, love, enjoyment, fascination, enthusiasm, and excitement, to embarrassment, frustration, anger, jealousy, fear, hurt, discouragement, discomfort, despondency, and distancing (Prus 1996, 174). It is these bodily experiences of interactions and associated emotional responses that allow people to make sense of their world and associated interactions with their surroundings, behaving in a way that is both appropriate and meaningful to themselves and to others.

Identifying influential factors

For interactions between humans and animals to have taken place both humans and animals had to exist. However, the existence of humans and animals in the past was not uniform: different species and groups of humans, and different animal species existed. For instance, in Europe during OIS3 two species of humans (Neanderthal and AMH) and a number of different groups of humans (Mousterian, Micoquian, Bohunician, Szeletian – Neanderthals; Aurignacian and Gravettian – AMH) existed (van Andel *et al.* 2003a). Also, a large array of different faunal species existed in Europe during this period, including Artiodactyla such as elk, red deer, reindeer, fallow deer, roe deer, giant deer, auroch, bison, chamoix, ibex, musk ox, saiga antelope, and wild boar. Perissodactyla such as horse, rhino, mammoth, straight-tusked elephant, existed together with Carnivora such as lion, leopard, wild cat, lynx, spotted hyena, wolf, red fox, Arctic fox, brown bear, badger, stoat, weasel, polecat, marten, wolverine, and otter (Musil 2003, 189-190).

Moreover, although the existence of humans and animals in the past was not homogenous in terms of human species and group and animal species, this also changed over time. For instance, although Neanderthals existed in Europe from around *c.* 300,000 years ago (Gamble 1999, 174), evidence of Mousterian occupation, largely equivalent to the presence of Neanderthals as a whole, is absent by at least 23,000 Cal. BP, and most probably earlier. Micoquian occupation is dated to between about 56,000 Cal. BP and 32,000 Cal. BP (Marks and Monigel 2004, 67-68; Musil 2003, 171), Szeletian presence is demonstrated for periods between 45,000 Cal. BP and 37,000 Cal. BP, and evidence of Bohunician occupation is largely documented after 47,000 Cal. BP and before 43,000 Cal. BP. The first signs of AMH, present in the form of the Aurignacian, can be found in Europe by 47,000 Cal. BP, and although AMH remain in Europe continuously after this period, the presence of Aurignacian occupation is largely absent by about 23,000 Cal. BP. Evidence of Gravettian occupation can be found in Europe after 38,000 Cal. BP, but their presence diminishes by about 21,000 Cal. BP (van Andel *et al.* 2003a, 32). Also, although a large number of faunal species existed in the past, many of these became extinct over time. This is the case with species such as mammoth, woolly rhino, giant deer, and spotted hyena, which all became absent in Europe by about 10,000 years ago (Martin and Steadman 1999, 20; Stewart *et al.* 2003, 114).

Interactions between humans and animals, however, were not simply a matter of human and animal existence and changes in their existence over time: humans and animals had to exist in similar places for interactions to take place. The spatial distribution patterns of humans and animals, however, were not uniform. For instance, although Neanderthals occupied regions across most of Europe during OIS3, they were particularly associated with areas such as the coastal regions of Italy, and the Atlantic shores of Portugal, northern Spain and Atlantic France (Davies *et al.* 2003, 191). Although Mousterians were largely associated with the overall distribution patterns of Neanderthals, Micoquians existed mostly in areas between Belgium and Romania, concentrated particularly in regions such as southern Germany and Poland (Allsworth-Jones 1986, 47), but also present in places such as Ukraine, Crimea, and Moravia (Czech Republic) (Meignen *et al.* 2004, 51). Bohunicians were mostly associated with Central Europe, in particular Moravia (Czech Republic), as was the case with Szeletians (Kozłowski 2004, 18) who occupied areas such as Moravia, west Slovakia, and central and eastern Hungary (Svoboda 2004, 35). AMH occupied much of the same regions of Europe as Neanderthals, but in contrast to Neanderthals also occupied areas such as the forelands and plains of the Pyrenees and Alps, and for the most part, the distribution of Aurignacians and Gravettians can largely be equated to the overall distribution of AMH (Davies *et al.* 2003, 192, 193). Also, in terms of faunal species in Europe during OIS3, there was a gradual north-south/east-west change, with species such as mammoth, woolly rhino, red deer, reindeer, auroch/bison, saiga and musk ox being common in areas such as the central and northern regions of Europe, but totally absent in places such as the central Mediterranean region (Stewart *et al.* 2003, 115-119). Moreover, although the distribution patterns of humans and animals were spatially heterogeneous, they were also temporally variable. For instance, prior to *c.* 60,000 years ago, Neanderthal occupation of Europe was largely located below 45° latitude; following this period, though, occupation extended to areas above 50° latitude, and between about 60,000 years ago and about 30,000 Cal. BP gradually became more focused in areas such as south-western France and the Iberian Peninsula (van Andel *et al.* 2003a, 37). This is similarly the case for AMH: prior to 43,000 Cal. BP Aurignacian occupation was largely absent from areas such as France, but by about 29,000 Cal. BP these areas, in particular south-western regions of France, became densely occupied by these people (van Andel *et al.* 2003a, 41-43). Also, faunal species such as reindeer, largely absent from regions such as south-western France up until about 29,000 Cal. BP, became increasingly common in these areas after this period (Champion *et al.* 1984, 63).

One of the most important factors governing the spatial distribution patterns of both humans and animals in the past was their use of, and association with, specific site locations. Binford (1980, 5) showed that settlement sites used by modern-day hunter-gatherers are vital in governing their everyday spatial activities, as is the case with the G/wi San hunter-gatherers of Africa who repeatedly use site locations, centring their foraging activities around their residential site and moving to and from these places on a daily basis. Moreover, archaeological data such as hearths and remnants of structures indicate that such places were also important for past people, and the associated presence of hunted animal remains and stone tools brought to these sites from other locations in the landscape indicate that people made regular trips to and from these places. This is the case at the site of Combe Grenal in France, occupied by Mousterians during OIS3, where large numbers of stone tools have been found alongside the remains of hearths, burnt material, remains of hunted animals, and post holes (Mellars 1996, 292-308). Dense deposits of Micoquian-type stone tools, indicative of repeated settlement activity, have been recovered from the site of Buran-Kaya III (Crimea) (Marks and Monigal 2004, 67). This is similarly the case at the Bohunician site of Stránská skála (II-III) (Moravia) and the Szeletian site of Szeleta where large numbers of associated stone tools have been recovered (Allsworth-Jones 1986; Kozłowski *et al.* 2009). Repeated use of the sites of Willendorf II (Austria) and Geissenklösterle (southern Germany) by Aurignacian people is also apparent (Svoboda 2004, 47), and at the Gravettian sites of Höhle Fels, Geissenklösterle, and Brillenhöhle in Germany stone tools have been found alongside bone tools, decorative objects, mobile art, and hearths (Scheer 2000). In terms of faunal association with specific site locations, animals such as foxes and wolves often use places such as the base of a tree, a hollow log, or cave as den sites to bring up their young (Dell'Arte and Leonardi 2008, 169; Joslin 1967, 286; Smith *et al.* 1992). Hyenas also use places such as caves as breeding dens and lairs (Tilson and Henschell 1986, 177), wolverines often use snow tunnels or fallen trees as dens for maternal purposes (Magoun and Copeland 1998, 1313), lynx commonly use hollow trunks, bushes or thickets as natal dens (Fernández *et al.* 2002, 1). Also, Doncaster and Woodroffe (1993, 88) with reference to badger territories in Britain, made the point that the spatial distribution patterns of associated den sites can determine the location, shape, and size of badger territories.

However, the frequency and duration with which such sites were used by humans and animals, and so the degree to which these places governed their associated spatial distribution patterns, was not uniform. For instance, deep sequences of hearths and ashy lenses associated with Mousterian occupation of sites such as Kebara (Levant) indicate repeated occupation of the same specific site location (Lieberman and Shea 1994, 306). On the other hand, according to Mellars (1996, 267) a small number of stone tools found together with the bones of a single mammoth found in the Carrière Thomasson region of France in association with Mousterian occupation during OIS3, may be indicative of a stop-over site possibly used only once. Also, Meignen *et al.* (2007, 160-162) have commented that the duration of occupation at the site of Umm el Tlel (Layer VI3b'1) (El Kown basin, Levant) by Mousterians during OIS3 was relatively short, and Mousterian occupation of Farah II (Negev, Levant) during the same period was no longer than about a few weeks. Archaeological evidence associated with Gravettian occupation at sites such as Willendorf (Austria), however, point to much longer periods of prolonged occupation (Musil 2003, 178). Also, in the case of animals, some faunal species such as reindeer simply do not use specific site locations on a regular basis. In cases where sites are used by animals, site fidelity, the tendency to return to a previously occupied site (Switzer 1993, 533), can vary tremendously. Spotted hyenas inhabiting the desert region of Namibia (Africa) have been shown to have particularly high site fidelity, regularly returning to the same site location (Tilson and Henschel 1986, 177-178), whereas other species such as racoons can shift sites on a daily basis (Gehrt 2003, 618). Also, recent studies undertaken on wolves in the Central Canadian Arctic have shown that 13 out of the 15 wolves studied returned to within a 25 km radius of a previously used site, but only 2 actually returned to the same site location (Walton *et al.* 2001, 867-876). Also, some animals use single site locations for months on end, as is the case with species such as Arctic ground squirrels that hibernate during the winter periods (Carl 1971), but other species such as racoons may use a site location for only a single night (Gehrt 2003, 618).

Human and animal association with specific site locations was not the only factor governing their spatial distribution patterns: both humans and animals used offsite locations, and in both cases the most important factor governing their association with offsite locations was the procurement of resources. Higgs and Jarman (1975, 5) commented, 'a major determining factor in the animal world is the relationship between populations and resources, and we may assume that the same factor is of similar importance for human behaviour'. One of the most important resources procured by past humans was lithic raw materials to make stone tools, as is the case in Europe during OIS3 where lithic raw materials were used extensively by both Neanderthals (Mousterian, Micoquian, Bohunician, and Szeletian) and AMH (Gravettians and Aurignacians) (Bar-Yosef and Pilbeam 2000; Brantingham *et al.* 2004). Another crucial resource for past humans was animals. Mellars (1996, 193) commented that the first requirement of any biological species is to secure adequate food supplies, and the fact that animals were one of the most important food resources for humans is demonstrated by the presence of large numbers of faunal remains such as horse, reindeer, bison, and mammoth found at a number of archaeological sites throughout Europe during OIS3 (Chilardi *et al.* 1996, 557; Hardy 2004, 547; Patou-Mathis 2009, 449; Svoboda *et al.* 2005, 214; Zilhão 2001, 599). For animals, association with particular habitats is governed mostly by dietary habits and the distribution of associated resources. Tilson and Henschel (1986, 179) have shown that the spatial distribution patterns of spotted hyena groups inhabiting the desert region of Namibia were governed largely by the distribution patterns of food resources. Dell'Arte and Leonardi (2008, 168) commented that different patterns of the use of space by red fox depend on the distribution and availability of food patches, and Gough and Rushton (2000, 199) made the point that the location of food is an important factor in governing the distribution patterns of animals such as badgers, polecats, and wolverines. Moreover, according to Musil (2003), although the distribution patterns of food may have been important in governing the distribution patterns of animals, the most important factor controlling the distribution patterns for both carnivorous and herbivorous animals is vegetation.

However, although the procurement of resources governed the offsite distribution patterns of both humans and animals, the places associated with the procurement of such resources would not have been uniform. For instance, at the same time that a number of different lithic raw materials were used by past humans to make stone tools, the distribution of such materials is highly variable (e.g. Gamble 1986, 336; Mellars 1996, 141, 166) and governed by a range of factors including the character of underlying geological formations, distribution and depth of overlying soil, and the effects of wind, rain, and rivers (Růžičková *et al.* 2001). In a similar vein, at the same time that past humans hunted a large range of fauna (Adler *et al.* 2006; Anikovich *et al.* 2007; Boyle 2007;

Grayson and Delpech 2003; Niven 2007; Musil 2003), and the distribution of such fauna is largely governed by their diet and vegetation, the dietary habits of animals can vary significantly according to body size, gut type, capability to obtain food, available feeding time (Senft *et al.* 1987, 792), and season (Fedriani *et al.* 1999, 141-142), and the distribution of vegetation can vary according to geology, topography (elevation, slope and aspect) (Barbour *et al.* 1999, 387; see also Barbour *et al.* 1999, 390; Huntley and Allen 2003, 81; Petrik and Wild 2006, 226; Pokorný 2004), hydrology (Brown 1997, 17; Gremmen 1982, 62; Richardson 2000, 455; Zelený and Chytrý 2007, 217), species type (Barbour *et al.* 1999, 320; Gamble 1986, 98; http://www.rook.org; Willson *et al.* 2008, 299), and long-term climatic change (Müller *et al.* 2003, 244) and seasonal change (Tockner *et al.* 2002, 266; Velichko and Zelikson 2005, 142). Moreover, in the case of humans and their procurement of resources such as animals, it was not only the distribution of individual species that was important, but also overall species diversity (Gamble 1986, 42; Jochim 1976, 16).

Humans and animals, though, did not have universal access to resources, but were governed by factors such as home range. Home ranges are the areas in which people and animals acquire necessary resources to carry out their biological requirements for life (Burt 1943). In the case of humans, this can also be understood as the 'site catchment', and be considered as the area around a site which people use to acquire resources for survival. Home range for animals can be thought of as the total area that an animal population covers in a year (Martin 2000, 22). The home ranges of humans and animals, however, are not uniform but are governed by factors such as diet, productivity of the environment, demography and social affiliation, energy, and the character of the landscape. For instance, an increase in the reliance on meat and a less productive environment is likely to result in an overall increase in home range (Lindstedt *et al.* 1986, 415; Roebroeks 2003, 105). Harestad and Bunnell (1979, 389) have shown that carnivores generally have a larger home range in comparison with omnivores and herbivores of the same mass living in the same environment. Walton *et al.* (2001, 867) have shown that male wolves in the Canadian Arctic had average home ranges of about 63,058 km^2, whereas female wolves in the same environment had a home range of about 44,936 km^2. The home range of female hyenas studied by Boydston *et al.* (2003, 1006) varied according to their association with den-dwelling cubs: those with den-dwelling cubs had the smallest home range extents. Also, Kelly (1995, 133) commented that the distance from a residential camp at which a human forager can procure resources is limited by the amount of energy expended when retrieving resources. Most modern-day hunter-gatherer groups can collect food up to about 5 km from the camp, and Binford (1980, 6) demonstrated that foraging trips undertaken by the !Kung San were largely no more than about 5 km. Similarly, for the Pitjandjara of western-central Australia, a walk of greater than 5 km is sufficient to induce 'grumblings' (Tindale 1972, 245). Lindstedt *et al.* (1986, 415) made the point that an animal's home range is associated with its body mass and metabolic rate. Also, Higgs and Vita-Finzi (1972, 33) made the point that as energy expenditure is affected by the character of the environment, the shape and spatial extents of the home range will be offset accordingly, with broken or difficult terrain distorting the shape of a home range.

However, although human and animal access to resources is governed by their associated home range and this is governed by factors such as diet, productivity of the environment, demography and social affiliation, energy, and the character of the landscape, the impact of these factors on human and animal home range would also have varied according to factors such as faunal species, climate and season, sex, topography, hydrology, and vegetation. For instance, the dietary habits of an animal (e.g. carnivorous or herbivorous) are normally governed by its associated species grouping. More productive environments and thus smaller home ranges are likely to exist in warmer climatic and seasonal periods. Female animals are more likely to nurture their young and so more likely to have smaller home ranges. Larger faunal species generally have larger metabolic rates, and so greater home ranges, as is the case with some larger species of gazelle (e.g. *Gazella granti*) which have a home range of about 290 km^2/year, but other smaller species (e.g. *Gazella thomsoni*) have a home range of about 142 km^2 (Martin 2000, 23). Increases in slope are likely to correlate with increases in energy expenditure and may account for as much as 90% of overall energy costs, but the relationship between slope and energy expenditure is unlikely to be linear (Bell and Lock 2000, 88; Llobera 2000, 67, 70). Areas of either liquid (e.g. rivers, floodplains) or frozen (e.g. snow) water are likely to increase energy expenditure, and in some cases may act as virtual barriers for human and animal movement (Garshelis 2000, 137; Gillings 1995, 70; Velichko and Zelikson 2005, 143). Vegetation is likely to affect energy expenditure with increases in density generally resulting in increased energy costs; however, vegetation is likely to be responsible only for about 10% of overall energy expenditure (Connolly and Lake 2006, 215; Llobera 2000, 67). Moreover, the impact of the environment on energy expenditure is likely to decrease during cold climatic and seasonal periods in comparison with warmer times, as floodplains and possibly rivers become frozen or semi-frozen, and as vegetation becomes sparser; however, energy expenditure also increased due to more snow fall occurring at lower levels during these periods.

Humans and animals, though, were not simply associated with discrete areas in the landscape: humans and animals (i.e. land mammals) moved through the landscape using pathways. One of the most important factors governing the routes of human and animal pathways is energy. For instance, modern-day animals such as cattle often take the most energy-efficient pathways (Ganskopp *et al.* 2000, 179), reindeer prefer to follow routes of least topographic resistance (West 1997, 32), and this is also probably the case for humans (Rockman and Steele 2003). However, it is not simply energy constraints that would have governed the particular routes taken by humans and animals in the past: humans, and no doubt to some extent animals, would also have been governed by other factors such as knowledge, learning (Rockman 2003), memory (Golledge 2003), social networks (Kelly 2003, 51), visibility and proximity to water (Steele and Rockman 2003).

Interactions, however, were not simply a matter of the spatial and temporal distribution patterns of humans and animals, but were also governed by human ability perceptually to recognise animals. This in turn would have been governed by factors such as human perceptual capabilities, the character of the environment, weather, animal perception and associated predator/anti-predator behaviour, animal appearance, and animal population density. Clearly, interactions cannot take place unless humans are able perceptually to recognise animals, and Cant and Temerin (1984, 342) commented that human visual and auditory senses can be hampered by the character of the environment: closed environments, and small-scale topographic relief and foliage density can affect the visual and auditory field. Ingold (2000d, 251) commented that the Umeda (Papua New Guinea) inhabit an environment of dense and virtually unbroken forest, and as a result walk with their eyes to the ground, listening for game instead of looking for it. Rainfall may hamper visual, olfactory and auditory perception, limiting sensory evaluation of the environment (Mithen 1990, 37; see also Gell 1995, 235; Ingold 2000d, 251). Also, many animal species are able to detect the presence of humans before humans are able to detect the presence of animals. Biesele (2001, 69) made the point that in thick vegetation animals hear and smell a person long before a person is able to detect the presence of the animal. As a result, although animals might be in human perceptual range, whether or not humans are able to perceive animals may depend on the animal's subsequent and associated predator/anti-predator behaviour: prey animals such as gazelle, zebra, wildebeest are more likely to flee if they identify the presence of humans, whereas animals such as lions might be more prone to attack (Griffin 1992, 58). Also, an animal that is particularly large or contrasting in appearance to the surrounding environment, loud, smelly, or which congregates with other animals is likely to be easier to perceive than a small, quiet, odourless and lone animal, and generally speaking, increases in animal population density are likely to result in an increase in the potential for human perceptual recognition of animals.

However, factors affecting human perceptual recognition of animals can vary according to the age of the human individual, topography, climate and season, faunal species, the social affiliation and behavioural state of the animal, idiosyncratic features of animals, overall animal population structure, and proximity to the centre of the home range. For instance, human perceptual capabilities can change significantly over time: eyesight can vary tremendously between humans, and this is particularly the case with different-aged individuals (Collins and Britten 1924, 3192). Hilly or mountainous regions and warm climatic or seasonal periods are likely to create more closed environments, and weather can change with climate and season. The perceptual capabilities of animals can change between species, and predator/anti-predator behaviours of animals can vary according to their association with young and feeding habits: female animals can become particularly defensive if accompanying young, and this is also the case for some animals, particularly carnivores, when feeding (Domico and Newman 1988, 171, 165). Also, the appearance of an animal can change over time. Guthrie (2005, 76) commented that individuals of any species vary. Reindeer, for example, are most rounded with fat after a summer of ample food, and dominant reindeer bulls reduce their food intake during the October rut to an extent that their abdomens become almost concave. Also, horses have two different pelage coats per year: the summer coat is short and of a different colour to that during the winter, which is shaggier (Guthrie 2005, 74). Furthermore, the tendency for animals to congregate in groups can change seasonally: although gazelle normally form herds, female gazelle tend to become solitary just before birthing (Martin 2000, 24). Population density of most animal populations is normally higher towards the centre of the home range, but as the age and sex of animals can affect an animal's home range, population density changes according to the overall population structure of the species. However, in most cases the population structure of an animal species is not uniform. Data on the ratio of male to female gazelles in a case study undertaken by Martin (2000, 25) indicated that, although roughly equal proportions of males and females are born, by adulthood there are fewer males than females in the populations. Also, in most stable animal populations there are normally more juveniles than older animals (Klein and Cruz-Uribe 1984, 56; Martin 2000, 25).

Overall, interactions and the frequency of interactions between humans and animals were governed by their spatial and temporal distribution patterns and human ability perceptually to recognise animals. Moreover, these were governed by a number of other factors including human species and groups, animal species, use of specific site locations, vegetation, geology, topography, hydrology, animal diet, human hunting and lithic raw material procurement practices, energy, home range, animal demography, chronology, climate, seasonality, human and animal perception, animal social affiliation, animal characteristic (appearance, sound and smell), and animal population structure and density.

Approaching influential factors

Although it is acknowledged that interactions and the frequency of interactions between humans and animals are governed by a range of factors, it is not my intention to address every issue here. Instead, I focus only on those issues most relevant for this book. Also, one of the most evident factors to arise from the previous discussion is that to understand potential interactions between humans and animals, it is important to understand their spatial distribution patterns and so necessary to undertake some kind of spatial analysis. Therefore, I also discuss 'tools for undertaking spatial analyses'. The issues discussed are:

- identifying humans, animals, and their chronological context;
- identifying human and animal sites, human hunting and lithic raw material procurement practices, human use of animal remains and/or depictions of animals, animal demography, diet, appearance, perception, social affiliation, and predator/anti-predator behaviour;
- tools for undertaking spatial analysis;
- climate, geology, topography, palaeohydrology (rivers and floodplains), palaeovegetation, and energy;
- mapping human and animal site locations, human and animal home ranges, human pathways, lithic raw material outcrops, animal habitat and species diversity, and animal population density;
- identifying interactions and the significance of interactions for past humans.

In most cases, where archaeological residues have been found and published they are normally done so within the context of associated human species or group (e.g. Neanderthal – Mousterian, Micoquian, Bohunician, and Szeletian; and AMH – Aurignacian and Gravettian). This is the case with van Andel *et al.* (2003a), who have documented the presence of humans in Europe during OIS3 on the basis of associated published records of human fossil remains and stone tools and their associated production techniques. Thus, by using such data it is possible not only to identify the presence of past humans, but also to associate species or groups of humans with particular archaeological remains. However, one problem is that fossil remains of humans are rare, and so human affiliation with archaeological remains mostly has to be done in terms of stone tools and associated production techniques. Although archaeologists are generally agreed about human species and groups and their affiliation with different types of stone tools, this is not universal. For instance, though most archaeologists agree that the Aurignacian is associated with modern humans (Svoboda 2004, 30), Vishnyatsky and Nehoroshev (2004, 96) commented that we still do not know who was responsible for the origins the Aurignacian, and Patou-Mathis (2009, 453) commented that craftsmen who made Mousterian and Micoquian industries were undoubtedly Neanderthals, but it is still unknown who made Szeletian stone tools: they could have been Neanderthals or AMH. Some doubt also exists over whether Neanderthals or AMH were responsible for the production of Bohunican stone tools, as there is little or no human fossil evidence to indicate direct association one way or another. However, in the case of both Szeletian and Bohunician stone tools, the general consensus seems to be that they were both produced by Neanderthals rather than AMH (Allsworth-Jones 1986, 210, 217; Banks *et al.* 2008; Bocquet-Appel and Demars 2000, 544; d'Errico *et al.* 1998, S37). Another problem is that using such an approach to affiliate human species and groups with archaeological remains provides a rather culture-historic perspective of the past, i.e. it seems to give the impression that discrete peoples were moving around the landscape in isolated groups. However, this is unlikely to have been the case. For instance, evidence that groups of people were not isolated from each other is apparent in the presence of Aurignacian bone points found at the Szeletian sites of Dzerava skala (Slovakia) and Oblazow (Poland) (Kozłowski 2000, 99). Moreover, there is still a large amount of uncertainty about interactions between past humans. Bar-Yosef (2000, 137) commented that the Szeletian emerged from the Mousterian, but others believe that the Szeletian arose from the Micoquian (see Kozłowski 2000, 87). Also, Bar-Yosef (2002, 370) made the point that the Bohunician stone tool production method lacks any relationship to earlier Mousterian industries, and Svoboda (2004, 48) commented that Bohunician and Szeletian stone tool industries evolved in a parallel manner, with Bohunician stone tool production techniques possibly influencing Szeletian production methods. Evidently, associating past humans with archaeological remains in this way has its problems; however, the fact that such an approach is widely used within existing archaeological literature means that as long as caution is exercised this is probably the most sensible and logical approach to take. Chapter 4 uses relevant published literature documenting associated fossil remains (Neanderthal and AMH) and stone tools and their associated production techniques (i.e. Mousterian, Micoquian, Szeletian, Bohunician, Aurignacian, and Gravettian) as a means for associating archaeological remains with past humans, and this format is continued throughout the whole of the book.

Musil (2003, 171-181; 1980a; 1980b; 1980c) has published a comprehensive overview of faunal remains recovered from archaeological and palaeontological sites in Europe during OIS3, documenting the presence of individual animal species. By using such data some insights can be gained into the presence of different faunal species alive in the past. However, one problem is that these data may be biased and not truly representative of the actual fauna that existed. For instance, Goebel (2004, 166-168) has documented the presence of faunal remains discovered at archaeological sites throughout Siberia in association with different environmental conditions (Alpine, steppe, forest-steppe, forest, and tundra) during OIS3. However, although the remains of a number of different species (e.g. hare, wolf, fox, hyena, woolly mammoth, horse, woolly rhino, bison, gazelle, red deer, and roe deer) are reported as being found in association with Alpine, steppe, forest-steppe and forest conditions, only two species (Arctic ground squirrel and reindeer) are found in association with tundra conditions. Although it is likely that changes in the environment manifested changes in the presence of different faunal species, it is also likely that the apparent lack of species associated with tundra conditions is related to taphonomic circumstances: animal remains deposited in tundra conditions are generally less likely to survive in comparison with those deposited in more favourable environments (e.g. Alpine, steppe, forest-steppe, forest). Moreover, even if faunal remains are preserved and documented, in some cases it is simply not possible to distinguish between species. Martin (2000, 14) commented that it is often difficult to make species identifications from archaeological material. As a result, animal remains, and the associated species to which they belong, are often grouped: Musil (2003, 189) documented the presence of auroch/bison and Goebel (2004, 168) reported the presence of horse/ass. Clearly, there are some issues regarding the use of published data and their use in gaining insights into the presence of past fauna. However, the fact that animal remains are often found in association with archaeological and palaeontological sites, details of which are commonly published, suggests that this is probable the most sensible approach to take, and so in Chapter 3 and Chapter 4 I use published literature to identify the presence of relevant faunal species.

In most cases where human and/or animal remains have been found they are normally published with associated chronometric dates, mostly the result of science-based techniques such as conventional radiocarbon dating – C14, radiocarbon dating using accelerator-mass-spectroscopy – AMS-C14, Thermoluminescence – TL, Uranium-series – U-Series, and Electron Spin Resonance – ESR (Gamble 1999, 272). For instance, van Andel and Davies (http://www.esc.cam.ac.uk/research/research-groups/oistage3) have published nearly 2000 scientifically derived chronometric dates associated with human occupation of Europe during OIS3. Also, as part of the same project, Stewart *et al.* (2003, 103, 104) commented that a comprehensive database was constructed of western and central European mammalian fossils dated between *c.* 60,000 and 20,000 years ago. Other researchers have used a similar approach. For instance, Carbonell *et al.* (2000) have used a series of scientifically derived published chronometric dates to gain insights into the chronological character of human occupation of the Iberian Peninsula during OIS3, and Mellars (2000) has used this approach to get some understanding of the chronological character of human occupation of France during the same period. Also, in cases where scientifically derived dates are not available, often archaeologists and palaeontologists have dated human and animal remains according to sedimentary deposits and associated past climatic change (e.g. Würmian scale) (e.g. Musil 1980a; 1980b; 1980c). However, it is necessary to be cautious here. For instance, at the same time that many of the published scientific dates are obtained using radiocarbon dating methods, there are problems with using radiocarbon dates to date samples older than about 45,000 years. Kozłowski (2000, 99) made the point that there are not many radiometric dates from the period 50,000 and 40,000 years ago, and most that are available indicate minimum age of samples. Van Andel *et al.* (2003b, 27) commented that beyond 45,000 years there are serious calibration problems, and error margins associated with dates can range between 4000 and 8000 years. Also, although scientifically derived dates are now available to anchor the beginning and end of periods of past climatic episodes, using sedimentary deposits and associated changes in past climate to comprehend the chronological context of humans and animals in the past is generally not accurate: in most cases, it is the major climatic changes that are apparent in sediment layers, and these can often span tens of thousands of years. Moreover, in some cases (e.g. caves) it is difficult to be confident about the sedimentary context to which human or animal deposits belong: sediments are often mixed as a result of repeated fluvial action. Thus, it is wise to be cautious when using such approaches. However, the fact that so many scientifically derived dates are now available, particularly for the period of OIS3, suggests that using such data is the most sensible approach, and in cases where scientifically derived data are not available, establishing a chronological context is probably best achieved using associated sedimentary deposits and past climatic change (e.g. Würmian scale). In Chapter 4, I mostly use available scientifically derived (radiocarbon) dates to establish the chronological context of past humans and animals, and where this is not possible I use the Würmian scale together with associated sedimentary deposits and past climatic change.

A number of archaeologists have published details regarding the use of important sites by past humans on the basis of archaeological evidence such as remnants of hearths, structures, stone tools, animal remains, and remains of artistic behaviours (e.g. figurines and ornaments) (Davies *et al.* 2003; Gamble 1999; 1986; Vishnyatsky and Nehoroshev 2004). However, one of the main problems with this is that archaeological evidence and associated literature may not evenly represent the settlement strategies of past humans. For instance, archaeological remains, particularly those associated with open-air sites, are often destroyed due to environmental factors and may therefore be less likely to survive in comparison with other sites and related archaeological material located in caves or rock shelters. Vishnyatsky and Nehoroshev (2004, 80) commented that there are no late Middle and early Upper Palaeolithic open-air sites located in the Russian Plain north of 52° where glaciation is thought to have been extensive. Also, even in cases where evidence of human use of specific site locations is apparent, in many cases this is manifest in the form of secondary archaeological data such as stone tools and remains of animal remains, and direct evidence, i.e. remnants of hearths and structures, is often not found. However, although the presence of stone tools and animal remains demonstrate the use, and in some cases repeated use of, particular locations by people, the abundance and presence of such data, particular organic material such as animal remains, can vary enormously and be governed by factors such as related taphonomic circumstances, meaning that it can be difficult to be sure the extent to which the site was used by past humans. Clearly, when using published literature documenting the use of site locations by past humans it is necessary to be cautious; however, as large amounts of relevant published literature exists documenting the presence and location of sites used by past humans, using these data probably offers the most promising and efficient way of gaining insights into past people's use of site locations. In Chapter 4, I use available literature to identify important site locations occupied by past humans.

Important site locations occupied by past animals can also be identified using published literature: researchers commonly publish records of sites repeatedly occupied by animals. For instance, Quam *et al.* (2001, 389-390) have documented the presence of fox, hyena, and wolf remains associated with Valdegoba Cave, Spain, and suggest that the site was regularly used by these animals as natal dens or hibernation refuges. Diedrich (2009a; 2009b) has documented the presence of hyena remains recovered from caves in Germany, commenting that these animals made regular and repeated use of these places during the Late Pleistocene period. Moreover, in some circumstances researchers have published quantitative details regarding the abundance of animal remains found at sites (Musil 1980b, 36) and these can be used to estimate the significance of a site for a particular species. However, it is necessary to be cautious here: animal remains associated with open-air human sites, for example, are likely to represent human hunting activities, and are probably not indicative of the natural ecological and associated spatial distribution patterns of animals (Musil 2003, 167). Also, even in cases where animal remains are found at sites not occupied by humans, it is possible that the remains were brought to the site by some other means: in caves in Alaska, limb bones of bears were brought into caves by other carnivores (Sattler 1997, 680), and Diedrich (2009a, 361) commented that in the Perick caves, northwest Germany, evidence of chewing and cracking of lion bones indicates that associated carcasses were brought into the cave by cave hyena. Furthermore, animal remains are likely to have undergone significant post-depositional disturbance as a result of factors such as freeze-thaw actions, and trampling, gnawing, chewing, and digestion by carnivores (Hesse and Wapnish 1985, 25-26; Wojtal 2007, 9). As a result, published details regarding the number of associated animal remains found at a particular site may not accurately represent the degree to which the site was used. However, although it is clear that a significant amount of caution must be exercised when using such an approach, published accounts documenting the presence of animal remains at site locations are relatively common, and so using these data is one of the best ways to identify specific site locations important for past animals. In Chapter 4, I identify important site locations occupied by relevant faunal species using published literature, based mainly on the apparent abundance of animal remains found.

Published literature documenting the examination of faunal remains found at archaeological sites can provide useful insights into which animal species were commonly hunted by past humans. Patou-Mathis (2000, 386) commented that the study of animal bones provides evidence of game acquisition, Musil (2003, 189) has published a comprehensive list of faunal remains hunted by humans in Europe during OIS3, and Meignen *et al.* (2004, 53) have published descriptions of animals commonly hunted by Mousterian at the site of Molodova (Ukraine) during this period. Patou-Mathis (2009, 449) has recorded the presence of hunted faunal remains associated with Micoquian deposits at the site of Buran Kaya III (Crimea), and Svoboda *et al.* (2005, 214) have recorded the dominance of hunted mammoth remains at the Gravettian site of Kraków Spadzista (Poland). Also, Goebel (2004, 190) has used published literature regarding the abundance of hunted faunal remains found at early Upper Palaeolithic sites in Siberia as a means of determining the apparent significance of different species for past humans and their associated hunting practices. However, one problem with this is that such data simply may not be available. Wojtal (2007, 10) commented that bones can become weathered by physical and chemical agents, and can be completely destroyed with 5 years, and Mellars (1996, 193) commented that animal bones can become dispersed entirely from the archaeological context. This can partially be overcome by using faunal remains from analogous sites. Goebel (2004, 190) makes generalised statements about the hunting practices of early Upper Palaeolithic humans in Siberia during OIS3 based on faunal remains found at only a few sites. Also, although animal remains may be available detailed zooarchaeological analysis may not have been undertaken. According to Goebel (2004, 190), many early Upper Palaeolithic sites in Siberia, for example, have preserved fauna, but so far only a few have been subjected to detailed analysis. Also, even in cases where details of faunal remains associated with specific sites have been published, in many cases only a general overview is provided (e.g. Chilardi *et al.* 1996, 557; Hardy 2004, 550; Meignen *et al.* 2004, 53), and specific details regarding the exact number of faunal remains found are not presented. This can sometimes make it difficult to achieve a full and comprehensive understanding of the hunting practices of past humans. Evidently, using published literature to understand the hunting practices of past humasn can be troublesome; however, faunal remains are some of the most commonly found data associated with archaeological sites, and where present provide direct insights into the hunting practices of past humans. In Chapter 4, I use published literature documenting the remains of hunted fauna found at relevant archaeological sites, and where such data are not available, suitable analogous data are used.

Often the types of lithic raw materials used by past humans and the frequency at which they were used are commonly documented by archaeologists (e.g. Oliva 2005). However, one problem with using such data is that in many cases the frequency of lithic raw material use is documented only in general terms. This can be overcome, at least to some extent, by focusing on the most commonly used raw material as a means for understanding the lithic raw material practices of past humans. Although it is not possible to gain a full understanding of all the lithic raw materials used using this approach, in most cases lithic raw materials found in association with prehistoric hunter-gatherers are dominated by a single lithic raw material type. As a result, this approach at least provides some potential for gaining insights into the most dominant lithic raw materials used by past humans. In Chapter 4, I use published literature to gain insights into lithic raw materials used by past humans, focusing in particular on the most commonly used raw materials.

Details regarding the use of animal remains by past humans are commonly published. For example, Holgueras (2009) has documented Mousterian use of ungulate limb bones to make retouching tools at the site of Axlor (Vizcaya, Spain) during OIS3, Adams (2009, 436) has recorded the use of animal remains by Szeletians at Szeleta Cave, and in association with Aurignacian occupation at the caves of Peskõ and Istállóskõ (Bükk Mountains, Hungary), and Conrad and Bolus (2002, 345) have documented the presence of mammoth ivory used by Aurignacians at Höhle Fels (Germany) to make beads. However, one problem with using such data is that although it is generally possible to identify whether or not past humans made use of animal remains, it is often difficult to establish the frequency at which this occurred: most publications simply comment on the presence or absence of use of animal remains without providing associated quantitative details. One way to overcome this is to categorise past human use of animal remains, not in precise quantitative terms, but rather according to broad-ranging categories (e.g. low, medium, and high frequency). Although this does not offer the ideal solution, it does provide some means of using the available literature in a way that still provides insights into the frequency of past human use of animal remains. In Chapter 4, I use published literature to further understanding the use of animal remains by past humans, and the results are categorised into three broad-ranging groups (low, medium and high frequency).

Published literature also documents the frequency at which past humans depicted animals (e.g. rock art or sculpted figurines). For instance, a number of archaeologists have documented the presence of animal depictions associated with archaeological sites in Europe occupied by past humans during OIS3 (e.g. Bahn and Vertut 1988). One problem with this is that although sometimes the frequency of animal depictions is recorded, in many instances this is not the case. However, in the same way that past human use of animal remains can be grouped into broad-ranging categories, the same approach can be taken here. In Chapter 4, I use published literature to reveal the frequency of past human depictions of animals, and the results are categorised into three broad-ranging groups (low, medium and high frequency).

Insights into the number of males and females in past animal populations can be gained by examining their teeth. Teeth are some of the most commonly preserved parts of an animal (Hillson 2005, 284-285); and as some animals are sexually dimorphic and this is commonly manifest in the size of their teeth (Stiner *et al.* 1998, 84), by measuring teeth it is possible to get some understanding of the relative number of males and females within a population. In most cases, this is done by taking the maximum dimensions of the crown of the tooth (the mesiodistal crown diameter or length, and the buccolingual diameter or breadth) normally to within 0.1mm (Hillson 2005, 260, 263). However, caution has to be exercised here. Klein and Cruz-Uribe (1984, 41) commented that where there are differences in the size of particular elements of an animal due to sexual dimorphism, often the difference is subtle and the size of male and female parts of the body can overlap considerably. This point is similarly made by Hillson (2005, 268, 269) who commented that although dimorphism may be manifest in tooth size, the level of dimorphism can vary widely: tooth size can vary between members of one sex, and within and between populations. Another problem is that teeth can be difficult to measure consistently: teeth are inconsistent in shape, often rounded in form, without flat surfaces or right angles, and have few easily definable reference points (Hillson 2005, 260). As a partial solution to this problem, Hesse and Wapnish (1985, 76) suggest that measurements of teeth should be taken more than once. Also, even if a comprehensive understanding of the number of male and female animal remains can be obtained, the fact that the living population can sometimes be biased in favour of one particular sex means that further caution must be exercised when making deductions about living populations. However, it is unlikely that discrepancies in the number of male and female animals within a living population compared with the number manifest in their remains are likely to be significant (Klein and Cruz-Uribe 1984, 56; Martin 2000, 25). Clearly, when using such an approach it is wise to be careful. However, the fact that teeth are some of the most commonly found parts of an animal within archaeological and palaeontological sites, and that a number of animals do demonstrate sexual dimorphism manifest in the size of their teeth, suggests that this approach is probably the most logical to take. In Chapter 7, I measure teeth associated with specific site locations to gain insights into the relative numbers of male and female animals within a particular population.

Teeth can also provide insights into the age profile of an animal population (Hesse and Wapnish 1985, 76), and the most commonly used approach to achieve this is tooth-wear analysis. Tooth-wear analysis involves examining teeth, normally molars, for wear: since teeth become more worn as the animal gets older, by analysing the degree of tooth wear it is possible to gain an understanding of the relative age of an animal at death. In most cases, the most suitable and most commonly used teeth for constructing age profiles are cheek teeth, and in particular the permanent molars M1, M2, M3 (Klein and Cruz-Uribe 1984, 53). Often, the teeth are divided into broad-ranging categories, e.g. juvenile, prime aged, and old (Stiner *et al.* 1998); although in some cases, teeth have been used to determine the precise year of death (Debeljak 2007, 478). However, it is necessary to be cautious when using this approach. Tooth wear in animals is not necessarily consistent and depends on factors such as the overall morphology of the tooth crown, the disposition, height and depth of the cusps and fissures on the tooth, the area of occlusal surface, thickness of the enamel, defects within the tooth, chewing mechanism and diet, and age of the animal (Hillson 2005, 214, 216). Another problem is that it is unlikely that the living animal population is directly proportionate to the age profile manifested in the teeth. In normal circumstances where the animal population dies as a result of attritional processes, i.e. population members die of starvation, accidents, predation, endemic disease, and other routine attritional mortality factors, the very young and old are most affected (Klein and Cruz-Uribe 1984, 56). As a result, their teeth are likely to be more frequently found. Furthermore, the survival rate of teeth associated with different aged animals may not be the same. For instance, the teeth of very young animals may be less common as the enamel caps have not yet been fortified with dentine, and so are unlikely to survive to the same extent as teeth from older animals (Stiner *et al.* 1998, 81). Evidently, caution has to be exercised when using tooth-wear analysis to obtain insights into the age profiles of living animal populations, particularly as regards teeth associated with young animals. In Chapter 7, tooth-wear analysis is used to gain insights into the relative ages of animal populations.

It is possible to reveal insights into the dietary habits of past animals by undertaking scientific stable isotope analyses on their remains, along with using already published scientific (e.g. stable isotope and tooth abrasion) data, and available literature documenting the dietary habits of relevant modern-day animals. For instance, there are two stable isotopes that are of particular interest for revealing diet in animals (carbon and nitrogen). Carbon allows for insights regarding the particular vegetation that an animal has consumed (i.e. C3 or C4 plants) and whether the diet includes sea foods. Nitrogen provides insight into the degree to which an animal was carnivorous, or their trophic level, with higher levels indicating a more carnivorous diet and lower levels a more herbivorous diet (Peterson and Fry 1987, 300). Also, a number of archaeologists (e.g. Bocherens *et al.* 2007; Pinto *et al.* 2005) have published the results of stable isotope analysis and tooth abrasion studies on past animal populations, and these can be used together with newly obtained stable isotope data to gain further insights into the dietary habits of past animals. Furthermore, studies on the dietary behaviours of analogous modern-day species, such as those published by Garneau *et al.* (2008) and McLoughlin *et al.* (2002) can be used to substantiate or refute results obtained through other sources. However, it is wise to be cautious when using this approach. For instance, in order to be able to retrieve stable isotope samples from bones it is necessary that there is sufficient collagen remaining within the bone; however, collagen in bones slowly degrades and undergoes replacement, therefore sometimes making it difficult to obtain samples with sufficient collagen, and this is particularly problematic in older bones (Peterson and Fry 1987, 312). A further point is that stable carbon and nitrogen isotope values obtained from animal remains do not necessarily represent a direct correlation with diet: there are numerous factors that can affect nitrogen stable isotope values in animals, such as the age (Keeling and Nelson 2001), metabolic processes, hibernation, dormancy, starvation, (Bocherens *et al.* 2006; Stiner *et al.* 1998), and weaning (Bocherens *et al.* 1994). Also, using the dietary habits of modern-day species as a direct analogue for past animals can be problematic: even animals of the same species living in similar environments can have different dietary habits. In this book, an attempt was made to undertake a stable isotope study on faunal remains; however, the initial pilot study indicated that collagen levels in the animal bones were generally not sufficient to warrant a more comprehensive investigation, and so this was not take any further (Appendix 2). In Chapters 3 and 5, insights into the dietary habits of past animals are gained using a combination of published scientific studies and available literature regarding the dietary habits of analogous contemporary faunal species.

It is possible to gain some understanding of the appearance of past animals by using published literature documenting descriptions of their remains, associated prehistoric art, and the appearance of contemporary animals. Illustrators of prehistoric animals such as Rinaldino (see Diedrich 2008, 830) use fossil remains, prehistoric art, and knowledge of the anatomical features of contemporary animals as a means for understanding and depicting past animals (Rinaldino pers. comm.). Guthrie (2005, 53-55) commented that we can compare portrayals in prehistoric art of animals such as horses, red deer, ibex, reindeer, and chamoix to the appearance of their modern-day analogues, and in cases where animals are now extinct, bones, skulls, horns, antlers, and tusks assist in helping to understand the appearance of such animals. However, although it is possible to gain some understanding of the appearance of past animals using such data, caution must be exercised. Although animal remains are useful for understanding the appearance of animals, especially in cases when the species is now extinct, the fossil record rarely preserves details of external appearance, and may be a poor clue to the actual appearance of the animal when alive. For example, living bison species show much more variation in their coat patterns and hair colour than in their skeletons (Guthrie 1990, 114). Also, modern-day species do not necessarily have the same appearance as animals of the same species alive in the past. Guthrie (2005, 55) commented that some prehistoric art makes it evident that extinct animals such as giant deer, woolly rhino, woolly mammoth, steppe bison, and reindeer did not look the same as their nearest living analogues. Clearly, there are problems in attempting to understand the appearance of past animals using such an approach; however, by using published literature associated with animal remains, prehistoric art, and contemporary animals in combination probably represents the most reliable way of gaining at least some understanding of the appearance of past animals. In Chapter 3, I use these data to gain an understanding of the appearance of past animals.

Some understanding of the perceptual mechanisms of past animals can be gained using published literature documenting the perceptual mechanisms of analogous contemporary animals. A number of researchers have published details regarding the perceptual capabilities of modern-day faunal species (Burton 1970; Nachtigall *et al.* 2007). Thus, by using these data some comprehension of the perceptual mechanisms of animals alive in the past can be gained. The main problem with this approach, though, is that there is no way of knowing if the perceptual mechanisms of past animals were the same as those belonging to similar animals alive today. However, although this is evidently the case, the perceptual mechanisms of most individual species are largely similar. For instance, dogs normally have an acute sense of smell and particularly sensitive hearing, and this was probable also the case for close relatives of these animals (e.g. wolf) alive in the past. Therefore, in Chapter 3 I use published literature to gain an understanding of the perceptual mechanisms of past animals.

It is possible to gain some understanding of the ethology (e.g. social affiliation and predator/anti-predator behaviour) of past animals by using published literature on analogous contemporary species. Berger *et al.* (2001, 131) commented that elements of modern-day animal behaviour can be used as a direct analogue for past animals. However, the process of using modern-day animal behaviour as a means for understanding the behaviour of animals alive in the past, especially those without direct living relatives, can be difficult. Berger *et al.* (2001, 131) made the point that reconstructing the behaviour of extinct mammals can be especially challenging in the absence of living relatives. A similar point is made by Martin (2000, 13), who commented that there are difficulties in directly imposing modern data onto the past and in assuming species-specific behaviour. However, although it is acknowledged that using the ethology of modern-day species to understand associated behaviours of past animals, especially extinct animals, is troublesome, so long as caution is exercised and contemporary data are used as a guide rather than as a transposition of past animal ethology, then using such an approach is likely to provide useful and meaningful insights into the social behaviours of past animals. In Chapter 3, I use published literature regarding the ethology of contemporary animals as a means for understanding the social affiliation and predator/anti-predator behaviours of past animals.

Of the best tools for undertaking spatial analyses are GIS (Geographic Information Systems). GIS are computer dependent technologies that involve three main components: software, hardware, and people. They provide a range of tools that help people interact with and understand spatial information (Connolly and Lake 2006, 11, 15; Heywood *et al.* 1998, 19), allow for different (digital and non-digital) spatial data to be overlaid and utilised, and new data synthesised. Today, there are numerous GIS software packages available, some of which are free to use (GRASS GIS) (http://www.grass.itc.it), and other more commercial packages such as Idrisi (http://www.clarklabs.org), ArcGIS (http://www.esri.com), and MapInfo (http://www.mapinfo. com) (Connolly and Lake 2006, 15). GIS work by using spatial (locational) and non-spatial (descriptive) data, representative of some element of the real world (Connolly and Lake 2006, 14). Spatial data are provided to GIS using spatial data models (vectors and rasters) (Gillings and Wise 1990, 10; Heywood *et al.* 1998, 40). Vector spatial data models use pairs of co-ordinates, in the form of points, lines, or polygons (areas), to locate objects in GIS, and are particularly good for handling information relating to discrete objects, but are not good at representing more complex and fluid spatial entities such as vegetation (Aldenderfer 1996, 4; Gillings and Wise 1990, 10, 11, 32; Heywood *et al.* 1998, 50). Raster data models use cells or pixels to display spatial data, with each cell assigned a value representative of the status of that thing at that location (e.g. altitude). Raster data are good for dealing with continuously varying spatial entities such as vegetation or elevation, but are not so good at dealing with discrete, bounded things such as buildings (Connolly and Lake 2006, 27; Gillings and Wise 1990, 12). One of the most common types of raster data models used is a Digital Elevation Model (DEM) (Gillings and Wise 1990, 34). One of the main advantages of using GIS is that they allow the user to deal with and depict large amounts of spatial data, and most importantly, allow new spatial data to be easily synthesised (Connolly and Lake 2006, 16, 17). However, GIS do have their disadvantages: they can be particularly time-consuming to learn, and the software can sometimes be expensive to obtain. GIS have also been criticised as simply being a means of producing pretty maps, technologically deterministic, and results obtained using GIS are often considered as being representative of the 'truth' (Connolly and Lake 2006, 1; Fisher 1999, 8; Gaffney and van Leusen 1995, 377; Kvamme 1995, 7, 123). Also, when using GIS it is necessary to undertake a number of procedural tasks in order to ensure that spatial data are consistent and results obtained are reliable. Spatial data must be correctly georeferenced, i.e. a uniform projection method (e.g. UTM – Universal Transverse Mercator Geographic Coordinate System) and co-ordinate system (e.g. WGS84) must be used, and all data must be at the same resolution and scale (Gillings and Wise 1990, 13-16). Here, GIS software package ArcGIS9.2 is used from Chapter 5 onwards, and all data are projected using the same projection method (UTM) and mapped using the same coordinate system (WGS84).

There are a number of ways in which researchers have been able to gain insights into past climates e.g. deep-sea cores, volcanic cores, and marine sediment (van Andel 2003, 13-15); however, one of the most common is to use published stable isotope ratios ($\delta^{18}O$ – $^{18}O/^{16}O$) associated with the Greenland ice core (GISP2). The Greenland ice core provides a picture of the last 100,000 years of climate change from the perspective of the Greenland ice cap: as the local air temperature changes, so too do the $\delta^{18}O$ values within the ice. As a result, ice cores, and associated $\delta^{18}O$ levels within the ice, provide a useful proxy for understanding the character of the climate in the past. Moreover, results of these data and summaries of their associated impact on past climates are commonly published (van Andel 2003, 12). The usefulness of the Greenland ice core has been acknowledged by Mellars (1996, 25) who commented, 'through detailed studies of the oxygen-isotope ratios in the cores it has been possible to identify climatic oscillations between 25,000 and 60,000 BP', and van Andel (2003, 13) made the point that the Greenland ice core has a near annual resolution for the same period. Also, researchers such as Barren *et al.* (2003) have used $\delta^{18}O$ values obtained from the Greenland ice core to reconstruct the climate of Europe during OIS3. However, some caution has to be exercised. It is unlikely that past climates were spatially homogenous: the climate in Greenland is unlikely to have been the same as the climate in and around the Mediterranean region for example. Mellars (1996, 26) commented that many of the shorter changes, apparent within the Greenland ice core, are not apparent in many of the pollen sequences across Europe. However, available palynological data suggest that large-scale climatic changes evident with the $\delta^{18}O$ values of the Greenland ice core are representative of large-scale changes in climate throughout Europe. Van Andel (2003, 16) remarked that although short-term oscillations may not be evident within palynological records, long-term temperature trends evident in the Greenland ice core agree well with pollen-based climate records from many parts of Europe. Also, Mellars (1996, 26) commented that the longer and major climatic episodes evident in the Greenland ice core are clearly represented in many pollen sequences, such as those from La Grande Pile and Les Echets in eastern France. Thus, in Chapter 4 insights into past climates are gained using published data associated with Greenland ice core (GISP2) $\delta^{18}O$ values and related published summaries on past climatic change, and in some cases published palynological literature are used as corroborative data.

Overviews and graphical representations of the main geological formations associated with different regions of Europe are commonly published (e.g. Golonka and Picha 2006, 54; Suk *et al.* 1984, 17; Svoboda *et al.* 1996, 202) and these can be used as a guide for understanding the character of the major geological formations that existed in the past. Although the main geological formations of a region are not static over time, such changes occur over millions of years (Putman 1964, 432), and so are unlikely to be of significance within the time-scale of this project. In Chapter 4, I use published literature to gain insights into the character of regional geology.

Using a Digital Elevation Model (DEM) together with GIS is a good way of obtaining some understanding of the topography of a region: digital elevation data for most regions of the world are freely available online (http://seamless.gov.us), and these can be easily incorporated directly into most GIS. Moreover, some GIS (e.g. ArcGIS9.2) provide semi-automated algorithms that enable the user to derive associated slope and aspect maps directly from a DEM, and will normally complete this procedure in a matter of minutes or even seconds. Derived topographic maps can be classified according to broad-scale elevation, slope, and aspect categories and displayed visually. Archaeologists have acknowledged the usefulness of GIS and DEMs for understanding and visually representing variation in topography (Wheatley and Gillings 2002, 123). However, some caution does have to be exercised. Connolly and Lake (2006, 101) commented that the accuracy of the DEM is essential if it is to be used for spatial analysis; however, in many cases the resolution of a DEM is such that many of the topographic features apparent in the actual landscape being represented are not apparent in digital format. DEMs of Europe provided by USGS (http://seamless.gov.us) typically have a resolution of 90 m x 90 m, and as a result are unlikely accurately to represent areas of the landscape that have significant and acute changes in topography. It is possible to overcome this problem to some extent by using data with a higher resolution, that is if such data are available; however, increases in the resolution of data normally result in increases in the computational time of the computer, but this also depends on the processing capabilities of the computer and the overall surface area of the region represented by the digital data. In most cases, a compromise is reached between the resolution of the DEM and computational time. Also, DEMs acquired from online sources often contain errors (i.e. pixels with no data), and these can have serious consequences for the results of the overall project and subsequent inferences.

'No-data' cells can be replaced with pseudo values in most GIS by reclassifying the erroneous cells based on the values of surrounding cells. This is not an ideal solution as the pseudo values will inevitably not be a true representation of actual elevation values; however, replacing the 'no-data' cells with pseudo values, albeit estimated, is a better solution than leaving the erroneous cells without data. Also, when using DEMs to derive topographic elements such as slope and aspect, it is important that the geographical extents of the original DEM are at least 10% larger than the extents of the target study region: many operations in GIS rely on surrounding cells (Wheatley and Gillings 2002, 120); however, where surrounding cells are absent (i.e. at the edge of the DEM) this can mean that erroneous results are obtained. According to Connolly and Lake (2006, 91), the 'edge effect' as it is known, can affect up to 10% of the outer region of the digital map. In Chapter 5, insights into topography are gained using a DEM, at least 10% larger than the target study area, obtained from online sources (http://seamless.gov.us). Erroneous 'no-data' cells are reclassified and assigned pseudo values based on surrounding cell values, and slope and aspect algorithms are used together with a DEM to derive slope and aspect models.

Insights into the geographical distribution of palaeochannels (ancient rivers) can be gained using a DEM together with GIS and paper and online maps. In some GIS (e.g. ArcGIS9.2) it is possible to derive drainage channels using a DEM and an in-built algorithm (Drainage Network). Wheatley and Gillings (2002, 121) made the point that modules to undertake simple simulations of hydrological processes are available in several commercial GIS that can be used to estimate the course of ancient rivers and stream networks. Moreover, when drainage channels are derived in this way they are sometimes presented with associated inter-connectivity values, representative of their inter-connectedness to each other, and so to some extent representative of the relative size of the derived drainage channel. Using present-day and historic maps of the study region, together with digitally derived drainage channels and associated inter-connectedness values, it is possible to identify and isolate drainage channels that correspond with the most significant rivers in the study area: digitally derived drainage channels can be filtered according to the inter-connectedness value by interrogating the GIS until remaining drainage channels are visually similar to the main rivers depicted on paper maps. It is necessary, however, to be cautious here. First of all, there is no guarantee that the routes taken by modern-day rivers take the same route as ancient rivers: over time, both the direction of a river and its flow rate can change significantly. Furthermore, there is a high degree of user discretion involved in this approach: visual comparisons are made between digitally derived drainage channels and river channels on paper maps, and so it is inevitable that some degree of error will occur. Given the evident problems, however, this approach does represent a good way of gaining an approximate understanding of the geographical distribution of major palaeochannels associated with a particular region. Thus, in Chapter 5 palaeochannels are derived using the approach described.

Together with the Buffer algorithm available in ArcGIS9.2, already derived palaeochannels can be used to comprehend the geographical distribution of palaeofloodplains. Artificial buffer zones of a set width (e.g. 10 km) around the derived palaeochannels can be created using the Buffer algorithm, outside of which the presence of floodplains are not considered a possibility. The mapped buffer zones are interrogated using GIS to determine their elevation in comparison with the height of the associated palaeochannel, plus an estimated value to take into account the depth of the palaeochannel when flooding. Buffer regions that are higher than the proposed height of the flooding palaeochannels are considered as potential dry areas, and other areas are considered as potential floodplain regions. Potential palaeofloodplain areas are further interrogated to determine their inter-connectedness, in terms of water flow, to the related derived palaeochannel. Any regions that are not connected to the associated palaeochannel are not considered potential floodplain regions. Similar approaches have been used by archaeologists such as Gillings (1995, 72) and Rhodda (2005). However, there are some problems with this approach. For instance, it does not take into account factors such as the local soil conditions and associated drainage potential, existing level of the water table, and seasonality and associated changes in the extents of the floodplains. However, although it is acknowledged that this approach is by no means comprehensive, it does represent a good way of gaining a general understanding of the overall distribution and extents of ancient floodplain regions. In Chapter 5, palaeofloodplains are derived using the approach described above.

An understanding of the geographical distribution of palaeovegetation can be accomplished using published literature documenting the regional distribution of palaeovegetation associated with climatic endmembers (i.e. extreme warm and cold climatic periods), together with palynological data, and newly created digital topographic and palaeohydrological (palaeochannels and palaeofloodplains) maps, and GIS. A number of publications document the regional character of past vegetation, as is the case with Huntley and Allen (2003) who outlined major vegetation types in Europe for extreme warm and cold climates during OIS3. Also, a number of researchers have published palynological data documenting the character and abundance of vegetation associated with more specific geographical locales (Svobodová 1991a; 1991b). Using both sets of data it is possible to gain an understanding of the character of the vegetation on a regional and more local scale. Also, GIS and topographic and palaeohydrological maps can be used to identify 'neutral regions', i.e. areas unlikely to greatly affect the composition and abundance of the generic vegetation-type associated with warm and cold periods. All other regions can be assigned hospitality coefficients based on their expected hospitality/hostility towards different types of vegetation. Using this approach, it is possible to produce palaeovegetation maps for summer and winter periods associated with warm and cold climates. It is, however, necessary to be cautious. Modelling palaeovegetation using climatic endmembers in such a binary fashion does not take into account the character of the palaeovegetation during other intermediate periods: it is probable that for most of the time the character of vegetation in the past was probably neither 'warm' nor 'cold' but somewhere in between. Also, the spatial distribution of vegetation would not have been characterised by discrete bounded areas as is depicted when using such an approach, but would have been 'fuzzy' without clear-cut spatial divisions. Clearly, although the approach outlined above will not result in an exact facsimile of the vegetation that existed in the past, it does represent one way of gaining a general understanding of the character of past vegetation during contrasting climatic conditions, and provides some means of taking into account local topographical and hydrological effects on vegetation. In Chapter 4, I use available literature to gain insights into the general palaeovegetation of the study region during warm and cold climates, and use palynological reports associated with specific, spatially relevant locales to corroborate these data. In Chapter 5, general palaeovegetation types associated with warm and cold climates are offset using the approach described above to take into account topographical and hydrological effects on vegetation, with palaeovegetation maps being produced for summer and winter periods during warm and cold climates.

Friction maps or 'cost of passage maps' provide values corresponding to different areas of a region that represent how difficult they are to traverse (Connolly and Lake 2006, 215), and these can be used to understand the energy expenditure experienced by past humans and animals. By using GIS together with already created digital slope, palaeohydrology, and palaeovegetation maps, numerical values can be assigned to areas in accordance with the energy costs that each element of the environment is expected to demand when being traversed. By using the Weighted Sums algorithm available in ArcGIS9.2, it is possible geographically to combine friction values associated with each element (slope, palaeohydrology, and vegetation) in accordance with their expected overall weighting in terms of energy expenditure (e.g. slope 90% and vegetation 10%). Thus, using this approach it is possible to produce friction maps for summer and winter periods during warm and cold climates that represent to some extent the energy expenditure for humans and animals when moving through the landscape. There are, however, a number of issues with this approach. For instance, it is possible that past humans used a means of transportation other than walking (e.g. sledges and boats). However, although it is acknowledged that this may have been the case, there are no existing archaeological data within the context of Europe during OIS3 demonstrating this. Another problem is that this approach does not consider the fact that the impact of the environment on human and animal energy expenditure would probably not have been the same. However, this approach is not designed to provide bespoke details of absolute energy expenditure values for humans and animals: friction maps are designed to provide insights into the *relative* energy costs incurred by humans and animals. As the relative impact of the environment on human and animal energy expenditure is similar (e.g. going up a slope is likely to increase energy expenditure for both humans and animals), friction maps therefore provide a good way to understand the general and relative energy expenditure experienced. In Chapter 5, friction maps are created for summer and winter periods during warm and cold climates using the approach described above.

Published data together with GIS can be used to map human and animal site locations. For instance, van Andel *et al.* (2003a, 53-56) have published a comprehensive series of coordinate locations associated with sites repeatedly occupied by humans and animals in Europe during OIS3. Moreover, coordinate location can be cross-referenced using online digital maps (e.g. http://www.mapy.cz). Once coordinate locations of human and animal sites have been obtained they can be converted to UTM format using an online coordinate converter (http://www.fcc.gov/mb/audio/bickel/DDDMMSS-decimal.html), and stored digitally using Microsoft Excel. The coordinate locations can then be imported directly into GIS and used to produce vector point locations representative of the case study site locations in real space. However, some caution does have to be exercised. For instance, published coordinate locations of sites may be inaccurate or inconsistent as is the case with the coordinate locations published for Kůlna Cave (Moravia, Czech Republic) (van Andel *et al.* 2003a, 53; and Davies *et al.* 2003, 207). However, by cross-referencing published site coordinate locations with other sources of geographical data such as online maps, any errors can normally be accounted for and corrected. In Chapter 5, human and animal site locations are digitally mapped using a combination of published sources (van Andel *et al.* 2003a) and online digital maps (http://www.mapy.cz).

An understanding of the distribution patterns of lithic raw materials commonly procured by past humans can be achieved using published articles and reports together with topographic, topological (http://www.mapy.cz), and geological maps (http://mapy.geology.cz/website/ne_tisk/viewer2.htm), and GIS. Often, the locations of lithic raw materials commonly used by past humans are described in reports and articles (Bahn 1982, 249; Burke 2006, 519; and Zilhão 2001, 604), and in some cases, details regarding the specific locations of lithic raw materials are available online (http://www.flintsource.net). Such data can be used together with available geological, topographical and in some cases topological maps to identify the probable location of lithic raw material outcrops. Coordinate locations can be logged and stored digitally using Microsoft Excel and subsequently imported into GIS, where they can be used to create either point or polygon (area) locations, representative of the lithic raw material outcrop in real space. Other archaeologists have used a similar approach. For instance, Mellars (1996, 142, 143, 161) has used a published geological cross-section of the central Perigord region of south-western France, together with published accounts of detailed prospection of the region, to identify areas where Neanderthals and AMH procured lithic raw materials. However, one problem with this is that although the location of lithic raw material outcrops may be described within existing literature, the description of the location may be vague and made only with reference to general topographic, hydrological, or topological features. In these cases, although a general idea of the source location can be achieved by cross-referencing the description with relevant maps of the area, because the description is made in general rather than specific terms (i.e. specific coordinate point locations are not provided), a precise digital representation of the location of the outcrop is not possible. Also, Mellars (1996, 144-145) commented that it is not always possible to locate the precise location of lithic raw materials, as in many cases raw materials are not distinctive and are either mostly uniform over large areas, or show localised and erratic varieties within a single source that make exact provenancing impossible. Another point is that the lithic raw materials may not have been procured from the same location as the original geological outcrop: rocks can be moved to secondary or tertiary locations by the effects of weathering and rivers (Růžíčková *et al.* 2001). In Chapter 5, locations of relevant lithic raw material outcrops are obtained using a combination of published literature, together with associated topographic, hydrological, topological, and geological maps. Coordinate locations are obtained, stored digitally using Microsoft Excel, and imported into ArcGIS9.2 where they are used to map the locations of lithic raw material outcrops in point or polygon format.

Together, Least-cost Surfaces, already mapped site locations, friction maps, GIS, and published literature can be used to gain some understanding of the home range of past humans and animals. Least-cost Surfaces is an algorithm available in ArcGIS9.2 that creates a model of the landscape in which each part of the surface is assigned a cost that effectively represents effort needed to reach that point from another predetermined point (e.g. site location) (Bell and Lock 2000, 86). Using the already created friction maps, derived least-cost surfaces can be offset according to the expected energy costs incurred when moving away from the site location. Also, a number of researchers have published details regarding the home range extents of contemporary and recent hunter-gatherers and different animal species (Binford 1980; Kelly 1995, 133; Walton *et al.* 2001), and these can be used to define maximum radii of the derived least-cost surface areas. In the case of animals, home ranges can be further offset according to demographic affiliation, and climate. Similar approaches have been used by archaeologists such as Bailey *et al.* (1983) and Hunt (1992, 286) who have mapped the home ranges of past humans, and Boydston *et al.* (2003) who have used GIS to map the home range of hyenas according to their particular demographic species grouping. One problem with this approach, however, is that in the case of animals there is no way of knowing if the site around which the home range is based is at the centre or at the periphery of the home range (Miracle pers. comm.). This can be overcome simply by doubling the radius of the expected home range: even if the animal site is located on the periphery of the home range area, it can be no more than a diameter's distance from all other regions of the home range (Fig. 2.1). In Chapter 6, home ranges for past humans are derived using the approach described above. In the case of animals and their associated home range, a similar approach is used in Chapter 7; however, home ranges are derived for different demographic groups, offset for different climatic periods, and the site around which the home range is calculated is considered to be at the periphery rather than at the centre of the home range.

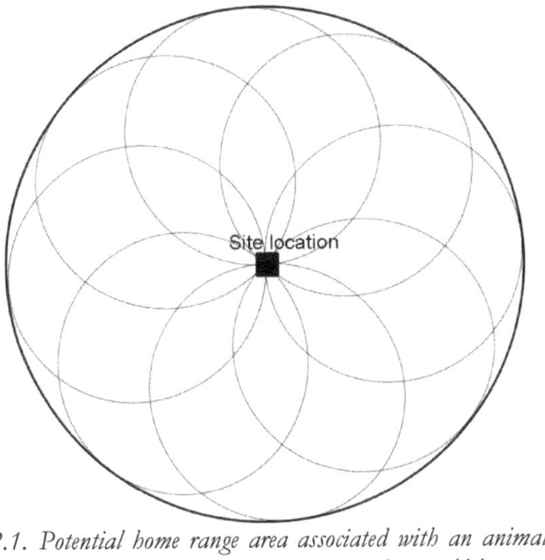

Fig. 2.1. Potential home range area associated with an animal site. Small circles are individual home range areas that could be associated with a site location, and the large circle represents the overall area that could potentially be associated with the same site location.

Insights into the pathways taken by past humans can be achieved using Least-cost Corridors, and already mapped site locations and lithic raw material outcrops. Least-cost Corridors is a semi-automated computer algorithm available in ArcGIS9.2 that calculates multiple return routes (i.e. corridors) from one place to another within a particular set energy requirement. For instance, if it takes 100 units of energy to move from point A to point B using the most energy-efficient pathway, and a corridor width of 12% is allowed for, then the outer perimeter of the derived least-cost corridor represents a return journey which takes up 112 units of energy. Thus, by mapping least-cost corridors between site locations and lithic raw material outcrops it is possible to get some understanding of the pathways taken by past humans (Fig. 2.2). However, one of the main problems with this approach is that it only considers energy and does not take into account factors such as knowledge, learning (Rockman 2003), memory (Golledge 2003), social networks (Kelly 2003, 51), visibility and proximity to water (Steele and Rockman 2003). This point is similarly made by Bell and Lock (2000, 87) who commented that these sorts of approaches do not take into account cultural attractions or deterrents that may affect movement across the landscape. However, although this approach does not consider such factors, energy inevitably had a significant impact on the movement and associated pathways taken by past humans, and this approach at least provides some means of understanding this. In Chapter 6, human pathways are mapped using Least-cost Corridors between already mapped human settlement sites and lithic raw material outcrop locations.

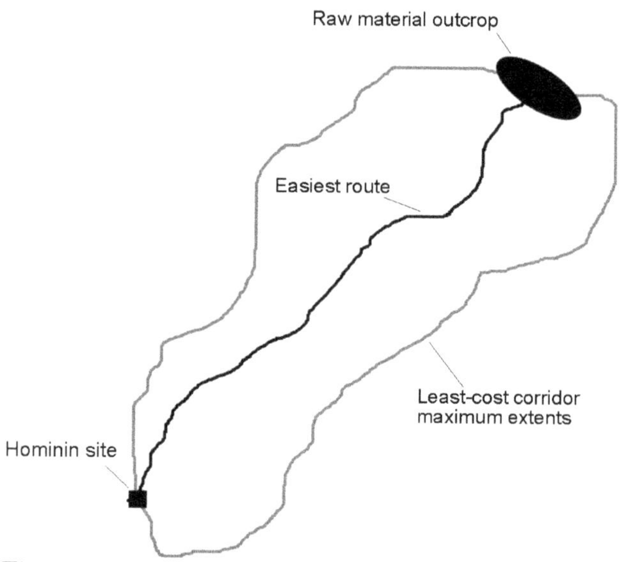

Fig. 2.2. Hypothetical least-cost corridor, showing a derived least-cost easiest route between two points (Hominin site and Raw material outcrop), and maximum extents of the associated derived least-cost corridor.

The preferred habitats of past animals can be mapped using already obtained dietary data, together with published literature regarding habitat preference of relevant contemporary animals, together with GIS, topographic, palaeovegetation, and palaeohydrology maps. For instance, by using the expected dietary habits of past animals, along with published accounts of habitat preferences based on topographic and hydrological features, and the already created digital palaeovegetation, topographical, and palaeohydrological maps it is possible to interrogate GIS and identify and highlight preferred habitat regions associated with specific faunal species. A similar approach has been used by Burke *et al.* (2008, 899) who have used published ecological studies of the modern-day Asiatic ass and its associated habitat preference, together with GIS, and digital slope and palaeovegetation maps to model the distribution patterns of now extinct species of horse (*Equus hydruntinus*) in the Crimean Peninsula during the Middle Palaeolithic. However, it is necessary to be cautious. Musil (2003, 167) made the point that 'it is not always possible to compare the present ecological adaptations of individual species with those of past species'. Although it is acknowledged that using the ecological behaviours of contemporary species is not an ideal solution for understanding the preferred habitats of past animals, it does at least provide one way of gaining some understanding of the habitats that past animals were associated with. In Chapter 5 and Chapter 7, the preferred habitats of relevant faunal species are mapped using this approach.

Some understanding of the prey species diversity of past animals can be obtained by overlaying preferred habitat distribution maps for each relevant faunal species. This can be done easily within ArcGIS9.2 by assigning a value of '1' to areas associated with each animal species habitat. Contemporaneous fauna, i.e. those animals associated with the same climatic and seasonal periods, can then be geographically summed to produce aggregate maps corresponding to the number of species associated with each location. A similar technique has been used by Steele and Rockman (2003) who have mapped the species diversity of modern vertebrate species in Wyoming (North America). One of the most obvious problems with this approach, however, is the largely uniform manner in which species diversity is represented. The geographical distribution of animals is, inevitably, spatially and temporally heterogeneous and as a result such maps are unlikely to accurately represent the realities of associated species diversity. However, although this is acknowledged it is not possible to recreate such spatial and temporal dynamism, and the best that can be done is to predict general patterns of diversity. In Chapter 5, species diversity maps are produced for relevant faunal species for summer and winter periods during cold and warm climates.

Population densities of animals alive in the past can be derived using already published literature associated with analogous animal populations, and expected home range surface areas. A number of researchers have published details of the population demographics of modern-day animals (Craighead and Mitchell 1982; and Knight and Eberhardt 1985), and so by dividing the expected relative population numbers associated with each demographic group by the related home range surface area, some understanding of the relative population densities can be achieved. However, there are some problems with this approach. For instance, home ranges associated with different demographic groups are likely to overlap, and so the relative densities are likely to be cumulative. Another point is that changes in the surface area and overall extents of the home range due to energy constraints are likely to change the relative densities of associated animal populations: as the surface area of the home range decreases due to increased energy expenditure, relative population density will increase. In Chapter 7, relative animal population densities are derived using relative population numbers associated with different demographic groups of analogous modern-day faunal species, and these are divided by the surface areas of associated home ranges. Changes in home range surface area due to energy constraints are not taken into account.

The frequency of interactions between past humans and animals can be derived by reconstructing human and animal distribution patterns using the approaches outlined above, and overlaying climatically and seasonally contemporaneous results and identifing areas of overlap. Moreover, potential areas of interaction occurring close to human sites, in areas with high species diversity, high animal population density levels, and in preferred faunal habitat regions are considered as manifesting relatively high frequency interactions. Interactions occurring in regions further from human sites, in areas with low species diversity, low animal population density levels, and in regions outside of preferred faunal habitat areas are considered as correlating with relatively low frequency interactions. Thus by using this approach it is possible to identify where and when (climate and season) potential interactions are likely to occur, and the potential frequency of these interactions in broad terms (e.g. low, medium, and high frequency). However, one obvious problems with this is that it does not consider other factors that affect interactions between humans and animals and human ability to perceive animals, such as the character of the environment (topography, vegetation, and weather), and characteristics of animals like their appearance, smell, the sound they make, whether or not they congregate in groups or spend time alone, and their predator/anti-predator behaviour. However, although it is possible to take some of these factors into account by employing GIS and other computer modelling software to undertake tasks such as visibility analysis and to create virtual reality models of human perception, this is beyond the scope of this project. Thus, in Chapter 8 I identify the frequency of potential interaction using the approach outlined above.

By assigning quantitative values to the expected potential frequency of interaction, and different types of interaction contexts based on the expected manifestation of apparent super-human and human-like qualities of animals, some understanding of the significance of interactions for past humans can be achieved. Where interactions occur, assigned values associated with the expected frequency of interaction and manifestations of particular types of behaviour are multiplied. The results are grouped into broad categories (low, medium, and high), and can be thought of as representative of the potential that interactions with animals offered past humans for establishing a sense of relational cohesion, and thus to some extent, indicative of the significance of those interactions for past human sense of identity. The results can be compared with associated human use of relevant animal remains and depictions of associated animals. If expected opportunities for relational cohesion through interactions are inversely correlated with the frequency of use of animal remains and depictions of animals, then interactions can be considered as playing a central role in the lives of past humans: relational cohesion is established through regular interactions, and the use of animal remains and depictions of animals are used to compensate for a lack of relational cohesion arising through infrequent interactions. On the other hand, no difference in the use of animal remains and depictions of animals compared with the expected opportunity to establish a sense of relational cohesion through interaction, indicates that such interactions are not important in establishing a sense of relational cohesion and so are not important in associated past human sense of identity. One of the main problems with this approach, however, is that although most past humans used animals remains to some extent, they did not all depict animals. For instance, there are no indications that Neanderthals depicted animals in any way. This raises the question of whether such an approach is made on a level playing field, i.e. did all past humans manifest their understanding of the world in material objects in the same general way. If this is not the case then apparent difference in the significance of interactions for past humans, manifest in their use of animal remains and/or depictions of animals, may simply reveal differences in world views, and may not be to do with how often they interact with things. However, this does not provide a reason not to undertake such comparisons: if a lack of animal depictions coincides with a localised lifestyle, for example, then this might actually corroborate the working hypothesis, and suggest that people living a local life are embedded in their world to such an extent that they interact with important animals on a daily basis, and therefore have no need to depict animals: they are governed primarily by their interactions with the perceived world, and not by memories of the perceived world. Another issue is that quantifying past human sense of identity in such a way tends to portray past humans and their understanding of the world in over-simplistic terms. However, such an approach is not meant to provide an absolute understanding of past humans and their associated sense of identity, but instead is intended to provide some general insights into the degree to which interactions were involved in past humans and their understanding of the world. In Chapter 9, I use the approach described above to gain some understanding of the significance of interactions for past humans and their sense of identity.

Summary and conclusions

In this chapter, I have attempted to highlight the main issues governing human interactions with animals, and the frequency of those interactions. I then attempted to discuss how the issues most relevant for this book can be approached, problems associated with these approaches, and where in the remaining part of the book such approaches are applied. In the next chapter, I discuss the target animal species chosen for this book (cave bears), and examine those issues most important for understanding its potential interactions with humans, the frequency of those interactions, and the significance of those interactions for the people that encountered it.

Chapter 3: Cave bears (*Ursus spelaeus* Rosenmüller 1794)

Introduction

The objective of this chapter is to provide an overview of cave bears. I discuss the history of cave-bear research, cave-bear phylogeny, evidence for their existence in the past, their geographical and chronological distribution, relevant aspects of their ecology, biology, physiology, ethology, and evidence of human use of their remains and depictions of cave bears.

Cave bears

The first records of cave bears were made by Central European cave explorers during the 15th century, the first drawings of cave bears were made in 1673 and 1676 (Fig. 3.1), and the first cave-bear monograph was published in 1774 by Esper. In 1794 Rosenmüller, with the help of J. Chr. Heinroth, published his doctoral thesis, looking specifically at the skull of the cave bear, in 1794 he identified cave bears as a separate bear species, and in 1804 subsequently published his report on the post cranial bones of cave bears (Rosendahl *et al.* 2005, 89). During the 19th and early 20th centuries, cave-bear research was mostly focused on attempting to identify the cave bear as a religious, cult-like figure for Neanderthal people. The presence of cave-bear remains, sometimes found in unusual positional contexts, together with a number of human-made stone tools, led many researchers during this period to believe that specialised cave-bear hunting and veneration was taking place (Kurtén 1976, 9, 10). This is the case with Bächler, who in 1921 and 1940 published his findings of apparently aligned and encased cave-bear bones found at Draconloch Cave in Switzerland. According to Bächler, the deposition and alignment of these bones was the result of human behaviour (Kurtén 1976, 83). Towards the middle and latter part of the 20th century, researchers became more interested in trying better to understand issues such as the origins and extinction of cave bears, their biological characteristics, the environment in which they lived, and their geographical and chronological distribution. Kurtén published articles (1955; 1958) and a book (1976) looking specifically at these issues, and in 1980 Musil (1980a; 1980b; and 1980c) published an extensive geographical and chronological overview of cave-bear finds. Over the last 15 years or so, the majority of research has been disseminated through journal articles, most of which have been associated with the annual Cave Bear Symposium. The research focus of these articles has varied tremendously from genetics and cave-bear dietary and hibernation habits (e.g. Bocherens *et al.* 2007; Hofreiter *et al.* 2002) to cave-bear-human relations, the latter of which have focused mainly on presenting evidence of cave bears being hunted by humans rather than any kind of cave-bear veneration (e.g. Münzel and Conrad 2004; Stiner 1999; Wojtal 2007). Other important publications include Gargett's (1996) spatial investigation of cave-bear remains found at Pod hradem Cave in the Czech Republic, an edited volume by Nagel (2005) that includes several important papers on cave bears, and other works such as Miracle's (2007) publication on the faunal remains of Krapina Cave (Croatia), which contains a substantial and detailed overview of cave-bear remains recovered from this site.

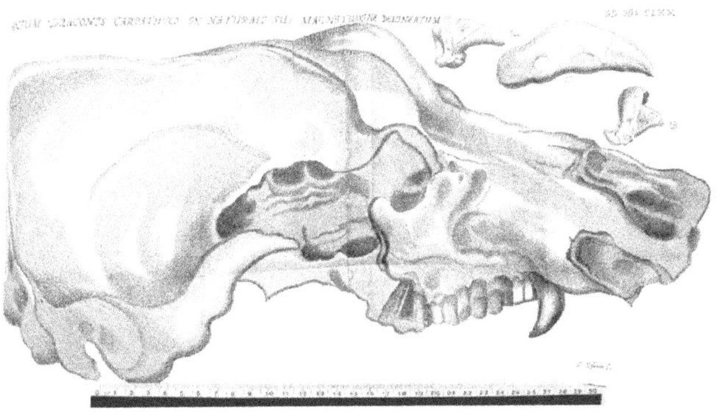

Fig. 3.1. 17th-century picture of a cave-bear skull. The heading reads "Skull of a dragon from the Carpathian Mountains in natural size". (After Paterson Hain and Vollgnad 1676, reproduced by Rosendahl et al. 2005, 93.)

The cave bear belongs to the terrestrial order Carnivora, and to the second superfamily Arctoidea (dog branch), which includes dogs (Canidae), racoons (Procyonidae), stoats and weasels (Mustelidae), and bears (Ursidae) (Ward and Kynaston 1995, 35-36) (Fig. 3.2). The cave bear can be traced back to around 5 million years ago to *Ursus minimus*. At around 2.5 million years ago, *Ursus minimus* evolved into *Ursus etruscus*, and 1.5-1 million years ago *Ursus etruscus* diverged into three separate lineages, *Ursus thibetanus* and *Ursus americanus*, *Ursus arctos*, and *Ursus sivini*. *Ursus thibetanus* and *Ursus americanus* are equivalent to the present-day Himalayan and American black bears, and *Ursus arctos*, also known as the grizzly bear in North America and Canada, presently occupies regions of Europe and Northern America. *Ursus sivini* is the direct antecedent of the cave bear, and evolved into *Ursus deningeri* about 700,000 years ago, and about 400,000 years later into the cave bear (Kurtén 1976, 37-45) (Fig. 3.3).

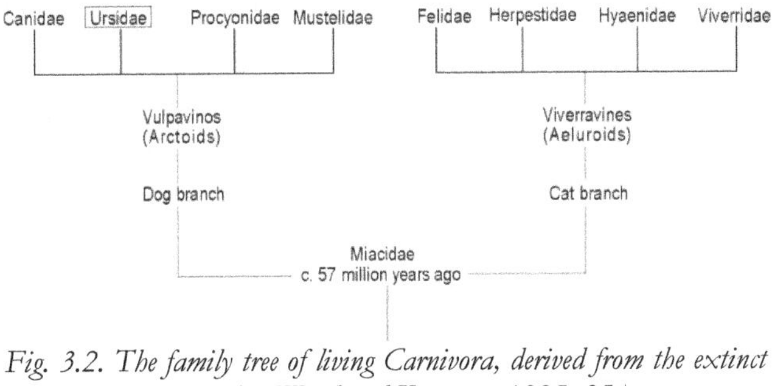

Fig. 3.2. The family tree of living Carnivora, derived from the extinct family Miacidae. (After Ward and Kynaston 1995, 35.)

Cave bears occupied most of Eurasia, from northwest Spain across central Europe to the Urals, and from

Belgium and the Harz region of Germany to as far south as Italy and Greece, and as far southeast as Crimea (Pacher and Stuart 2008, 2). Although there are claims that cave-bear remains have been found in Britain (Kurtén 1976, 61), it is likely that these are the remains of *Ursus deningeri* and not *Ursus spelaeus* (Rosenmüller 1794) (Pacher and Stuart 2008, 2). Most evidence indicates that cave-bear remains originate from deposits dating from *c.* 300,000 years ago (Gargett 1996, 38) to *c.* 10,000 years ago (Kurtén 1976; Musil 1980a; Musil 1980b; Musil 1980c). Pacher and Stuart (2008, 2), however, have argued that much of these data, in particular more recent material, are most probably associated with discrepancies due to the stratigraphic provenance and the misidentification of brown bears for cave bears. According to Pacher and Stuart (2008, 191), results from a range of old and new radiocarbon dates indicate that cave bears became extinct about 24,000 ^{14}C yr BP (*c.* 27,800 Cal. BP). Although Pacher and Stuart's (2008) work does suggest that cave bears may possibly have become extinct earlier than previously thought, they have sampled only a relatively small number of cave-bear remains, and to date the large majority of data still indicate that cave bears became extinct around 10,000 years ago (Musil 1980a; Musil 1980b; Musil 1980c).

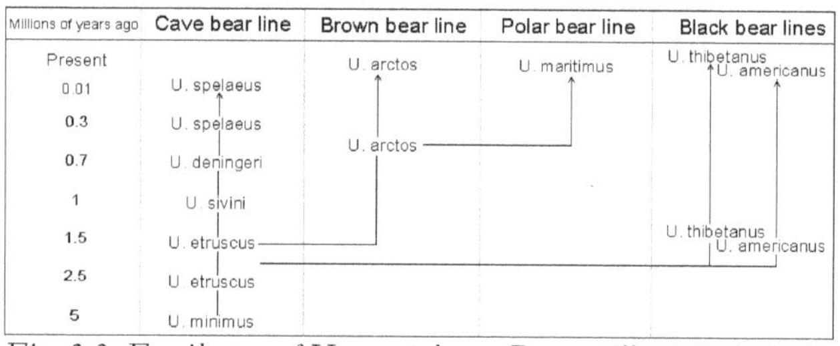

Fig. 3.3. Family tree of Ursus spelaeus (Rosenmüller 1794). (After Kurtén 1976, 45.)

Most of the data demonstrating the presence of cave bears in the past are cave-bear remains (bones and teeth); these are mostly recovered from cave sites, and largely thought to be the remnants of cave bears that died a natural death while hibernating (Stiner *et al.* 1996, 304). This is the case at sites such as Chauvet Cave in France, where a large number of cave-bear remains have been recovered (Chauvet *et al.* 1996, 51), and at the cave site of Dragon Cave in Switzerland, where the remains of between 30,000 – 50,000 individual cave bears have found (Kurtén 1976, 74). In some cases, cave-bear remains are associated with open-air sites, as is the case at sites such as Biedensteg near Bad Wildungen in northern Germany (Diedrich 2006), but these represent only a very small proportion of the overall number of cave-bear bones found. Where cave-bear remains are found, these are normally biased in favour of remains associated with young and old cave bears. Stiner *et al.* (1996, 309) commented that it is likely that the remains of juveniles and old adults are likely to be relatively more common than prime-aged adults as they have higher mortality rates. Also, Debeljak (2007, 478) showed that the age profile of cave-bear teeth found at Mokrica Cave in northern Slovenia demonstrated relatively more young and older cave-bear teeth compared with prime-aged adult teeth. However, cave-bear remains are not the only data indicating the presence of these animals in the past. Trace fossil evidence, such as footprints, scratch marks, polished rock surfaces (*Bärenschliffe*), bear dens, and kidney stones also indicate the possible presence of cave bears (Rosendahl and Döppes 2006, 241). Bear footprints have been found in the French cave sites of Chauvet Cave, claw marks have been discovered at sites such as Toirano Cave in Italy (Kurtén 1976, 11), and *Bärenschliffe*, thought to be the result of cave bears rubbing along the walls of caves (Fig. 3.4), have been found at Vogelherd Cave, Bärenhöhle, Charlottenhöhle, Hohle Fels, and Kleine Scheuer, all located in the Swabian Alb, Germany. Also, apparent cave-bear dens, areas dug out by cave bears for sleeping and hibernating, have been found in Neu-Laubenstein-Bärenhöhle in the Chiemgau Alps in Germany, and at Jubiläumshöhle and Groβeklingerberg Höhle in the Franconian Alb in Germany (Rosendahl and Döppes 2006, 246).

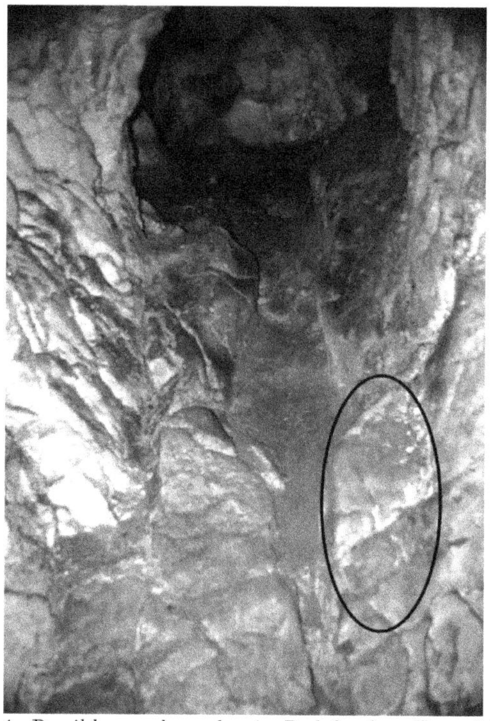
Fig. 3.4. Possible cave-bear den in Pod hradem Cave (Moravia, Czech Republic) with suspected areas of Bärenschliffe highlighted. (Photograph taken by author.)

Although cave bears are taxonomically classified as carnivores, most data indicate that they were probably omnivores, with the vast majority of their diet made up of vegetation, such as roots, tubers, and deciduous plants (Bocherens *et al.* 1997, 370; Pinto *et al.* 2005). This is similarly the case with some extant bear species: the diet of modern-day brown bears in Cantabrian Spain is largely made up of grass, pulpy fruits, dry fruits, and other vegetable matter, but they also consume mammals and insects, and have been known to kill large animals such as cows (Pinto *et al.* 2005, 603). Cave bears probably also had the ability to adapt their diet depending on their surroundings. Richards *et al.* (2008, 600, 602-603) commented that cave bears were carnivores, herbivores or omnivores depending upon the availability of food resources, body size requirements, competition, seasonality and temperature. Modern-day brown bears can vary their diet from almost completely vegetarian, to being mostly carnivorous, and eating mainly fish and/or ungulate meat (Richards *et al.* 2008, 602-603). Also, Japanese black bears have been shown to vary their diet throughout the year, relying more on fruits during the summer months, and nuts during the autumn months (Takahashi *et al.* 2008, 472). In some cases, cave bears also had the capability and inclination to eat other cave bears: Quilés *et al.* (2006, 927) have commented that there is a high degree of cannibalism apparent on cave-bear bones recovered from the cave site of Peştera Cu Oase in Romania.

Cave bears, like most stable mammal populations, would probably have had a living demographic that was biased towards younger cave bears, with possibly more adult female cave bears than males, with about 1/3 of the female population pregnant at any one time, and about another 1/3 with yearlings as is the case with modern-day brown bears inhabiting Yellowstone National Park (Knight and Eberhardt 1985, 327-334) (Fig. 3.5) (Table 3.1). The cave-bear population would have been mostly sedentary, inhabiting largely the same areas of landscape for most of their lives. Strontium stable isotope analysis of cave-bear teeth from Slovakia indicate that cave bears underwent little or no migration during their lifetime (Ábelová 2006, 126). Also, cave bears probably hibernated for most of the winter months (Nelson *et al.* 1998, 177), most likely for a period of up to about six months, between the end of November and the beginning of May, as is the case with most contemporary brown and black bear species (Dahle and Swenson 2003b, 663; Mitchell *et al.* 2005, 164). In some cases, however, cave bears may have become active during hibernation periods: some modern-day black bears have been known to exit their dens during the winter months after being disturbed by humans, and some bears can become active during the winter months if they have been unable to store sufficient food supplies during the summer (Goodrich and Berger 1994, 105). In these circumstances, however, bears are unlikely to have remained outside of their den for a substantial period of time, and if they did venture out, would probably have remained within the local vicinity of the den (Stiner *et al.* 1996, 307). During the summer period, when cave bears would have been active in the landscape, their distribution was probably mostly governed by the distribution of suitable foodstuffs: studies of grizzly bears within the Canadian Arctic demonstrated that their distribution corresponds well with the spatial and temporal availability of food in the landscape (McLoughlin *et al.* 2002, 106), and Garneau *et al.* (2008, 48) showed that black bears move in accordance with the growth of preferred plant foods. The summer was probably split into two main periods; a mating period and a post-mating period. The mating period probably began as soon as cave bears emerged from hibernation, around the beginning of May, and ended close to July, and the post-mating period most

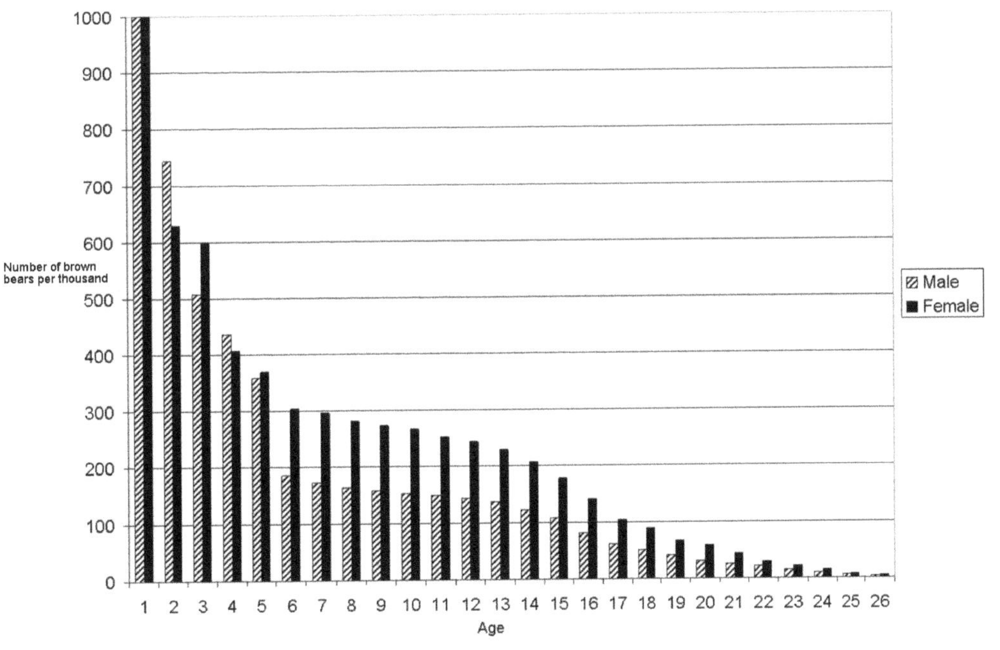

Fig. 3.5. Living age profile of male and female brown bears from Yellowstone National Park, showing the number of individuals alive per thousand brown bears over the course of a twenty-six year lifespan. (After Craighead and Mitchell 1982, 539.)

All females	Lone females	Females with yearlings (1-2 years olds)	Females with cubs	All males	Lone males
50% of total population	33% of all females - 16.67% of total population	33% of all females - 16.67% of total population	33% of all females - 16.67% of total population	50% of total population	50% of total population

Table 3.1. Expected demographic make-up for a cave-bear living population, based on modern-day brown bear populations, assuming 33% pregnancy rate, and 50% survival rate for cave-bear cubs. (After Knight and Eberhardt 1985, 327-334.)

likely continued until the beginning of the hibernation period, around October time, as is the case with modern-day brown bears (Wilson and Ruff 1999, 158).

The home range of cave bears probably varied significantly as a result of sex, age, social affiliation, mating habits, and climate. Male cave bears probably had larger home ranges than females, as is the case with modern-day black bears studied in North America: the home ranges of male black bears are consistently a quarter to a third larger than those of female black bears (Powell *et al.* 1997, 14), and the home range of male brown bears from two regions in central and northern Scandinavia was shown to be larger than associated female brown bears: on average, male brown bears had a home range 3-4.9 times larger than female brown bears (Dahle and Swenson 2003a, 329-333; see also Preatoni *et al.* 2005, 195). Modern-day female brown bears with cubs and yearlings have been shown to have home ranges consistently smaller than lone females, with the home range of females with cubs between 25% and 76% smaller than those of brown bears with yearlings (Dahle and Swenson 2003b, 664). Also, oestrous female brown bears often use larger (up to 62%) home ranges during the mating period than during the post-mating period, and male brown bears often reduce their home range during the post-mating season by up to 53%. Conversely, female brown bears with cubs commonly reduce their home range during the mating season, and can have up to a 64% smaller home range during this period than during the post-mating season, and the home range of female brown bears with yearlings remains relatively constant both during and after the mating period (Dahle and Swenson 2003b, 660-666; Preatoni *et al.* 2005, 196) (Fig. 3.6). Furthermore, as vegetation is likely to be more abundant and of a higher quality during warmer climatic periods, cave bears would have been able to reach their calorific and nutrient requirements within a smaller home range than during colder climatic periods when food was scarcer and of a lower quality. The average home range of modern-day adult black bears in south-eastern North America can be in the tens of square kilometres, whereas in the more barren landscape of Labrador in Canada, black bears can have home ranges in hundreds of square kilometres (Powell *et al.* 1997, 14). Also, male brown bears studied within arctic-tundra conditions had a home range of 8000-3500 km^2, and female brown bears had a home range of 2500-1200 km^2, but in warm-boreal conditions male brown bears had home ranges of 2400-800 km^2 and females had home ranges of 450-300 km^2. Moreover, changes in climate are likely to impact more on the home range of male cave bears than female bears: the maximum difference between the home range of studied male brown bears inhabiting arctic-tundra and warm-boreal conditions was 6100 km^2, whereas the maximum difference in the home range of female brown bears inhabiting similar environmental conditions was only 1450 km^2 (Dahle and Swenson 2003a, 333) (Fig. 3.7) (Table 3.2).

Cave bear gait was probably that of an ambling motion, with the hind and fore feet on the same side moving together, creating a side-to-side rolling action, the same as modern-day bears (Palliser 1859, 181). Cave bears were also capable of moving at fast speeds and able to demonstrate motor skills with significant agility. Modern-day brown bears are able to run almost as fast as a horse, and are able to rise on to their hind feet when reaching for food or offering combat, and can sit upright with the hind legs extended while keeping the forelegs free (Storer and Tevis 1996, 44) to undertake particularly agile tasks with their paws (Mills 1919, 253).

Cave bears were comparable with, if not bigger than, the largest of all extant bear species (Christiansen 1999, 93; Kurtén 1976, 25) and at their maximum probably weighed up to 1000 kg (Pacher and Stuart 2008, 2). They were sexually dimorphic, with males generally much larger than females, probably weighing about half as much (Kurtén 1976, 65-82), and these differences were manifested in particular anatomical features such as their canine teeth: male cave-bear canines are significantly larger than those of female (Baryshnikov *et al.* 2003, 111). Also, cave bears probably changed in size according to season, being much larger prior to hibernation than after hibernation: large brown bears may consume as much as 41 kg of food a day prior to hibernation, and gain around 2-3 kg of fat in the same period (Ward and Kynaston 1995, 104), resulting in 6-10 inches of fat being deposited under the skin (Domico and Newman 1988, 47). As a result, brown bears can weigh as much as double before hibernation in comparison with after the hibernation period (Ramsay 1993, 62; Ward and Kynaston 1995, 109). Like modern-day brown bears, cave bears had a dish-like face (Domico and Newman 1988, 6), probably short, rounded ears, large canine teeth, and a forehead that had a step-like profile (Fig. 3.8), rather than sloping as is the case with contemporary bears species such as brown and black bears (Kurtén 1976, 14). Individual cave bears, however, probably varied in their appearance: cave-bear skulls recovered from Dragon Cave in Switzerland varied significantly in morphological character, and have been

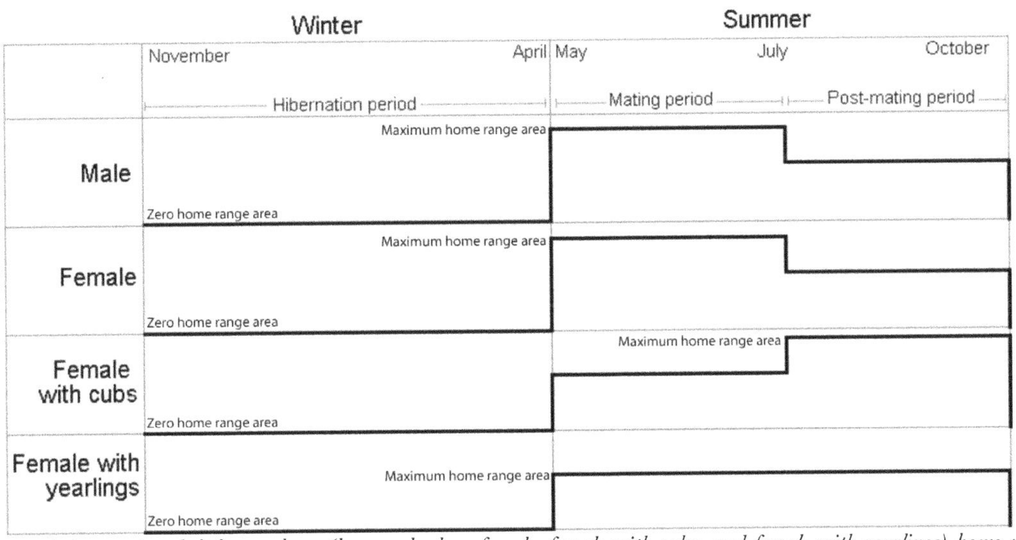

Fig. 3.6. Seasonal changes in adult brown bear (lone male, lone female, female with cubs, and female with yearlings) home range areas as inferred from contemporary brown bear data. (After Dahle and Swenson 2003a.)

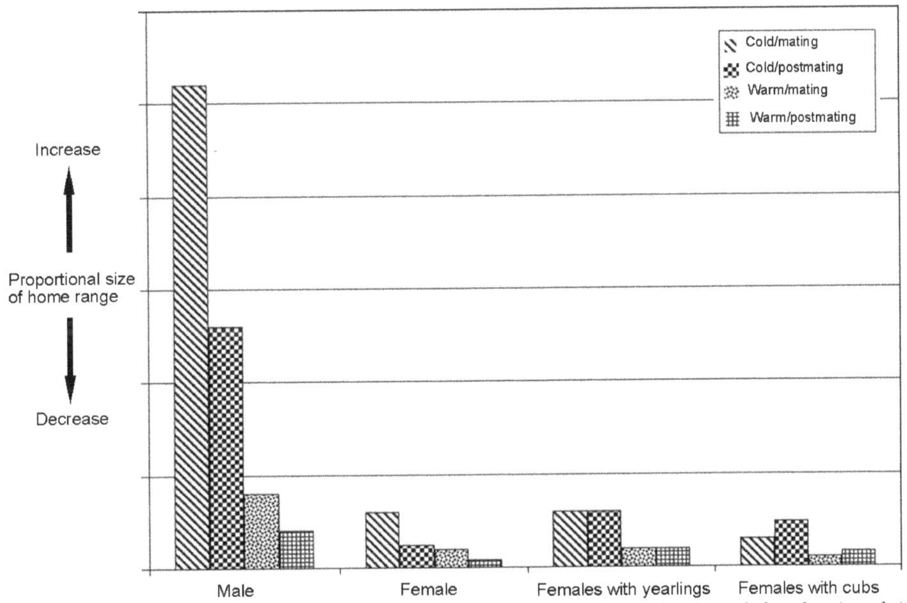

Fig. 3.7. Relative changes in adult brown bear (lone male, lone female, female with yearlings, and female wit cubs) home range areas, according to mating habits (i.e. mating/postmating season) and climate (i.e. cold/warm climates). Ratios are derived from figures given for brown bear home ranges by Dahle and Swenson 2003a.

described as large, small, domed, flat, elongated and greyhound-like, and short, broad, and pug-like in character (Kurtén 1976, 65). Similar to modern-day bears, cave bears had five toes on each foot, each toe equipped with a curving, non-retractable claw, as is the case with modern-day bears (Domico and Newman 1988, 6). Also, it is likely that the colour of cave bears was not uniform: the pelage of modern-day brown bears can be cream, tan, mousey, cinnamon, or golden yellow in colour (Barker 1956, 147; Storer and Tevis 1996, 35), and the coat of American black bears can be brown, reddish-brown, blue-black, and even white (Ward and Kynaston 1995, 55).

Cave-bear eyesight was probably relatively good, and at least as good as humans'. Bears generally have relatively good eyesight, and like humans and other animals such as dogs and elephants, have both rods and cones in the retinas of their eyes, and so are able to see in both light and dark conditions (Burton 1970, 88). They probably also had relatively good hearing, most likely as good as human hearing. Although there have been few studies undertaken on the hearing capabilities of bears, recent research on polar bears have shown them to possess the ability to hear sounds from a wide frequency-range, and with the ability to hear a wider and higher frequency range of sounds than humans (Nachtigall *et al.* 2007, 1121). The olfactory senses of cave bears, like modern-day bears, were most likely very good, and considerably more sensitive than human olfactory capabilities. Modern-day brown bears have been known to be able to detect food refuse buried under 2-3 feet of soil (Domico and Newman 1988, 166), and often utilise their acute

	Warm climate		Cold climate	
	Mating season	**Post-mating season**	**Mating season**	**Post-mating season**
Male	2500	1272	8000	4240
Female	450	283.5	2500	1575
Female with cubs	61.86	137.48	343.69	763.75
Female with yearlings	211.5	133.25	1175	740.25

Table 3.2. Home range areas of brown bears (km2), as inferred from changes in brown bear (male, female, female with yearlings, and female with cubs) home ranges, during mating and post-mating seasons, and cold and warm climates. (After Dahle and Swenson 2003a.)

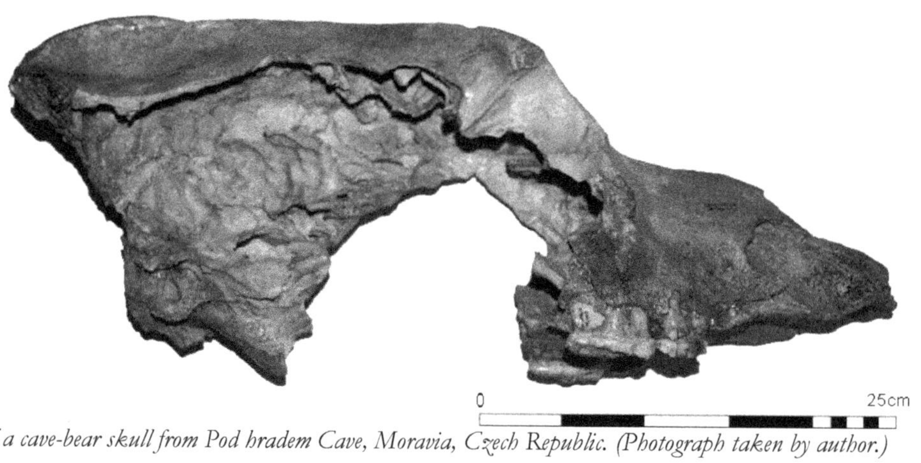

Fig. 3.8. Profile of a cave-bear skull from Pod hradem Cave, Moravia, Czech Republic. (Photograph taken by author.)

sense of smell to detect and communicate social information such as their sex and age (MacDonald 1985, 620; Ward and Kynaston 1995, 122).

Cave bears were probably mostly solitary animals. Records of most modern-day bear species suggest that they are generally secretive, spending large amounts of time by themselves (Argenti and Mazza 2006, 1552), and most recorded human encounters with brown bears have mainly been with lone bears (McMillion 1998). For some of the time, however, cave bears would have spent time in groups of two or more. Male and female cave bears would have needed to rendezvous in order to mate, and probably remained within each other's company for some time after: modern-day male brown bears are normally associated with females for up to a 10-week period during the mating period (Wilson and Ruff 1999, 158). Also, it is likely that young cave bears would have remained with their mothers, probably for a period of up to two or three years. According to Liden and Angerbjörn (1999, 1782), cave-bear mothers would have nursed their cubs for the first two winters, and young brown bears will typically remain with their mother for a period of up to three years. Also, brown bear sibling groups can remain together for a year or two after leaving their mothers, and some closely related modern-day brown bear females and their young have been known to form short-lived foraging groups. Groups of cave bears may also have been found in areas that were particularly rich in food. Recent observations of brown bears demonstrated that groups of up to 50 bears can accumulate in food-rich areas (Bunnell and McCann 1993, 91-92).

On the whole cave bears probably would have tried to avoid contact with people. Although polar bears regularly prey on humans (Derocher and Stirling 1993, 85), generally most modern-day bears will avoid contact with humans if possible, and would rather flee than attack people (Bunnell and McCann 1993, 90). In some circumstances, however, it is likely that cave bears would have become particularly defensive if met with the presence of a person. Modern-day brown bears can become defensive during periods of feeding, if accompanying young, or when surprised at close range, and may attack during such an encounter (Domico and Newman 1988, 171, 165). Most recorded attacks on humans by bears have been associated with the protection of offspring and surprise encounters (Derocher and Stirling 1993, 85), and recorded incidents involving brown bear attacks on people within North American national parks were mostly associated with female brown bears with young (Jope 1985, 32). Also, both modern-day black and brown bears can be effective predators, and if in a predatory behavioural state may attack humans without provocation (Derocher and Stirling 1993, 84): predatory brown bears can attack people, and American black bears can attack a human with the intention of eating them (Domico and Newman 1988, 171). If cave bears did intend to attack a person, whether defensively or offensively, it is likely that this would have been evident in their behaviour and appearance: when alerted, brown bears may rear up on their hind legs, and if aggravated may make woofing sounds or champ their teeth, and brown bears that are about to charge will lay their ears back (Domico and Newman 1988, 170).

There is little or no indication that cave-bear remains were used by humans to make objects or tools of any kind, but they were used on rare occasions for other purposes and sometimes depicted. Polished cave-bear penis bones have been recovered from caves in Germany, and the teeth of cave bears were used by humans to make pendants: cut and polished cave-bear canines have been recovered from the cave site of Deszczowa in Poland associated with Upper Palaeolithic Aurignacian deposits. An apparent 'flute', made from the shaft of a juvenile cave-bear femur, was found in Divje Babe I Cave in Slovenia in association with Neanderthals remains (Lau *et al.* 2004, 515); however, Morely (2006, 330) has shown that the holes are more likely the result of carnivore activity, and were probably not made by hominins.

What appears to be a depiction of a cave bear has been found on the interior walls of Chauvet Cave (Bocherens *et al.* 2006; Chauvet *et al.* 1996), and there are profiles of two cave-bear heads in Lascaux Cave and La Madeleine Cave, and engravings of what are thought to be cave bears in Combarelles Cave and Teyjat Cave, all of which are located in the Dordogne region of France. Engravings of possible cave bears have been found in the cave of La Colombière in Ain, France, and an engraving and painting of a cave bear has been found in a cave near Santander in northern Spain (Kurtén 1976, 92-94). Also, what appears to be a life-size headless clay sculpture of a cave bear was found in the cave of Montespan in the French Pyrenees, with the skull of a cave bear, now lost, found nearby (Kurtén 1976, 95-96), and carved ivory figurines have been found at sites such as Geissenklösterle in Germany resembling cave bears (Münzel 2001, 448).

Summary and conclusions

In this chapter, I have attempted to provide an overview of cave bears. I began by discussing the history of cave-bear research, their phylogeny, their geographical and chronological distribution, and evidence of their existence. I discussed important elements of cave-bear ecology, biology, physiology, and ethology, and finally discussed evidence of human use of cave-bear remains and depictions of cave bears. In summary, most evidence of cave bears originates from cave deposits in Europe associated with periods between *c.* 300,000 and 10,000 years ago. They were mostly sedentary animals, eating mostly vegetation, mating between April and June, and hibernating for much of the winter period. Although large animals, cave bears were probably able to move at fast speeds, were particularly dextrous creatures, with a particularly sensitive sense of smell. If confronted by humans it is likely they would have fled, but this would have depended largely on the context of the interaction. Although archaeological data indicate that humans utilised cave-bear remains and depicted their form, generally this is rare. In the next chapter, as well as discussing other important issues, I look in more detail at cave bears in relation to the case study context.

Chapter 4: Cave bears and hominins in the Czech Republic during OIS3

Introduction

The objective of this chapter is to provide an overview and essential background to the case study, and the case study sites chosen for this book. I first discuss the geographical and chronological study context, and the study region's geology, topography, hydrology, climate, and flora and fauna. I discuss hominins in the study region, looking at available evidence for their presence, geographical and chronological distribution, raw material procurement, hunting practices, use of animal remains, and depictions or sculptures of animals. I discuss cave bears, looking at evidence of their presence, geographical and chronological distribution, and demography. Finally, I provide an overview of the case study sites chosen for this project.

Background

This case study is focused in Moravia and Silesia, located in eastern Czech Republic, Central Europe, during OIS3 (*c.* 60,000 – 24,000 Cal. years ago). Bordered by Poland to the north and northeast, the Slovak Republic to the east and southeast, Austria to the south, and Bohemia (western Czech Republic) and Germany to the west, the study region measures about 250 km in an east-west direction, and about 180 km in a north-south direction (Fig. 4.1). OIS3 is preceded by OIS4 (*c.* 71,000 – 60,000 years ago) and followed by OIS2 (*c.* 24,000 – 13,000 Cal. years ago), and subsequently by the present-day climatic epoch of OIS1 or Holocene (*c.* 13,000 Cal. years ago – present) (Fig. 4.2).

There are three main units of geological structure within the study region; the Bohemian Massif, the Brno Unit, and the Western Carpathians (Suk *et al.* 1984, 17). The Bohemian Massif occupies the western and northern parts of the study region, the Brno Unit covers central, northern and southern areas of the study area in a cross-like

Fig. 4.1. Map of the study area and surrounding regions. (After http://www.europeetravel.com.)

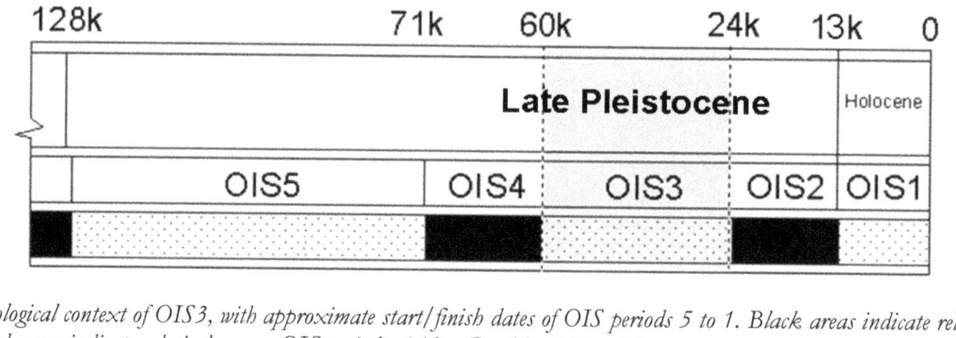

Fig. 4.2. Chronological context of OIS3, with approximate start/finish dates of OIS periods 5 to 1. Black areas indicate relatively cold OIS periods and dotted areas indicate relatively warm OIS periods. (After Gamble 1993, 43.)

formation, and the Western Carpathians are represented in the eastern, central and southern parts of the study region. Other geological areas of significance are karstic, limestone outcrop regions such as the Moravian and Štramberk karsts, which are home to more than 1000 caves (Musil 2007, 17) (Fig. 4.3).

The topography of the study region corresponds with overall changes in major geological formations. Low-lying (100-400 metres above sea level – m.a.s.l.) regions (e.g. Moravian Gate, Vyškov Gate and the Napajedla Gate) mostly follow the geographical areas of the Brno Unit, and the mid-elevated regions (400-700 m.a.s.l.) mostly correspond with areas of the Bohemian Massif and the lower regions of the Western Carpathians (e.g. Chřby Mountains). The highest regions (700-1300 m.a.s.l.) are located mostly towards the north of the study area, in the Jeseníký Mountains and the eastern regions of the study area on the northern edge of the Carpathian Mountains (Fig. 4.4).

The main rivers within the study region largely follow the topography, beginning in the smaller tributaries of the mid and higher elevated areas of the Western Carpathians and the Bohemian Massif, and flowing down towards the most southerly-central regions of the study area (Fig. 4.5). This is the case with the river Morava, which originates from smaller tributaries located in the Jeseníký Mountains, and flows through the low-lying areas of the northern and central parts of the study region, on through the Napajedla Gate and down towards the southern-central regions and on into Austria. The river Bečva begins in the Western Carpathians, flows east, and eventually joins the Morava. The river Dyje begins in the smaller tributaries of the Bohemian Massif in the extreme southwest of the study region, and flows east, joining the Morava at the most southerly point of the study area. Also, the river Svratka begins in the south-westerly regions of the Bohemian Massif, flows south and joins the Dyje about 20 km from the most southern edge of the study region. Although most rivers generally flow towards the south-central regions of the study area, some major rivers flow in a north-easterly direction into Poland. This is the case for the river Odra, which begins in the south-easterly regions of the Jeseníký Mountains, and flows south-east and then north-east through the Moravian Gate and on into Poland.

The present-day climate of the study region is mixed. The mean annual temperature is about 45°F (7°C), but temperatures can reach as high as 91°F (33°C) during the mid-summer, and as low as 1°F (-17°C) during the winter months. Annual precipitation normally ranges between 18 inches during the driest periods in February, and more than 60 inches during the wettest periods in July (http://www.southtravels.com). Most of the indigenous forests of the Czech Republic have now been artificially replaced by spruce woods, but more than 1/3 still remain. Indigenous forests typical for the Czech Republic include a mixture of oaks, firs and spruce, with the most common wild fauna including hare, marmot, otter, marten, mink, pheasant, partridge, wild boar, red deer, ducks and geese. Less common but also present are eagles, vultures and herons, and in some northern parts of Moravia wolves and brown bears can be found (http://www.czech.cz).

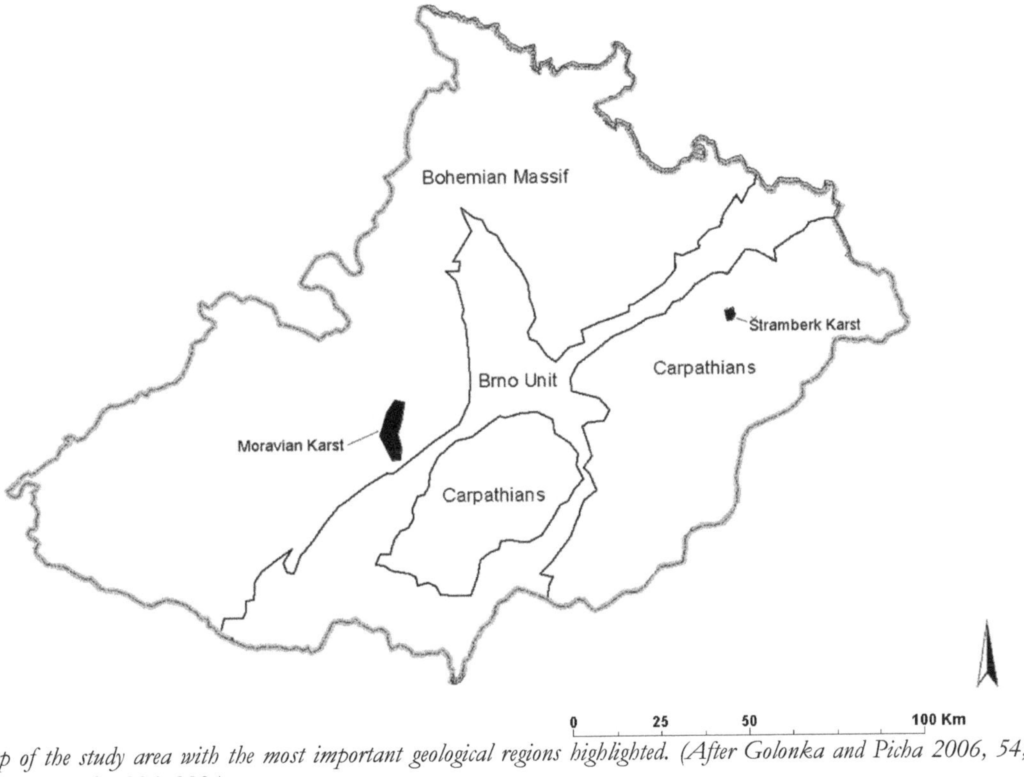

Fig. 4.3. Map of the study area with the most important geological regions highlighted. (After Golonka and Picha 2006, 54; Suk et al. 1984, 17; Svoboda et al. 1996, 202.)

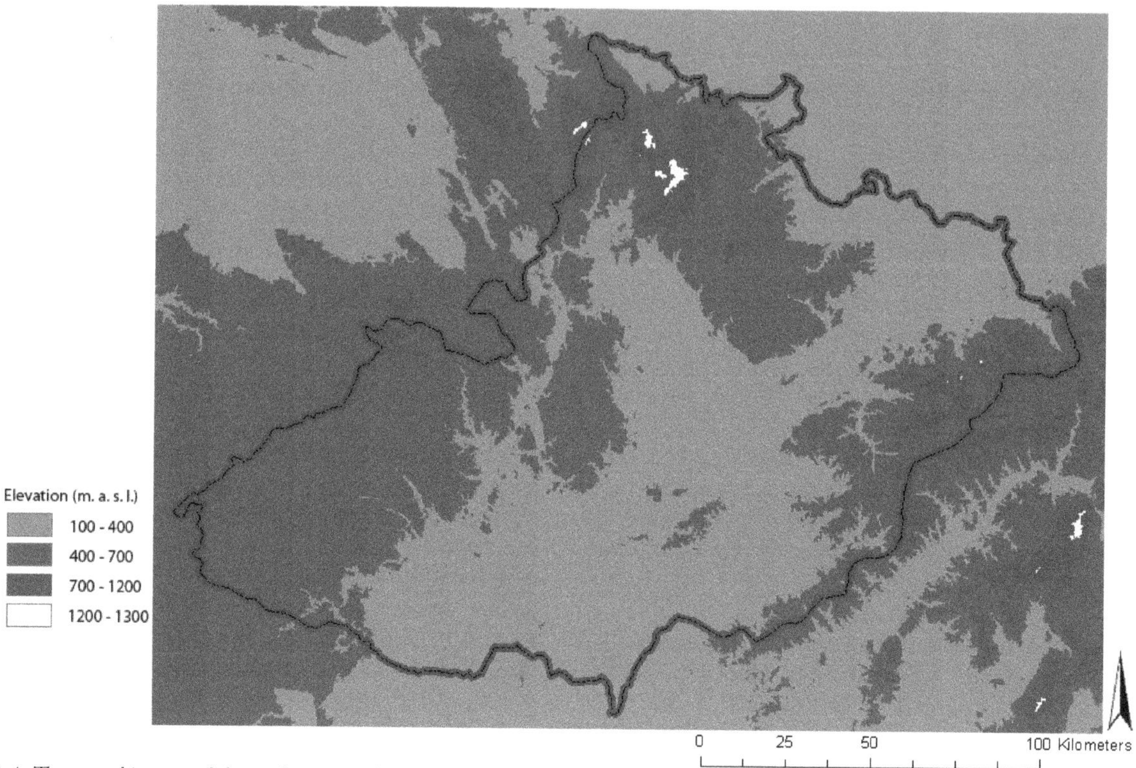

Fig. 4.4. Topographic map of the study region, showing elevation in metres above sea level (m. a. s. l.).

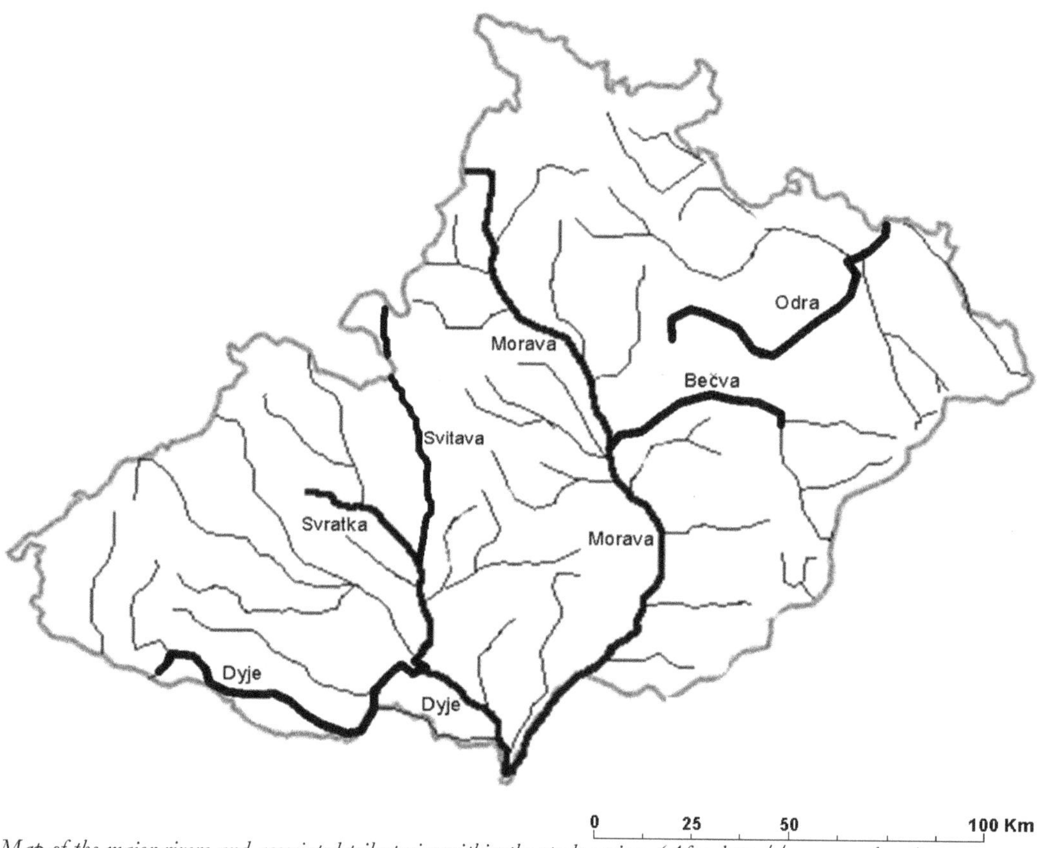

Fig. 4.5. Map of the major rivers and associated tributaries within the study region. (After http://www.czech.cz.)

In comparison with OIS4 and OIS2, OIS3 represents an overall warming of the climate (van Andel 2003, 16), and can be divided into three main climatic episodes: an initial warm phase (60,000 – 44,000 Cal. BP), a transitional phase (44,000 – 38,000 Cal. BP), and a later cold phase (38,000 – 24,000 Cal. BP). The initial warm phase included a sequence of fairly mild climatic events that were occasionally interrupted by brief and not very cold periods, during which time annual mean temperatures reached about 7°C (van Andel 2003, 11). The transitional phase consisted of many cold events that were separated by a few brief warm episodes, and concluded by a relatively warm period (Musil 2003, 170; van Andel 2003, 11); cold events during the transitional phase had average annual temperatures of around 0-2°C (van Andel 2003, 11). The last climatic phase of OIS3 was, on average, much colder than previous periods, and included episodes that were similar to that of the Last Glacial Maximum of OIS2 (Musil 2003, 170; van Andel 2003, 11). As the climate gradually cooled during OIS3, there was a series of cold (stadial) periods and warm (interstadial) periods that lasted between around 2000 and 5000 years. The main interstadials of OIS3, which were interrupted by cold stadial periods, were the Oerel (58,000–54,000 ^{14}C BP; 58,000–54,000 Cal. BP), Glinde (51,000–48,000 ^{14}C BP; 51,000–48,300 Cal. BP), Moershoofd (46,000–44,000 ^{14}C BP; 46,600–44,900 Cal. BP), Hengelo (39,000–36,000 ^{14}C BP; 40,600–38,000 Cal. BP) and the Denekamp (32,000–28,000 ^{14}C BP; 28,000–24,600 Cal. BP) (Allen and Huntley 2000, 117; van Andel and Tzedakis 1996, 494). OIS3 also experienced high frequency climatic fluctuations known as Dansgaard/Oeschger (D/O) events (Huntley and Allen 2003, 79). These climatic episodes represent relatively fast and dramatic changes in the climate, taking as little as 70 years to complete and sometimes manifesting as much as a 10°C temperature change (Burroughs 2005; Dansgaard *et al.* 1993). It is difficult to correlate D/O events with the pollen interstdaials listed above (Oerel, Glinde, Moershoofd, Hengelo, and Denekamp). Indeed, Fletcher *et al.* (2009, 7, 8) commented that the precise attribution of central and northern European terrestrial interstadials to particular D/O cycles is not possible. Although there were a series of long- and short-term warm and cold events during OIS3, as the climate was generally getting colder during this period, stadials and interstadials were generally warmer towards the beginning of OIS3, compared with the end of this period. The most significant point at which this difference was most pronounced was around 38,000 Cal. BP (36,000 ^{14}C BP), following which the climate became considerably colder (van Andel 2003, 11).

During prolonged cold periods of OIS3, the vegetation would have been characterised by a reduction in trees, with a predominance of herbaceous vegetation and a landscape of mainly grassland/steppe tundra (Huntley and Allen 2003, 99). This is demonstrated by pollen samples associated with cold climatic periods taken from Dolní Věstonice II in southern Moravia, Czech Republic, which suggest vegetation that was dry and steppic in character (Svobodová 1991b, 75-88). However, more favourable areas were also home to arboreal species: a number of arboreal taxa such as *Picea*, *Pinus cembra*, *Pinus sylvestris* and *Larix* have been shown to be able to withstand semi-permanent frozen soils and grow today in places such as Hungary where the air temperature and precipitation levels fall to similar levels as those during full glacial conditions (Willis *et al.* 2000, 208). Also, pollen data from Vyazivok in the Ukraine indicated that in sheltered regions trees such as *Pinus* would still have been present during extreme cold conditions (Rousseau *et al.* 2001, 354). In a similar vein, pollen data associated with cold climatic periods of OIS3 from Bulhary in Moravia, Czech Republic, indicate an arboreal pollen range of between approximately 30% and 50% (Svobodová 1991b, 84), and pollen data from areas close to Dolní Věstonice also suggest that *Pinus sylvestris*, *Pinus cembra*, *Picea abies*, *Larix europaea* and *Juniperus communis* existed with grassland-steppe vegetation during cold periods (Rybníčková and Rybníček 1991, 76; Willis *et al.* 2000, 209). During prolonged warm climatic conditions, the study region was probably dominated by either patchy woodland, or parkland conifer forest/savannah-like vegetation (Huntley and Allen 2003, 99), and deciduous trees such as oak, lime, hornbeam, beech, hazel, birch, willow and alder probably occupied more favourable areas (Mason *et al.* 1994, 49-50). This is corroborated by sediment deposits taken from Jablůnka in eastern Moravia, Czech Republic, which suggest that the vegetation consisted of trees such as larch, pine and pine limbus, birch and spruce, and shrubby herbaceous plants (Jankovská 2003, 192). Similarly, pollen records from Füramoos in Germany suggest an upsurge in Betula and Pinus coinciding with warm climatic conditions during OIS3 (Müller *et al.* 2003, 240). Even during the relatively colder, warm climate episodes of OIS3 (i.e. interstadial periods after 38,000 Cal. BP), pollen data indicate a park-type vegetation with groups of conifers and sparsely scattered deciduous trees (e.g. Pinus, Picea, Junipers, Larix, oak, hazel, and beech), but probably less densely populated than during the warmer interstadial periods prior to 38,000 Cal. BP (Mason *et al.* 1994, 50; Svoboda 1995, 292).

During cold climatic episodes, tundra and steppe-like conditions would have provided ideal environments for numerous cool-adapted herd species such as reindeer, wild horse, and steppe bison, as well as larger pachyderm species such as mammoth and woolly rhino (Mellars 1996, 51), but smaller carnivorous and herbivorous animals such as polar hare, foxes, grouse, snow owls, wolverines, wolves, hyena, and lion would also have been present (Oliva 2005, 58). Increases in temperature and the associated number of trees during warm climatic episodes of OIS3 would have corresponded with an increase in thermophilous, warm-adapted species such as red deer, roe deer, elk, boar, and aurochs inhabiting more favourable areas such as woodland regions (Musil 2003, 175). However, even during these warm climatic conditions, cold-adapted species such as mammoth, rhinoceros, reindeer and horse were still dominant within the study region (Musil 2003, 172, 173 and 182).

Overall, during OIS3 the study region experienced significant changes in climate, which were both long and short in time span. However, though prolonged warm and cold climatic episodes manifested contrasting floral conditions, this was less apparent in terms of the faunal species that inhabited the study region during this time: cool-adapted faunal species persisted during both cold and warm climatic periods.

Hominins

The presence of both Neanderthal and AMH hominins is evident in the study region during OIS3. Fossil remains of Neanderthal hominins have been recovered from sites such as Šipka Cave (Layer III) (Neruda 2006, 65; Svoboda *et al.* 1996, 234; Valoch 1965, 22-26) and Kůlna Cave (Neruda 2006, 65; Rink *et al.* 1996; Svoboda *et al.* 1996, 50; Valoch 1988). At the site of Mladeč I /II, AMH adult and child skulls have been found (Svoboda *et al.* 2002, 957-958), and at Pavlov I the burial of an AMH male has been discovered. The most complete evidence of human remains are associated with the AMH sites of Brno (I, II, and III), Předmostí, and Dolní Věstonice I and II (Svoboda *et al.* 1996, 56-67). Mousterian-type stone tools have been recovered from sites such as Šipka Cave and Švédův stůl Cave, and Micoquian stone tools have been recovered from the open-air site of Ráječko I (Oliva 1991a), as well as Kůlna Cave, Pekárna Cave, Drátenická Cave, Výpustek Cave (Svoboda *et al.* 1996, 87). Bohuncian-type stone tools have been found at sites such as Bohunice-kejbalý, Nová Hora, Brno-Líšeň and Ondrtice (Oliva 2005, 39; Svoboda *et al.* 1996, 100; Svoboda 1993, 27-29), and Szeletian lithics are associated with sites such as Vedrovice V (Oliva 2005, 32) and Pod hradem Cave (Oliva 1991b, 322). Aurignacian deposits are found in association with Stránská skála (II, IIa, IIIa, and IIIb) – the most significant of which are found in Layers 3 and 4 of Stránská skála IIa, where the largest Aurignacian stone tool assemblage in Moravia was excavated (Svoboda *et al.* 1996, 230-234) – Mladeč I /II (Oliva 2005, 51), Kvasice, Zdislavice, Milovice (Kroměřéž), Míškovice (Oliva 1993, 42), Vedrovice I, II, Kupařovice (Oliva 1993, 37; Svoboda *et al.* 1996, 101), Milovice (Oliva 1993, 39), Ondratice II, Brodek, Dobrochov, Určice (Oliva 1993, 41), and Pod hradem Cave (Oliva 1993, 41; Svoboda 1993, 29-30). Gravettian stone tools have been found at Dolní Věstonice (I, II and III) (Svoboda *et al.* 1996, 214), Pavlov (I and II), Milovice, Petřkovice, Předmosti, Boršice, Jarošov, Spytihněv (Svoboda *et al.* 1996, 132), and Kůlna Cave (Neruda and Valoch 2007, 71).

Many of the sites where hominin presence is demonstrated indicate settlement activity and evidence of repeated occupation. This is the case with Mousterian occupation of Šipka Cave where the remains of a hearth have been found associated with Neanderthals remains and large numbers of Mousterian stone tools, and at Kůlna Cave where large numbers of Micoquian stone and bone tools have been recovered (Neruda 2006; Svoboda *et al.* 1996, 215). At Bohunician sites such as Bohunice-kejbalý, although no direct evidence for settlement exists, large numbers of associated stone tools found at this site attest to repeated occupation (Tostevin and Škrdla 2006, 33), as is the case at sites such as Nová Hora, Brno-Líšeň and Ondrtice (Oliva 2005, 39; Svoboda *et al.* 1996, 100; Svoboda 1993, 27-29), and at the Szeletian site of Vedrovice V where more than 17000 stone tools have been found (Oliva 2005, 32). At the sites of Mladeč I /II and Stránská skála (IIa), large numbers of Aurignacian-type stone tools have been found together with a hearth, attesting to the repeated occupation of this site by these people (Oliva 2005, 51), and hearths have been found in association with Gravettian stone tools at the site of Petřkovice (Svoboda *et al.* 1996, 223), Předmostí (Svoboda *et al.* 1996, 223-229), Pavlov (Svoboda *et al.* 1996, 220-221), Dolní Věstonice (Svoboda *et al.* 1996, 209-214) and Milovice (Péan 2001, 331), with evidence also of built structures being found at Předmostí (Svoboda *et al.* 1996, 223-229; Valoch 1987a), Dolní Věstonice (I, II, and II) (Svoboda *et al.* 1996, 153, 209-214), Pavlov I (Svoboda *et al.* 1996, 220-223), and Milovice (Péan 2001, 331). However, although the repeated use of a number of sites is demonstrated, in some cases archaeological data indicate only very ephemeral occupation by hominins, as is the case at Pod hradem Cave, where evidence for both Szeletian and Aurignacian occupation has been recovered, and Kůlna Cave where archaeological remains indicate Gravettian occupation: in both these cases, archaeological data demonstrate only very rare or sporadic occupation by associated hominins (Neruda and Valoch 2007, 71; Oliva 1991b, 322).

The geographical distribution of hominins was mainly focused in an area of about 100 km x 100 km, centred in the south-central regions of the study area. The most commonly populated regions are areas in and around the Moravian Karst, and the river valley regions associated with the Morava river and its tributaries, and to a lesser extent the Dyje, Svratka, and Svitava rivers. Other regions also occupied by hominins, but much less often, are the Štramberk Karst and the river valley regions associated with the river Odra and associated tributaries. Mousterian and Micoquian Neanderthal sites are generally associated with caves of the karstic regions and in particualr the Moravian Karst, as is the case with Micoquian sites of Kůlna Cave, Pekárna Cave, Drátenická Cave, Výpustek Cave, Ráječko I; although the only Mousterian site that can be confidently associated with OIS3 (Šipka Cave) is located in the Štramberk Karst. Bohunician sites are mostly open-air, situated largely just south of the Moravian Karst in the Brno Basin (Bohunice-kejbalý, Nová Hora, and Brno-Líšeň, Stránská skála II-III) (Svoboda 2004, 41), and Szeletian sites, also largely open-air, are located within about a 30 km radius of the Moravian Karst. Most Aurignacian sites are open-air sites, and these are largely situated close to major, low-lying, open river valleys such as the Moravian and Napajedla gates (Oliva 2005, 48; Oliva 1993, 51; Svoboda 1995, 293). The major Aurignacian sites can be found along the Chřiby Mountains, along the bank of the

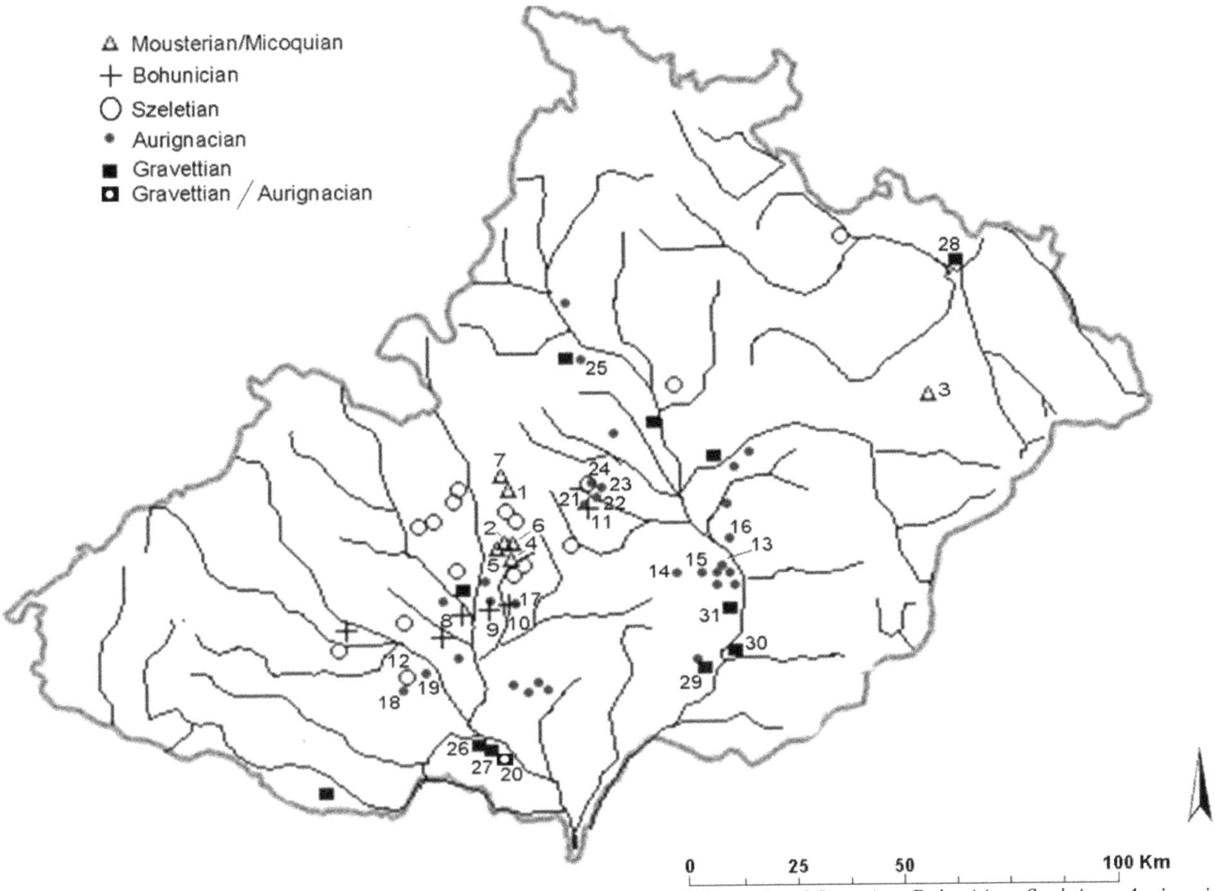

Fig. 4.6. Distribution map of main hominin sites and their associated group (Mousterian, Micoquian, Bohunician, Szeletian, Aurignacian, and Gravettian) within the study region occupied during OIS3. 1 = Kůlna Cave, 2 = Švédův stůl Cave, 3 = Šipka Cave, 4 = Pekárna Cave, 5 = Drátenická Cave, 6 = Výpustek Cave, 7 = Ráječko I, 8 = Bohunice-kejbalý, 9 = Nová Hora, 10 = Brno-Líšeň, 11 = Ondrtice, 12 = Vedrovice V, 13 = Kvasice, 14 = Zdislavice, 15 = Milovice (Kroměřéž), 16 = Míškovice, 17 = Stránská skála, 18 = Vedrovice I, II, 19 = Kupařovice, 20 = Milovice, 21 = Ondratice II, 22 = Brodek, 23 = Dobrochov, 24 = Určice, 25 = Mladeč I,II 26 = Dolní Věstonice, 27 = Pavlov, 28 = Petřkovice, 29 = Předmosti , 30 = Boršice, 31 = Jarošov, 32 = Spytihněv. (After Svoboda et al. 1996, 76, 100, 101, and 132.)

river Morava in eastern Moravia (Kvasice, Zdislavice, Milovice (Kroměřéž), Míškovice), and in the Napajedla Gate (Oliva 1993, 42). Other major regions of Aurignacian occupation are areas just south of the Moravian Karst (Stránská skála), Krumlovský les (Vedrovice I, II and Kupařovice) (Oliva 1993, 37; Svoboda *et al.* 1996, 101), the lower regions of the Svratka and Dyje rivers (Milovice) (Oliva 1993, 39), and central Moravia (Ondratice II, Brodek, Dobrochov, and Určice) (Oliva 1993, 41). The only cave site to demonstrate repeated Aurignacian occupation is Mladeč Cave, although this site is not associated with the major karstic regions but is located adjacent to the river Morava (Oliva 1993, 41; see also Svoboda 1993, 29-30). Gravettian sites are again mostly open-air, and largely associated with large, open river valleys. In comparison with the Aurignacian, there is less evidence for Gravettian sites, but as with the Aurignacian, Gravettian sites are focused on the river valley systems towards the south of Moravia (Dolní VěstonicePavlov, and Milovice), and the Moravian and Napajedla gates (Petřkovice, Předmosti, Boršice, Jarošov, Spytihněv) (Svoboda *et al.* 1996, 132) (Fig. 4.6).

Generally, evidence for the chronological distribution of hominins within the study region is sequential in nature. Neanderthals occupied the study region mostly before AMH. Mousterians and Micoquians occupied the study region largely before Bohunicians and Szeletians, and Aurignacians were mostly present before Gravettians. Neanderthals are generally associated with the warmest periods (i.e. interstadial periods prior to 38,000 Cal. BP), and AMH are mostly associated with the coldest periods of OIS3 (i.e. stadials after 38,000 Cal. BP). Sedimentary deposits associated with Mousterian occupation of Šipka Cave indicate associated occupation during at least one or possibly all of the prolonged warm periods of OIS3 (Neruda pers. comm.; Oliva 2005; and Valoch pers. comm.) (Fig. 4.7). Radiocarbon dates associated with Micoquian deposits of Kůlna Cave suggest associated occupation during the Oerel interstadial (Fig. 4.7; Table 4.1). Radiocarbon dates associated with Bohunician occupation (Stránská skála IIIa, Bohunice-kejbalý I and II, and Bohunice-cihelna) indicate associated occupation during the Moershoofd and Hengelo interstadials, and radiocarbon dates from the Szeletian site of Vedrovice V

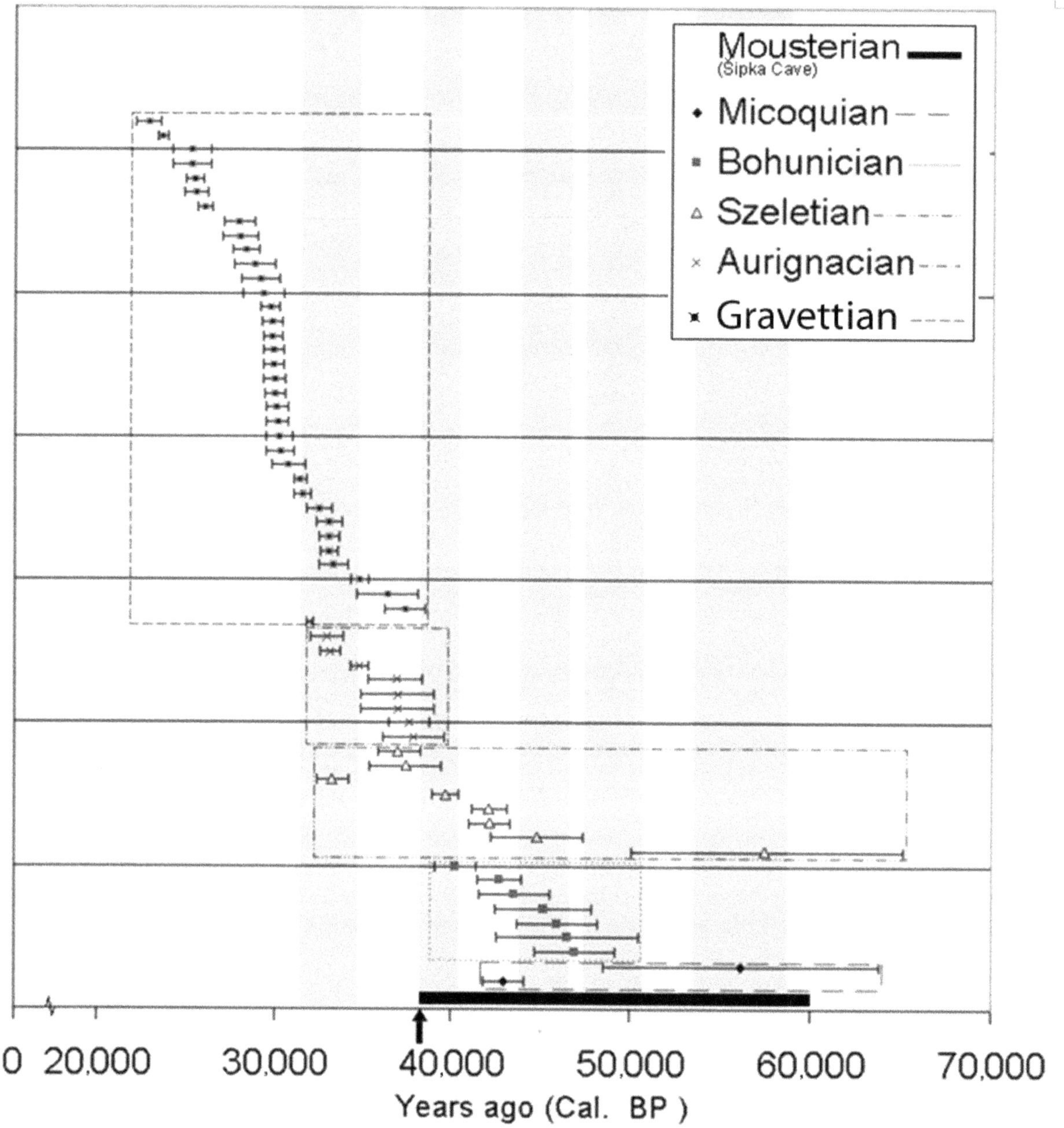

Fig. 4.7. Calibrated radiocarbon dates (using CalPal calibration program; see http://www.esc.cam.ac.uk/research/research-groups/oistage3/stage-three-project-overview) associated with the main hominin sites and their associated group (Mousterian, Micoquian, Bohunician, Szeletian, Aurignacian, and Gravettian) within the study region. Shaded regions are interstadial periods, and the black arrow indicates the point at which the climate deteriorates most significantly (i.e. approximately 38,000 Cal. BP). (After Allen and Huntley 2000, 117; http://www.esc.cam.ac.uk/research/research-groups/oistage3.)

Period	Site	Industry	Calibrated radiocarbon date			
Oerel	Kůlna Cave 7a	Micoquian	56143.00	7702.00	7702.00	
58000-54000 Cal. ka BP	Vedrovice V	Ancient Szeletian	57585.00	7576.00	7576.00	
54000-51000 Cal. ka BP						
Glinde						
51000-48300 Cal. ka BP						
48300-46600 Cal. ka BP	Bohunice-cihelna	Bohunician	46935.00	2212.00	2212.00	
Moershoofd	Stránská-skála IIIa	Bohunician	46491.00	3969.00	3969.00	
46600-44900 Cal. ka BP	Bohunice-kejbalý II	Bohunician	45949.00	2232.00	2232.00	
	Bohunice-kejbalý I	Bohunician	45145.00	2644.00	2644.00	
44900-40600 Cal. ka BP	Kůlna Cave 7a	Micoquian	42974.00	1152.00	1152.00	
	Stránská-skála III	Bohunician	43510.00	1981.00	1981.00	
	Stránská-skála III	Bohunician	42712.00	1275.00	1275.00	
	Vedrovice V	Ancient Szeletian	44781.00	2561.00	2561.00	
	Vedrovice V	Ancient Szeletian	42147.00	1147.00	1147.00	
	Vedrovice V	Ancient Szeletian	42144.00	999.00	999.00	
Hengelo	Bohunice-cihelna	Bohunician	40239.00	1144.00	1144.00	
40600-38000 Cal. ka BP	Vedrovice V	Ancient Szeletian	39684.00	748.00	748.00	
38000-34600 Cal. ka BP	Pod hradem Cave	Ancient Szeletian	37409.00	2004.00	2004.00	
	Pod hradem Cave	Ancient Szeletian	37056.00	1184.00	1184.00	
	Pod hradem Cave A	Aurignacian	37823.00	1720.00	1720.00	
	Pod hradem Cave A	Aurignacian	37601.00	1159.00	1159.00	
	Stránská-skála IIIb	Aurignacian	36942.00	2068.00	2068.00	
	Stránská-skála IIIb	Aurignacian	36942.00	2068.00	2068.00	
	Stránská-skála IIa	Aurignacian	36819.00	1503.00	1503.00	
	Dolní Věstonice I	Gravettian	37351.00	1189.00	1189.00	
	Dolní Věstonice I	Gravettian	36355.00	1727.00	1727.00	
	Dolní Věstonice	Gravettian	34802.00	476.00	476.00	
Denekamp	Vedrovice V	Ancient Szeletian	33311.00	878.00	878.00	
34600-31200 Cal. ka BP	Pod hradem Cave A	Aurignacian	33141.00	591.00	591.00	
	Milovice I	Aurignacian	32939.00	927.00	927.00	
	Pod hradem Cave A	Aurignacian	31936.00	227.00	227.00	
	Pavlov I	Gravettian	33315.00	822.00	822.00	
	Dolní Věstonice II	Gravettian	33081.00	484.00	484.00	
	Dolní Věstonice I	Gravettian	33079.00	594.00	594.00	
	Dolní Věstonice I	Gravettian	33059.00	755.00	755.00	
	Dolní Věstonice II	Gravettian	32476.00	737.00	737.00	
	Dolní Věstonice I	Gravettian	31495.00	521.00	521.00	
	Dolní Věstonice II	Gravettian	31367.00	377.00	377.00	
31200-22810 Cal. ka BP	Dolní Věstonice I	Gravettian	30698.00	949.00	949.00	
	Pavlov I	Gravettian	30223.00	790.00	790.00	
	Pavlov I	Gravettian	30181.00	730.00	730.00	
	Pavlov I	Gravettian	30068.00	640.00	640.00	
	Pavlov I	Gravettian	30048.00	628.00	628.00	
	Dolní Věstonice I	Gravettian	29954.00	580.00	580.00	
	Pavlov I	Gravettian	29895.00	598.00	598.00	
	Dolní Věstonice II	Gravettian	29856.00	555.00	555.00	
	Pavlov I	Gravettian	29824.00	561.00	561.00	
	Dolní Věstonice I	Gravettian	29778.00	533.00	533.00	
	Dolní Věstonice I	Gravettian	29734.00	568.00	568.00	
	Pavlov I	Gravettian	29653.00	539.00	539.00	
	Dolní Věstonice I	Gravettian	29277.00	1149.00	1149.00	
	Dolni Vestonice II	Gravettian	29131.00	1077.00	1077.00	
	Milovice I	Gravettian	28777.00	1176.00	1176.00	
	Pavlov I	Gravettian	28272.00	767.00	767.00	
	Dolní Věstonice III	Gravettian	27933.00	988.00	988.00	
	Dolni Věstonice II	Gravettian	27898.00	839.00	839.00	
	Milovice I	Gravettian	25971.00	398.00	398.00	
	Dolní Věstonice II	Gravettian	25485.00	654.00	654.00	
	Dolní Věstonice I	Gravettian	25418.00	495.00	495.00	
	Milovice I	Gravettian	25223.00	1041.00	1041.00	
	Milovice I	Gravettian	25223.00	1041.00	1041.00	
	Dolní Věstonice I	Gravettian	23604.00	300.00	300.00	
	Dolní Věstonice I	Gravettian	22810.00	692.00	692.00	

Table 4.1. Calibrated radiocarbon dates (using CalPal calibration program; see http://www.esc.cam.ac.uk/research/research-groups/oistage3/stage-three-project-overview) associated with the main hominin sites and their associated group (Mousterian, Micoquian, Bohunician, Szeletian, Aurignacian, and Gravettian) within the study region. Shaded regions are interstadial periods, and the black arrow indicates the point at which the climate deteriorates most significantly (i.e. approximately 38,000 Cal. BP). (After Allen and Huntley 2000, 117; http://www.esc.cam.ac.uk/research/research-groups/oistage3.)

indicate occupation took place during the Oerel, Moershoofd and Hengelo interstadials (Fig. 4.7; Table 4.1). Radiocarbon dates from the Aurignacian sites of Stránská skála IIIa and IIIb, and Milovice I suggest occupation during prolonged cold periods prior to the Denekamp interstadial, and radiocarbon dates associated with Gravettian sites (Dolní Věstonice I, II and III, Pavlov I, Milovice I) indicate occupation during prolonged cold periods following the Denekamp interstadial (Fig. 4.7 and Table 4.1).

The most common lithic raw materials used by hominins in the study region include flysch sandstone, Cretaceous hornstones, Krumlovský les chert, Stránská skála chert, erratic flint, and radiolarite (Valoch 1987b). Generally, Neanderthals tended to procure more local raw materials, and AMH procured more exotic raw materials. Mousterian occupants of Šipka Cave (Layers III and IV) (Valoch 1965, 22-26) utilised a range of materials (e.g. flint, radiolarite and quartz), but by far the most commonly used lithic raw material was flysch sandstone, located only a few hundred metres from the site (Svoboda *et al.* 1996, 93; Valoch 1987b, 263). This is similarly the case for Micoquians associated with Kůlna Cave (Layer 7a and Layer 6a): raw materials used include a wide variety of materials (Layer 7a – quartz 9.03%, dobra chert 2.92%, orthoquartz 2.28%, and Němčice chert 2.66%; Layer 6a – quartz 10.48%, non-distinguished chert 8%, orthoquartz 5.24%, dobra chert 2%, Němčice chert 2%, Olomačany chert 1.6%, Byči skala chert 0.4%, Krumlovský les chert 0.4%, and smoke rock crystal 0.4%) (Neruda pers. comm.), but by far the most commonly used is Cretaceous chert (Layer 7a, 77%; Layer 6a, 66.94%; Neruda pers. comm.), which is found in the river gravels of the Svitava located about 10 km west of Kůlna Cave (Svoboda *et al.* 1996, 93; Valoch 1987b, 265).

Bohunician-type stone tools are almost universally made of locally sourced Stránská skála chert, as is the case at the site of Bohunice-kejbalý where, although other materials (Krumlovský les chert and Cretaceous chert) are sometimes used (Svoboda *et al.* 1996, 207; Svoboda 1993, 32; Svoboda and Simán 1989, 294), the vast majority of stone tools are made using Stránská skála chert. Szeletian stone tools are made of mostly local materials, as is the case at Vedrovice V where 99% of the material comes from the local Krumlovský les chert deposits (Oliva 2005, 33; Svoboda *et al.* 1996, 237). This

	1952		1957	
	N°	%	N°	%
Flint	13015	96.5	7065	47.3
Radiolarite (red)	430	3.3	5088	34.1
Radiolarite (green)	Undetermined		2420	16.2
Moravian cherts	20	0.2	51	0.3
Obsidian	0	0	2	-
Rock crystal	0	0	2	-
Burnt in fire	Undetermined		305	2.1
Total	13485	100	14933	100

Table 4.2. Number and percentage of silicate raw materials recovered from the south-eastern (1952) and north-western (1957) excavations of the Gravettian site of Pavlov I, Moravia, Czech Republic. The most common are highlighted. (After Svoboda et al. *2000, 204.)*

	% 1952	% 1957
Mammoth	7.5	18.9
Wolf	12.5	14.6
Reindeer	10.1	15.1
Horse	4.6	9.0
Hare	18.5	19.2
Polar (Arctic) fox	16.9	13.9
Fox	12.7	3.2
Bear	1.6	0.6
Wolverine	4.4	2.3
Lion	0.5	0.3
Rhinoceros	-	0.9
Birds	8.3	1.7
Felids	0.7	-
Bovids	0.5	-
Red deer	0.3	0.3

Table 4.3. Percentage of animal bones recovered from the south-eastern (1952) and north-western (1957) excavations of the Gravettian site of Pavlov I, Moravia, Czech Republic. The most common are highlighted. (After Svoboda et al. *2000, 204.)*

is also the case with some Aurignacian sites such as Stránská skála (IIA and IIIA), where raw materials are made mostly of locally sourced Stránská skála chert (Svoboda *et al.* 1996, 231, 233; Svoboda and Simán 1989, 313-314); although at other Aurignacian sites, e.g. those located in eastern Moravia, in the Napajedla Gate, along the Chřiby Mountains and along the bank of the Morava, exotically sourced raw materials are used more often (Oliva 1993, 42). The vast majority of Gravettian raw materials used to make stone tools are imported (Valoch 1987b, 265): Gravettian stone tools found at Dolní Věstonice (I, II and III) and Pavlov I are largely (~ 90%) made of erratic flint; however, stone tools made using radiolarite are also common at Pavlov I (Přichystal *et al.* 1994, 31; Svoboda *et al.* 1996, 153; Verpoorte 2005, 79) (Table 4.2).

The remains of a large number of different faunal taxa discovered at hominin sites in the study region indicate that animals were commonly hunted by hominins during OIS3. Those most commonly found include large herbivores such as mammoth, reindeer, horse, bison, as well as smaller animal species such as wolf, hare, and fox (Musil 2003, 171-181). For the most part, Neanderthals hunted large herbivorous animals such as mammoth, reindeer, horse and less often bison, with those animal species hunted by AMH also including smaller taxa such as wolf, hare, and fox. Although faunal remains have been found associated with Mousterian deposits at sites such as Šipka Cave (Layer III and IV) (Valoch 1965, 22-26), no research has been undertaken to determine which of these are the remains of hunted fauna (Valoch pers. comm. and Neruda pers. comm.), but it is probable that their hunting practices were similar to those of Micoquian people. Remains of hunted fauna associated with Micoquian deposits of Kůlna Cave include mostly those of mammoth, reindeer, and horse (Musil 2003, 171, 180; Svoboda *et al.* 1996, 215-217). There have been few faunal remains recovered from Bohunician and Szeletian sites in the study area; however, it is probably that the hunting practices of these people were similar to those of Micoquians and Mousterians, hunting mainly large herbivores such as mammoth, reindeer, horse (Musil 2003, 171; Svoboda 1993, 32), and bison (Musil 2003, 172; Oliva 2005, 39). Of particular interest are suspected cut marks found by the author on a cave-bear bone recovered from the same layer (Layer 9) as Szeletian and Aurignacian stone tools at Pod hradem Cave, suggesting that Aurignacians and/or Szeletians may occasionally have hunted cave bears. Remains of hunted fauna associated with Aurignacian sites are also rare, but have been found at sites such as Mladeč I/II: numerous teeth of large herbivores such as horse have been found here (Oliva 2005, 51; Svoboda *et al.* 1996, 219), and Svoboda *et al.* (1996, 127) made the point that large herbivores such as mammoth, reindeer, and horse would probably have been important for Aurignacian people. In the case of Gravettians, large herbivores were also important, and one of the most significant was mammoth: large mammoth dumps have been found at Predmostí, Dolní Věstonice, and Milovice (Oliva 1988; Svoboda 2005). However, these were not the only fauna hunted by Gravettian people: at Dolní Věstonice and Pavlov hare, wolf, Arctic fox, mammoth, reindeer, and horse are among the most commonly found hunted fauna (Musil 2003, 176-177; and Table 4.3).

There is no apparent evidence that Mousterians (i.e. Šipka Cave, Layers III and IV) used animal remains to make tools or other objects, but split animal bones with cut marks, chips of mammoth ivory used as compressors, chips of retouched bone, a mallet of reindeer antler, and a fragment of a heavily use-abraded mammoth's rib have been found associated with Micoquian occupation of Kůlna Cave (Oliva 2005, 68; Rink *et al.* 1996, 981). The only evidence of Szeletian use of animal remains comes in the form of a harpoon made from the root of a mammoth molar (Oliva 2005, 51). Most evidence of Aurignacian use of animal remains in the study region comes from Mladeč I /II, where bone points have been found along with pendants made of pierced beaver, elk, horse, wolf, and bear teeth (Oliva 2005, 51, 87; Svoboda *et al.* 1996, 129). Teeth were also used by Gravettian people to make pendants, as well as mammoth bones and ivory to support and build structures, such as those found at the sites of Dolní Věstonice and Předmostí (Oliva 2005, 63; Svoboda *et al.* 1996, 209-229), and those used to make objects such as shovels, spoons, spatulas, hammer stones, and decorated headbands. Mammoth ivory was also commonly used at Gravettian sites as a palette for engravings, or for sculpting figurines of animals (Oliva 2005, 51-87; Oliva 2000, 145; Svoboda *et al.* 2000, 205; Svoboda *et al.* 1996, 221), and other objects such as mammoth, wolf, and cave-bear penis bones have been found in association with Gravettian sites (Absolon 1938, 45; Klíma 1963, 100). Many of the pendants made by Aurignacian and Gravettian hominins were made using the teeth of carnivores such as fox and bear more so than other animals, as is the case with the pendants found at Mladeč I/II and Dolní Věstonice (Oliva 2005, 51, 87; Svoboda *et al.* 1996, 129).

Depictions of animals have only been found in association with Gravettian occupation of the study region (Oliva 2005, 52). For instance, what appears to be a figurine of a bear's torso has been recovered from Pavlov I, along with a bear's head, near-complete figurines of bears, and figurines of lion, mammoth, and rhinoceros. Animals most often sculpted are carnivores such as bear, lions, wolves, and foxes (Svoboda *et al.* 1996, 167), and according to Bahn and Vertut (1988, 129) of 77 identifiable animal figurines discovered at Dolní Věstonice and Pavlov, at least 21 are thought to be bear, 9 lions, 5 wolves, and 3 foxes.

In summary, Neanderthals (Mousterian, Micoquian, Bohunician, and Szeletian) and AMH (Aurignacian and Gravettian) occupied the study region during OIS3. Their distribution was mainly associated with the central southern areas of the study region, and associated largely with geological features such as major karstic regions (Moravian and Štramberk karsts) and major rivers and associated valleys. Although in some cases particular sites used by Neanderthals and AMH might have been used rarely or even just once, Neanderthals and AMH, associated with all cultural groups, utilised settlement locations on a repeated basis. Occupation of the study area was generally sequential in nature, but there was significant overlap between Neanderthals and AMH and individual cultural groups. For the most part, Neanderthals occupied the study region during warm periods, and AMH occupied the study region during relatively cold climatic periods. Neanderthals largely procured locally sourced raw materials, and AMH procured the more exotically sourced, but this was not always the case, particularly for Aurignacians. All hominins hunted large herbivores, but Gravettians also commonly hunted smaller animals such as hare, wolf, and Arctic fox. Animal remains were used by both Neanderthals and AMH, with mammoth bone and ivory being the most commonly used. Neanderthals only used animal remains to make tools and rarely weapons (e.g. harpoon), whereas AMH used animal remains to make a variety of objects, including decorative objects, with carnivore canines being commonly used. Only Gravettian people depicted animals, and depictions of carnivores and in particular bears were most common.

Cave bears

Cave-bear remains have been recovered from sites such as Pod hradem Cave, the Sloup-Šošůvka caves, Výpustek Cave, Barová Cave, Švédův Stůl Cave, Kůlna Cave, and Šipka Cave (Musil 2007, 24; Musil 1980b; Fig. 4.8). For instance, more than 900 cave-bear skulls were found within Sloup-Šošůvka caves, remains of more than 300 individual cave bears were discovered in Pod hradem Cave, and a large number of cave-bear remains have been found in caves such as Švédův Stůl Cave, Barová Cave, and Šipka Cave (Layers III and IV) (Musil 2007, 24; Musil 1980b). In the case of cave-bear remains recovered from Layer III of Šipka Cave, some demonstrated evidence of burning and are associated with the fireplace where the mandible of a Neanderthal juvenile was found (Svoboda 2005, 70). Cave-bear remains have also been recovered from open-air sites such as Dolní Věstonice, Pavlov, and Předmostí; however, it is likely that these were brought to these sites by humans, and do not represent natural cave-bear spatial distribution patterns.

By far the largest majority of cave-bear remains originate from the Moravian Karst: all but one of the most important cave-bear sites (i.e. Pod hradem Cave, the Sloup-Šošůvka caves, Výpustek Cave, Barová Cave, and Švédův stůl Cave) are located within the Moravian Karst, and the only other important cave located outside of this area is Šipka Cave, situated in the Štramberk Karst. There has only ever been one radiocarbon date obtained from cave-bear remains associated with the study region, and this provided a date of more than 47,000 years ago (Pacher and Stuart 2008, 9). There are only a handful of

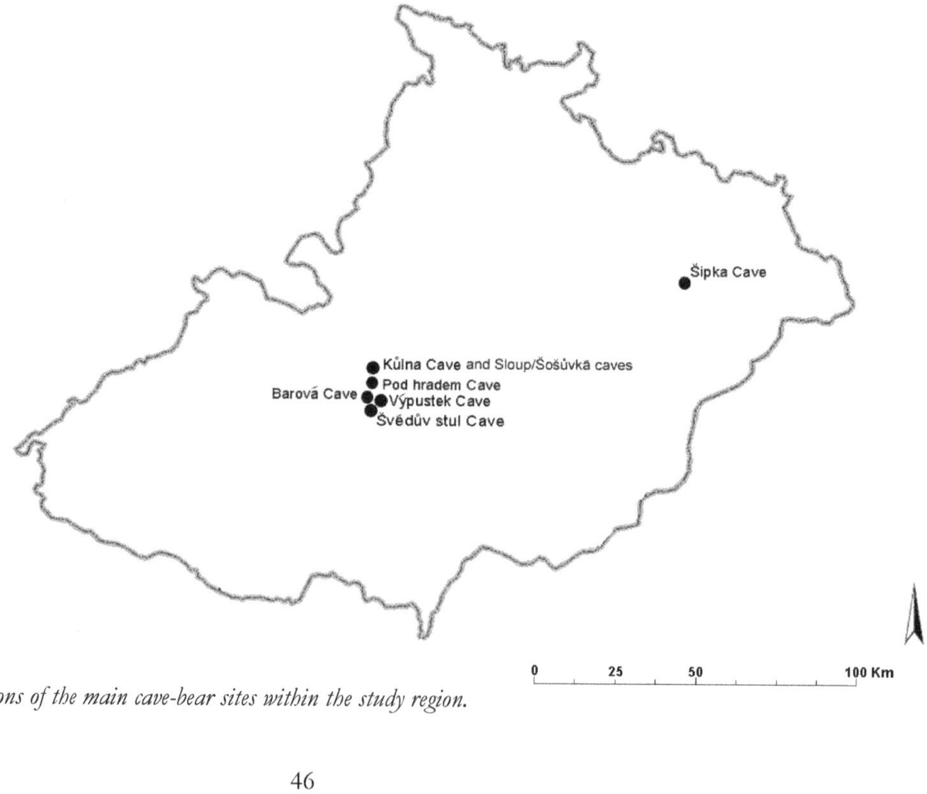

Fig. 4.8. Locations of the main cave-bear sites within the study region.

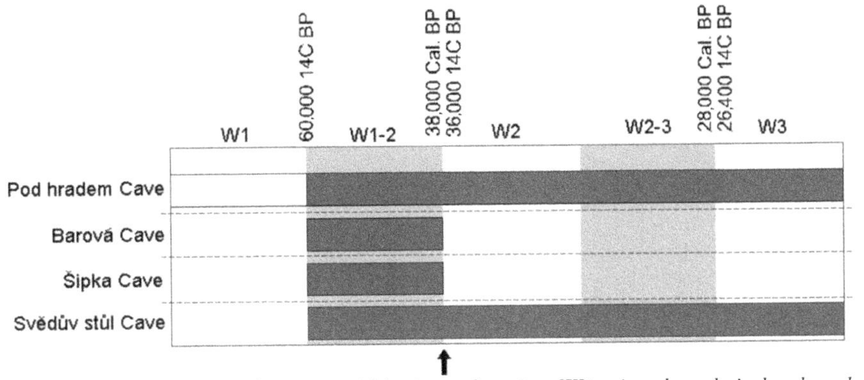

Fig. 4.9. Chronological comparison of major cave-bear sites within the study region. Würmian chronological scale and associated chronometric dates are shown, warm climatic periods are shaded, and cold climatic periods are un-shaded. Arrow indicates the point at which the climate becomes significantly colder (i.e. approximately 38,000 Cal. BP.). It should be noted that there are significant difficulties in determining the precise start/finish dates of the different Würmian episodes. Even if radiocarbon dates can be confidently associated with the beginning and end of each Würmian event, radiocarbon date error margins, particularly towards the upper radiocarbon date limit (i.e. approximately 50,000 Cal. BP; Cutler et al. 2004, 1127), can be in the order of several millennia.

Layer	Bone						
	Scapula	Mandibula	Humerus	Radius	Ulna	Femur	Tibia
5	Both present but no record of how many in each layer		1			1	1
6							
7							
8			1	1		1	1
9					1		
10						1	
11			1	1	1	4	1
12				1	1		7
13			2		1	2	
14							
15			1	1	3		6
16				1	1	1	
17			3	2	1	3	5
18			2	1	2	1	1
19							

Table 4.4. Number of cave-bear bones (number of individual specimens - NISP) recovered from Pod hradem Cave, Moravia, Czech Republic. (After Musil 1965, 68-76.)

Layer	M^1	M^2	M_1	M_2	M_3	Canine M/F	
5	3	2	12	7	7	1	1
6		1	4	3	3		
7		1	3	3	3		1
8	9	8	7	4	11	1	2
9	3		2		1		
10	4	1	4	11	5		
11	5	3	7	15	5		
12	4	11	14	21	9	1	3
13	10	12	7	13	8	7	6
14	5	7	4	4		2	5
15	8	17	18	21	16	11	8
16	5	10	9	9	3	4	3
17	12	15	11	16	16	9	6
18		5	11	18	11	4	6
19			3	2	2		2

Table 4.5. Number of cave-bear teeth (number of individual specimens - NISP) recovered from Pod hradem Cave, Moravia, Czech Republic. (After Musil 1965, 64.)

sites where cave-bear remains have been excavated from good chronological contexts (Švédův Stůl Cave, Pod hradem Cave, Šipka Cave, and Barová Cave). These suggest that cave-bear remains areassociated with all periods of OIS3, i.e. W1-2, W2, W2-3, and W3 (Musil 1980b), and are associated with both prolonged warm and cold climatic periods; however, by far the largest majority of cave-bear remains originate from deposits associated with prolonged warm climatic episodes (i.e. W1-2):
Švédův Stůl Cave, Pod hradem Cave, Šipka Cave, and Barová Cave all contain disproportionate numbers of cave-bear remains from deposits associated with this period (Fig. 4.9) (Musil 1980b). It is unlikely however that the apparent absence or reduction of cave-bear

Layer	MNI	% female	% male	% neonate-3/4 year	% 3/4 -1 1/4 year	% older than 1 year	% senile
5	20	50	50	16	33	33	16
6	7			100			
7	5			66		33	
8	25	50	50	40	30	30	
9	2						
10	18			50		50	
11	21			64	8	24	
12	37	75	25	55	11	22	11
13	21	53	46.6	37		62	
14	8	71.4	28.5	70	9	14	5
15	37	42.1	57.8	53		23	23
16	24	42.8	57.1	65	10	15	10
17	28	43.7	56.2	62	12		25
18	26	60	40				
19	4	100					

Table 4.6. Summary of cave-bear bones found in Pod hradem Cave, Moravia, Czech Republic. (After Musil 1965, 62-75.)

remains associated with prolonged cold periods is indicative of significant changes in cave-bear behaviour and/or mortality rates: it is more likely that differences in the particular taphonomic circumstances associated with cold or warm periods manifested such changes.

Cave-bear remains apparent in the case study demonstrate the presence of both sexes and different age groups, but the proportion of each is not consistent. For instance, at Pod hradem Cave, male:female ratios vary considerably from being largely in favour of females (75%) to being slightly in favour of males (57.8%), and the ages of cave bears associated with this site also vary significantly, but are mostly biased in favour of juvenile cave bears (Table 4.4, Table 4.5, and Table 4.6). Also, although there are no apparent records of the number and demographic nature of cave-bear remains recovered from Šipka Cave, the remains of a significant number of cave bears of both sexes and different ages are evident in museum collections (Musil 1965, 127-132) (Fig. 4.10 and Fig. 4.11). The remains of at least 21 individual cave bears were found at Barová Cave (Table. 4.7), and although there are no detailed published data regarding the sex and age of the cave bears, a general comparison of cave-bear teeth from this site and those from Pod hradem Cave has been made by Musil (1965, 59).

Case study sites

The hominin and cave-bear sites for this project are chosen on the basis that there exist sufficient data demonstrating repeated occupation by either hominins or cave bears, that hominin or cave-bear occupation can be confidently assigned to either prolonged warm or cold periods, and that access to relevant data is possible. In the case of hominin sites, first-hand examination of archaeological data is not necessary as the majority of research is undertaken via available literature. The hominin sites chosen are Šipka Cave, Kůlna Cave, Vedrovice V, Bohunice-kejbalý (I and II), Stránská-skála (IIa and IIIb), Dolní Věstonice (I, II, and III), and Pavlov I. Following a process of literature research, visits to the study region, speaking to relevant people (i.e. Prof. K. Valoch and Prof. R. Musil), and briefly examining relevant cave-bear remains, three cave-bear sites are chosen (Šipka Cave, Pod hradem Cave, and Barová Cave): large numbers of cave-bear bones have been recovered from all three sites, all three can confidently be associated with prolonged warm or cold period during OIS3, and access to the cave-bear remains from these sites is granted by Prof. K. Valoch of the Anthropos Museum, Brno, Czech Republic.

Šipka Cave is located on the northern slope of Kotouč, a Jurassic limestone hill near Štramberk, on the edge of the Carpathian Mountains at 440 m.a.s.l.. Access to the cave can be gained via a tourist path, and the cave is easily visible from the pathway on approach. The roof to the front part of the cave has now collapsed, which is likely to have taken place during OIS2 (Neruda pers. comm.), and so was most probably present during most of the study period. The cave is composed of one large chamber that subsequently divides into two separate corridors, Jezevčí Díra and the larger Krápníková corridor (Oliva 2005, 25). Four main layers were excavated from this site (I, II, III, and IV), with Layers III and IV corresponding to prolonged warm periods (W1-2) (Neruda pers. comm.; Oliva 2005, 25; and Valoch pers. comm.), and Layers I and II corresponding to cold episodes. Kůlna Cave is located in the northern part of the Moravian Karst at 464 m.a.s.l., around 30 km north of Brno (Oliva 2005, 22), and next to the Sloupsko-Šošůvka cave system (Neruda and Valoch 2007, 66; Rink *et al.* 1996, 889). The cave is located in a transitional area between the steep valley system of the Moravian Karst, and the largely open landscape towards the north of this area. The entrance to Kůlna Cave is approximately 20 m wide and 9 m high (Musil 2007, 25), and most visible if approached from the south. The cave is easily accessible via a gentle slope from the adjacent main road, and the interior of the cave is a large, almost horizontal, tunnel-like gallery about 92 m long (Svoboda *et al.* 1996, 215). Kůlna Cave was excavated by J. Wankel in 1881, by M. Křiž in 1881-1886, by J. Knies in 1887-1913, and subsequently by K. Valoch 1961-1976, and deposits span from the last interglacial to the Holocene (Rink *et al.* 1996, 890). Bohunice-kejbalý (I and II) is an open-air site located just south of Brno city. The site is situated in the area of Red Hill, an elevated locality on the western edges of the Brno basin. The site was initially discovered by an amateur archaeologist (Mr Klíma) who extracted artefacts from bulldozered trenches, which were subsequently examined and identified by Prof. Valoch (Tostevin and Škrdla 2006, 31, 33). Vedrovice V is located close to the village of Vedrovice, 40 km southwest of Brno on the eastern slopes of the southern part of Kromlovský les highland (Valoch 1993, 7), and has the best stratified and well-dated Szeletian finds in Moravia. Stránská skála (IIa and IIIa) is located close to the Jurassic Stránská skála limestone outcrop, located in the eastern regions of Brno. Dolní Věstonice (I, II and III) is located in the southern regions of the study area, about 500 m from the banks of the Dyje. Excavations have been almost continuous over the last century or so: between 1924 and 1938 excavations were undertaken by Absolon, and subsequently (1939-1942) by Bohmers, Žebra (1945-1946), Klíma (1947-1952, 1971-1979), and more recently by Svobodá (1990 and 1993) (Svoboda *et al.* 1996, 209). Pavlov I is located southeast of the complex of sites associated with Dolní Věstonice, a few hundred metres from the banks of the Dyje. Excavations at Pavlov I were carried out by Klíma (1952-1972).

Fig. 4.10. Cave-bear mandibles recovered from Layer III Šipka Cave. Scale resolution = 1cm. (Picture taken by author at Anthropos Institute, Brno, Czech Republic.)

Fig. 4.11. Cave-bear teeth recovered from Layer IV of Šipka Cave. Scale resolution = 1cm. (Picture taken by author at the Anthropos Institute, Brno, Czech Republic.)

Bone	Number of bones	Notes
Cranium	16	
Scapula	10	
Mandible	4	
Humerus	38	Numbers 3, 16, 30, 5, 19 and 20 epiphyses not connected
Radius	42	Numbers 9 and 22 epiphyses not connected
Ulna	20	Numbers 6 and 7 epiphyses not connected
Femur	23	Numbers 17 and 18 epiphyses not connected
Tibia	24	Numbers 16 and 17epiphyses not connected
Astragalus	18	

Table 4.7. Summary of bones recovered from Barová Cave, Moravia, Czech Republic. (After Musil 1965, 16-17.)

Pod hradem Cave is located about 4 km south of the northernmost tip of the Moravian Karst, close to the top of the west side of a steep-sided valley. The entrance to the cave is no more than around 1.5 m in height, with a width of approximately 2.5 m, and the cave interior is about 120 m in length and made up of one main chamber that separates into two further chambers (Musil 2007, 25). Although the site is relatively easy to access on foot, as there is a footpath leading from the valley floor that passes close by the cave, the cave itself is relatively difficult to find: it is located off the main footpath and behind the limestone rock that faces the adjacent footpath. The cave was excavated by Musil and Valoch in 1952, and the cave deposits are divided into 19 layers, spanning from W1 to W3. Layer 19 is associated with W1, which is prior to the period of interest for this study. Musil (1965, 59) suggested that Layers 18-8 are associated with W1-2; however, subsequent analysis by Kukla (1968) indicated that only Layers 17 and 18 can be associated with W1-2, and this is supported by the remains of typical warm-type species (*Sus scrofa*) associated with Layer 17. Also, it is likely that Layers 5, 7, 8, 9 and 10 are associated with cold periods, as the remains of typical cold-type species (e.g. *Alopex lagopus*) are found in these layers. Barová Cave is located in the central part of the Moravia Karst, on the north-eastern side of a steep-sided valley. Access to the cave is relatively easy, and can be found by following the footpath that leads up the valley wall near the entrance of Byči skála Cave. The entrance to the cave cannot be seen unless viewed from immediately in front of the cave itself: the cave entrance is about 1-2 m below the current level of the adjacent footpath/ledge. The entrance to the cave has been reconstructed to accommodate a security door, but the original width of the cave was about 2.5 m. Access to the cave was not possible during my visit because the security door was locked. The cave was excavated by Musil in 1965.

In summary, there are 9 case study sites chosen for this project, one of which (Šipka Cave) is associated with hominins (Mousterian) and cave bears, two (Pod hradem Cave and Barová Cave) are associated with cave bears only, and 6 (Kůlna Cave, Bohunice-kejbalý (I and II), Vedrovice V, Stránská skála (IIA and IIIA), Dolní Věstonice (I, II and III), and Pavlov I) are hominin sites only. All three cave-bear sites are located either in the Štramberk Karst (Šipka Cave) or Moravian Karst (Pod hradem Cave and Barová Cave). One hominin site is located in the Štramberk Karst (Šipka Cave), one in the Moravian Karst (Kůlna Cave), three to the south and south-west of the Moravian Karst (Bohunice-kejbalý I and II, Vedrovice V, Stránská skála IIA and IIIA), and two associated with the river Dyje at the very south of the study region (Dolní Věstonice I, II, and III, and Pavlov I). A comparative overview of the hominin and cave-bear case study sites is provided in Table 4.8.

Summary and conclusions

This chapter set out to provide an overview and the essential background for the case study and case study sites chosen for this book. I began by providing a background to the study region, looking at the geographical and chronological context of the case study, and associated geology, topography, hydrology, climate, flora and fauna. I discussed hominins and cave bears in the study region, and the case study sites chosen for this project. The next chapter builds on this chapter, and prepares the GIS digital framework that provides the basis for modelling hominins and cave-bear distribution patterns used for identifying potential interactions.

	All sites and individual sites		Climatic affiliation						
Cave bears		All	Warm and cold	Presence of male/female and different aged cave bears confirmed, but not uniform between sites or over time					
		Šipka Cave	Warm	Presence of male/female and different aged cave bears confirmed but exact numbers unknown					
		Pod hradem Cave	Warm and cold	Presence of male/female and different aged cave bears confirmed, and some details provided by Musil (1965)					
		Barorá Cave	Warm	Presence of male/female and different aged cave bears confirmed but exact numbers unknown					
Hominins				**General proximity of commonly used lithic raw materials**	**Most commonly hunted animals**	**Use of animal remains, or animal depictions**	**Animal depictions**	**Use of cave-bear remains**	**Depictions of cave bears**
	Mousterian	**All (Šipka Cave)**	Warm	Local	Unknown – but probably mammoth, horse and reindeer	Unknown	None	Unknown	None
		Šipka Cave	Warm	Local - flysch sand stone from Štramberk Karst	Unknown – but probably mammoth, horse and reindeer	Unknown	None	Unknown	None
	Micoquian	**All**	Warm	Local	Mammoth, horse and reindeer	Mammoth and reindeer used to make tools	None	None	None
		All (Kůlna Cave)	Warm	Local - Cretaceous chert	Mammoth, horse and reindeer	Mammoth and reindeer used to make tools	None	None	None
	Bohunician	**All**	Warm	Local and almost universally Stránská skála chert	Unknown – but probably mammoth, horse and reindeer	Unknown	None	Unknown	None
		Bohunice-kejbalý I and II	Warm	Local - Stránská skála chert	Unknown – but probably mammoth, horse and reindeer	Unknown	None	Unknown	None
	Szeletian	**All**	Warm	Local	Unknown – but probably mammoth, horse, reindeer, and bison	Largely unknown, but mammoth tooth harpoon	None	Unknown	None
		Vedrovice V	Warm	Local - Krumlovský les type chert	Unknown – but probably mammoth, horse, reindeer, and bison	Unknown	None	Unknown	None
	Aurignacian	**All**	Cold	Local and exotic	Mammoth, reindeer, horse	Bone tools and pendants made from teeth, in particular carnivore canines	None	Medium – possible use of cave-bear teeth for pendants	None
		Stránská skála IIA and IIIA	Cold	Local - Stránská skála chert	Unknown – but probably mammoth, horse and reindeer	None	None	Unknown	None
	Gravettian	**All**	Cold	Exotic	Mammoth, reindeer, horse, wolf, Arctic fox, hare	Animal remains used for tools, personal ornamentation	Sculpted animal figurines (mostly carnivores and in particular bear)	Low/medium	High
		Dolní Věstonice I, II, and III	Cold	Exotic - Erratic flint	Mammoth, reindeer, wolf, fox, hare	Animal remains used for tools, personal ornamentation	Sculpted animal figurines (mostly carnivores and in particular bear)	Low/medium	High
		Pavlov I	Cold	Exotic - Erratic flint and radiolarite	Mammoth, reindeer, horse, wolf, Arctic fox, hare	Animal remains used for tools, personal ornamentation	Sculpted animal figurines (mostly carnivores and in particular bear)	Low/medium	High

Table 4.8. Comparative overview showing general and site-specific details associated with cave bears and hominins in the study region during OIS3.

Chapter 5: Building the GIS framework

Introduction

The objective of this chapter is to construct a digital GIS framework in preparation for the next three chapters where hominin and cave-bear distribution patterns are mapped and potential interactions identified. Using GIS, I digitally map site locations, geological outcrops, topography, palaeohydrology (palaeochannels and palaeofloodplains), palaeovegetation, friction maps, and prey species habitat and diversity. In each case, I describe the methodology used and present and describe the results.

Mapping the GIS framework

Case study site locations are digitally mapped using coordinate locations obtained mostly from van Andel *et al.* (2003a, 53-56), and these are cross-referenced using an online digital map of the study region (http://www.mapy.cz). In the case of one site (Barová Cave), site coordinates are not readily available in published literature, and so these are obtained using only online digital map source data (http://www.mapy.cz). All site coordinates are converted into decimal UTM format using an online coordinate converter (http://www.fcc.gov/mb/audio/bickel/DDDMMSS-decimal.html), and stored digitally using Microsoft Excel. The coordinate locations are then imported directly into ArcGIS9.2, and used to produce vector point locations representative of the case study site locations in real space (Fig. 5.1). As the spatial characteristics of the case study sites are mostly self-explanatory and are presented in Chapter 4, they will not be described further here.

Lithic raw materials outcrops (erratic flint, Krumlovský les chert, Stránská skála chert, Cretaceous chert, flysch sandstone, and radiolarite) are digitally mapped using a combination of published literature and online digital maps. In the case of Krumlovský les chert, Stránská skála chert, Cretaceous chert, and flysch sandstone, general descriptions of the geographical location of these materials are obtained using available published sources, and these are cross-referenced using online digital geological, topographic, and topological maps (http://mapy.geology.cz/website/new_tisk; http://www.mapy.cz). In the case of erratic flint, the geographical extents are already mapped in a paper publication (Růžičková *et al.* 2001, 28), and so the paper map is digitally scanned to produce a raster image file that is imported into ArcGIS9.2, digitally traced to produce a vector polygon, and georeferenced using the coordinate locations on the original map. In the case of radiolarite, specific coordinate point locations are available at http://www.flintsource.net. These are converted to the relevant coordinate (UTM) format, stored digitally using Microsoft Excel, and imported into ArcGIS9.2 to produce vector point locations. The results of the mapped lithic raw material outcrops indicate that areas of Cretaceous

Fig. 5.1. Map of the study region showing site locations, and relevant geological outcrops.

chert are located towards the central-northern regions of the study area, about 10 km west of Kůlna Cave, and distributed in a north-south linear fashion over about 15-20 km. Stránská skála chert is located about 25 km south of the Cretaceous chert outcrops and within a very localised region, Krumlovský les chert is situated about 30-40 km southwest of the Stránská skála outcrop, again localised to a relatively small region, and outcrops of radiolarite are located on the south-eastern border of the study region, and restricted to three specific locations. Flysch sandstone is largely associated with the Štramberk Karst, located about 60 km north of the radiolarite outcrop, and erratic flint is located along most of the north-eastern border regions of the study area, covering an area of about 120 km in a northwest-southeast direction, and in some places more than 50 km in a northeast-southwest direction (Fig. 5.1).

The topography of the study region is mapped using a digital elevation model (DEM) of the study region (at least 10% larger in all directions than the target study area) obtained from the USGS online resource (http://seamless.gov.us) at a resolution of 90 m x 90 m. The DEM is checked for errors, and any errors found are isolated and given pseudo values based on the elevation values of surrounding cells. Slope and aspect models are derived from the DEM using slope and aspect algorithms available in ArcGIS9.2. These are classified into broad categories (elevation – < 400 m.a.s.l, > 400 m.a.s.l and < 700 m.a.s.l, > 700 m.a.s.l and < 1100 m.a.s.l, and > 1100 m.a.s.l.; slope – flat < 5°, gentle > 5°< 10°, medium > 10° < 30°, and steep > 30°; aspect – north-facing regions 337.5° - 67.5°, and all other regions 67.5°-337.5°; areas with a slope of less than 5° are considered flat and are not assigned an aspect value). The elevation of the study region is described in Chapter 3, and so here I describe the results associated with slope and aspect only. The flattest regions of the study area are located mostly in the north-easterly and central parts of the study area, and in particular in areas associated with the Brno Unit geological formation. Steeper regions occur in the central areas of the Bohemian Massif, in and around the Moravian Karst area, the most northerly parts of the Bohemian Massif, and the southern areas of the Jeseníký Mountains. Areas with the highest concentrations of steep regions occur most in the Carpathian Mountains, particularly towards the eastern parts of the study area (Fig. 5.2). North-facing aspects are located throughout most other regions, and in particular in the northern parts of the study area – towards the northern extremes of the Jeseníký Mountains – and areas of the Carpathian Mountains. East-, south-, and west-facing slopes are again located in most regions, with a high proportion existing in the south-easterly border regions of the Jeseníký Mountains, and some of the easterly areas of the Carpathian Mountains (Fig. 5.3).

Palaeochannels are derived using the Drainage Network algorithm available in ArcGIS9.2 and the already obtained DEM of the study region. The derived drainage channels are filtered by interrogating the GIS and using associated inter-connectedness values to identify and isolate drainage channels that closely match major river channels evident on paper maps. Associated floodplains are derived by assigning buffer zones of 10 km to the derived river channels using the Buffer algorithm available in ArcGIS9.2. Buffer regions are subsequently interrogated to determine their height in comparison with associated derived palaeochannels. Any buffer regions below the level of the associated palaeochannels, plus 2 metres to take into account an estimated depth of the river when flooding (see Rhodda 2005), are considered as potential floodplain areas, and other buffer areas are considered to be dry regions. The remaining potential floodplain regions are further interrogated to determine which zones are inter-connected in terms of water flow to the associated palaeochannels. All areas that are not inter-connected are considered as potentially dry regions, and the remaining areas are considered as potential floodplain regions. A general overview of the main palaeochannels river channels has been provided in Chapter 3, and so will not be described again here. Palaeofloodplain areas are associated mostly with the rivers Morava, and Dyje. Particularly apparent are the large floodplain regions related to the central and northern regions of the river Morava, areas towards the south-central parts of the study area where the Morava and Dyje meet, and areas east of this region where the Dyje meets with rivers Svitava and Svratka. Palaeofloodplain regions are also associated with the rivers Odra, Bečva, the northern and north-eastern parts of the Svitava, Svratka, but to a much lesser extent (Fig. 5.4).

Palaeovegetation is mapped using general palaeovegetation descriptions outlined in Chapter 4 (steppe/tundra for cold periods, and parkland conifer forests with grassland steppe for prolonged warm periods). These are assigned to areas considered as being neutral in terms of vegetation growth, i.e. areas that are not inundated with water, below 700 m.a.s.l. and that have flat, gentle or medium slopes. All other regions are considered to be, to greater or lesser extents, more or less hospitable for vegetation. More hostile regions generally correspond with increases in elevation, north-facing slopes, and steep-sloping areas, and more hospitable areas largely correlate with lower elevated regions, south-, east- or west-facing slopes, and flat, gentle, or medium sloping areas. Different types of vegetation are considered as being more or less able to withstand different degrees of hostility. Deciduous flora is thought of as being easily affected by changes in the general hospitality of a region, and is only assigned to more favourable regions; evergreen coniferous species are considered as less susceptible to changes in overall hostility, able to withstand relatively hostile (i.e. cold and exposed) conditions, and not associated with areas inundated with water. Grasses are thought of as being particularly durable and able to withstand all but the most arduous conditions, and along with herbaceous vegetation are considered as being particularly associated with wet

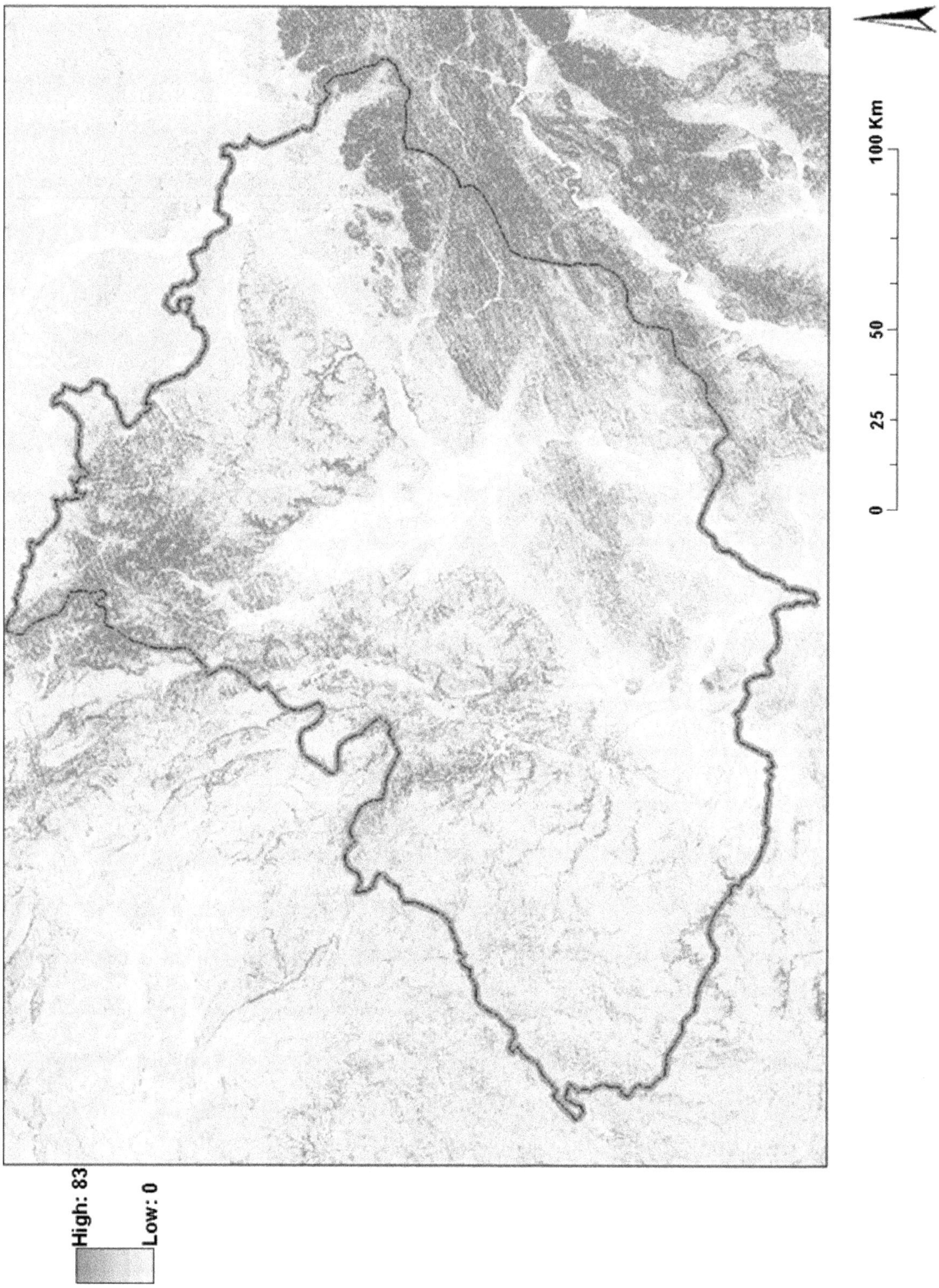

Fig. 5.2: Map of the study region showing derived slope (using ArcGIS9.2). Scale represents slope in degrees.

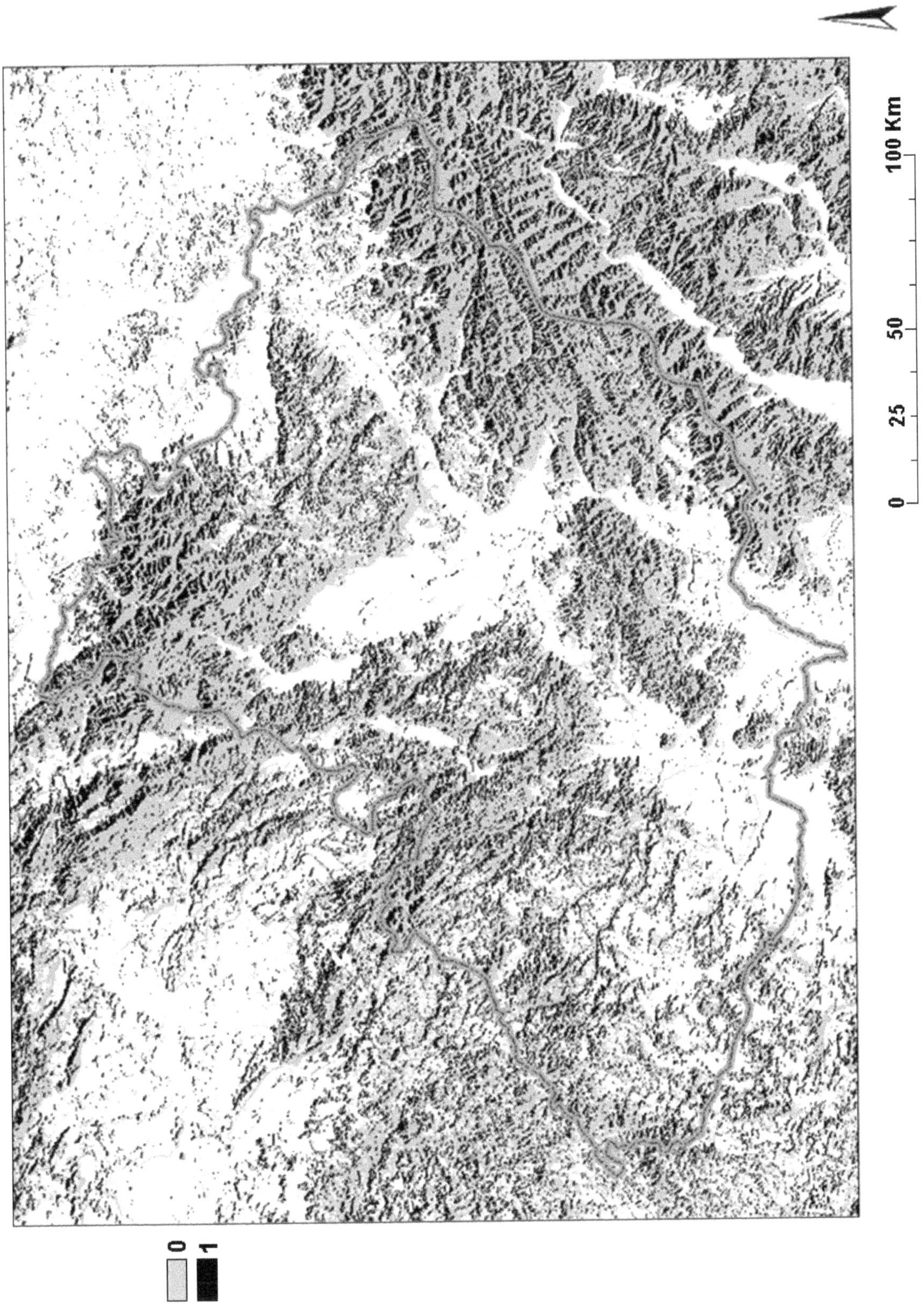

Fig. 5.3. Map of the study region showing derived aspect map (using ArcGIS9.2). 0 = north-facing slopes, 1 = east, south, and west facing slopes, and white areas are regions with no assigned aspect.

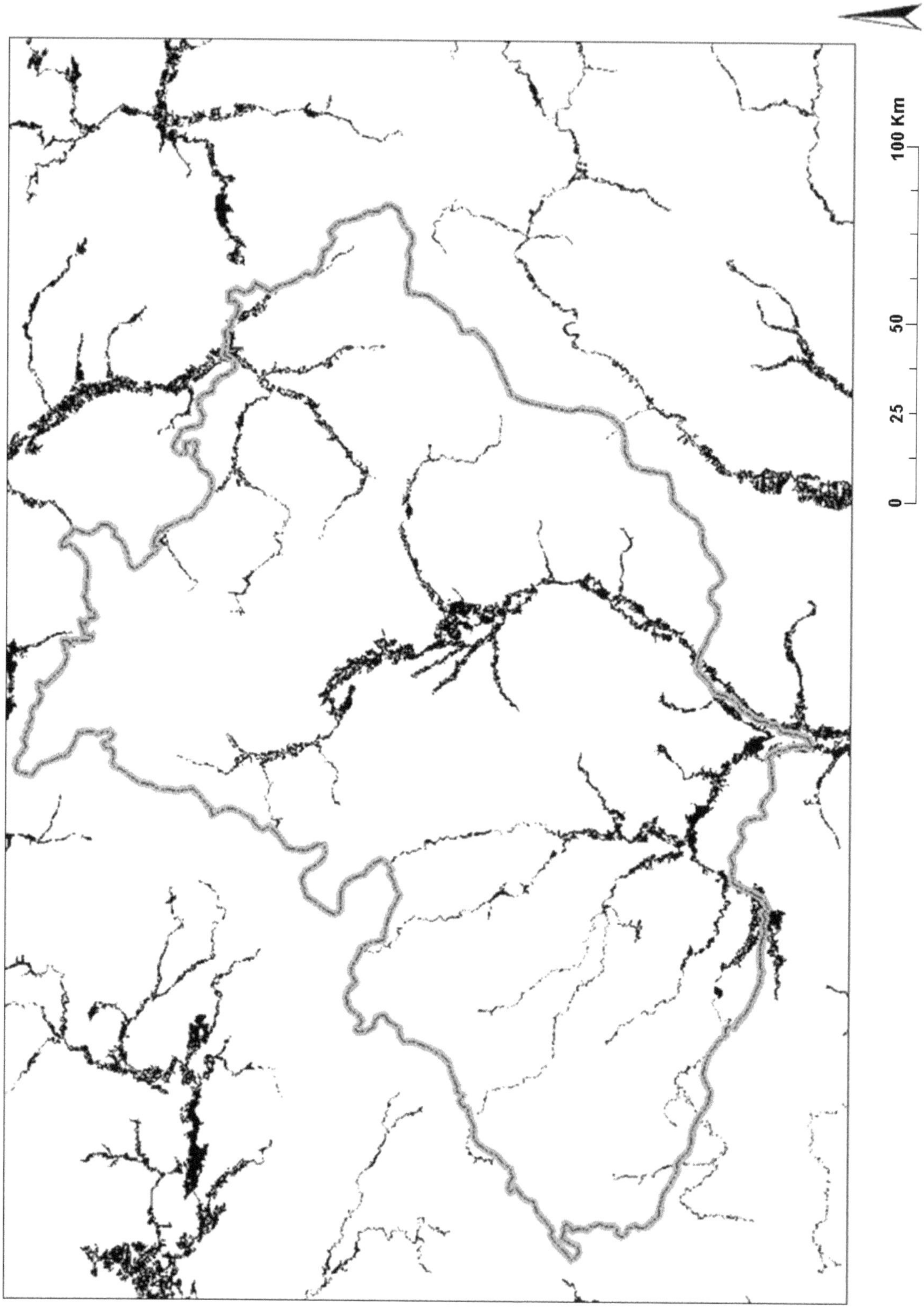

Fig. 5.4. Map of the study region showing derived palaeofloodplains (using ArcGIS9.2).

regions. In cases where vegetation is assigned to areas inundated with water, i.e. floodplains, topography is not taken into account: the presence of water is considered to be more influential in the growth and presence of associated vegetation than topographic factors, and so these regions are assigned appropriate vegetation types regardless of the topographic character of the landscape. For the most part, however, floodplain regions are located in low-altitude, flat areas, and so topography probably had little impact on associated vegetation in any case. During the winter months in warm climates and summer periods in cold climates, snow is assigned to all areas above 700 m.a.s.l., and during the winter months in cold climates, snow is also assigned to north-facing areas above 400 m.a.s.l. (Table 5.1; Table 5.2).

east-, and west-facing, steep-sloping areas grassland with stands of deciduous trees are most commonly found; in north-facing, medium-sloping regions, grassland steppe dominates with a few conifer stands; and in the most extreme areas only grassland exists. At very high altitudes (above 1100 m.a.s.l.) snow dominates during the winter months, and during the summer period grassland is most common, interrupted by bare rock on north-, south-, east-, and west-facing medium and steep slopes, with bare rock dominating on the most extreme north-facing steep-sloping areas. During cold climatic periods (Fig. 5.7 and Fig. 5.8), floodplain regions are mostly associated with steppe/tundra vegetation with herbaceous species and isolated stands of deciduous vegetation. At low altitude areas (100 – 400 m.a.s.l.), gentle slopes are mostly associated with steppe/tundra; in south-, west-,

	Vegetation type	
	Warm climates	**Cold climates**
Floodplains	Herbaceous plants and grasses with stands of deciduous trees	Steppe/tundra with herbaceous plants isolated deciduous stands
Hospitality coefficient		
+1	Grassland with conifer woodland and isolated stands of deciduous shrubs/trees	Steppe/tundra with pine and birch stands
0	Parkland conifer forest with grassland steppe	Steppe/tundra
-1	Grassland steppe with isolated stands of conifer trees	Grassland
-2	Grassland	Grassland/bare rock
-3	Grassland/bare rock	Bare rock
-4	Bare rock	Snow
-5	Snow	

Table 5.1. Palaeovegetation categories assigned to the study region. Hospitality coefficients for warm and cold climatic periods are not equivalent, i.e. -5 during warm periods does not equate to -5 during cold climatic periods.

Results of the reconstructed palaeovegetation associated with warm climatic periods (Fig. 5.5 and Fig. 5.6) indicate that floodplain regions are occupied mostly by herbaceous plant species with occasional stands of deciduous trees. At low altitudes (100 – 400 m.a.s.l.) the vegetation is mixed: gently sloping areas are dominated by parkland conifer forest with grassland steppe; in north-facing, medium sloping areas grassland/steppe with isolated stands of conifer trees are common; in steep, north-facing regions grassland dominates; and in south-, west-, and east-facing slopes, isolated stands of deciduous trees are present along with parkland conifer forests and grassland. At medium altitudes (400 – 700 m.a.s.l.), parkland conifer forest dominate in gently sloping areas, and in medium sloping, north-facing regions grassland steppes with isolated stands of conifer trees are most common; in north-facing, steep-sloping areas, grassland is most prevalent, and in medium and steep-sloping, south-, west-, east-facing regions parkland conifer forest and grassland steppe exist with a few stands of deciduous trees. At high altitudes (700 – 1100 m.a.s.l.), the landscape is covered with snow during the winter months; for the summer period, parkland conifer forest with grassland steppe exists in gently sloping areas, and in south-, west-, and east-facing, medium-sloping areas isolated stands of deciduous species exist together with parkland conifer forest and grassland steppe; in north-, and east-facing medium sloping areas steppe/tundra with some pine and birch are evident; and in north-facing, medium- and steep-sloping areas, grassland vegetation is interrupted by areas of bare rock. At medium altitudes (400 – 700 m.a.s.l.), gently sloping areas are covered mostly with steppe/tundra; medium-sloping, north-facing regions are populated with grassland and covered in snow during the winter months; south-, west-, and east-facing medium slopes are associated with stands of pine and birch, but are dominated by steppe/tundra vegetation; steep, north-facing areas are largely covered with snow during the winter, with grassland/bare rock existing during the summer months; and in areas of south-, west-, and east-facing steep slopes grassland and bare rock dominate. At high and very high altitudes (700 – 1100 m.a.s.l. and > 1100 m.a.s.l.) snow is always present.

Friction maps are derived using already created digital slope, palaeohydrology, and palaeovegetation maps. Friction values are assigned to each component (slope, palaeohydrology and palaeovegetation) in terms of how much each element is considered to affect human and animal movement. Friction values are geographically combined using the Weighted Sums algorithm available in ArcGIS9.2 to create aggregate friction values for summer and winter periods in warm and cold climates. Friction values associated with slope are given an overall weighting of 90%, and are created using the already derived digital slope map, and the following equation: C=tan s/ tan 1, where s=slope and C=cost. Friction values associated with vegetation are given an overall weighting of 10%, and are assigned different friction values on the basis of the relative energy expected to be expended when walking through different vegetation types. Generally, vegetation regions that coincide with floodplain regions are given the highest values, particularly during warm climatic and seasonal periods, areas that are thought to be more densely occupied with vegetation are given high friction values, and vegetation regions that coincide with snow-covered areas are also given high friction values. Also, high friction values are assigned to all major rivers independent of all other friction values. The results of the derived friction map associated with summer months during warm climatic periods (Fig. 5.9) indicate that the highest friction values are associated mostly with floodplain regions and to a less extent with some areas of the Bohemian Massif and the Carpathian Mountains. Relatively low friction values are mostly associated with areas of the Brno Unit, and some areas in the north of the study region in the Jeseníký Mountains, and isolated areas in the eastern regions of the Carpathian Mountains. Friction maps associated with winter periods during warm climates (Fig. 5.10) are largely similar to those during the summer months, but high friction values are apparent in some northern areas of the study area in the Jeseníký Mountains and some areas of the Carpathian Mountains. These areas largely correspond with areas of snow cover. Friction maps associated with summer months during cold climatic periods (Fig. 5.11) have the highest values in the Jeseníký Mountains, and some areas of the Carpathians. Areas with the lowest friction values generally correspond with floodplain regions, some areas in the Carpathian Mountains, and some regions of the central and northern areas of the Bohemian Massif. Regions with mid-range friction values are associated with areas mostly adjacent to floodplains and in areas corresponding to the Brno Unit. Friction maps associated with winter months in cold climates (Fig. 5.12) are largely similar to those for summer months during cold climates, except there is a higher concentration of high friction values in the northern parts of the study region (Jeseníký Mountains and Carpathian Mountains) corresponding with expected increased snow cover in these areas.

The distribution patterns of commonly hunted prey species (mammoth, wolf, horse, reindeer, bison, Arctic fox, hare) are mapped using available published literature, together with newly created palaeovegetation, and topographic maps. Regions associated with each faunal species are identified and highlighted by interrogating the relevant digital maps using ArcGIS9.2. This is mostly undertaken in terms of the preferred dietary habits of each faunal species, and the distribution of associated palaeovegetation types. Where relevant, I also attempt to take into account non-dietary preferences and accessibility. I briefly describe the preferred habitat of each faunal species, the regions they are assigned to for this project, and the results in terms of summer and winter periods during warm and cold climates.

Mammoth probably prefer tough, poor-quality grasses with woody plants and small forbs (Kurtén 1968; Stewart *et al.* 2003, 23), but are probably also associated with forest biotypes where they can browse for flora such as larch, birch, willow, sedge mosses and grasses (West 1997; Boyle 1990). River valleys are probably the principal source of food for mammoth, especially during cold periods, during which time herds are largely confined to these regions (Velichko and Zelikson 2005, 144, 149). During warm periods, floodplains are probably not able to support the weight of mammoths, as these areas are likely to be inundated with water. As a result, mammoth probably graze on higher, dry sites (Velichko and Zelikson 2005, 143-144), and are probably restricted to flat or gentle slopes (Boyle 1990; Kurtén 1968; Sturdy *et al.* 1997). For this study, during the summer months in warm climatic periods mammoth are confined to gently sloping areas below 700 m.a.s.l., with parkland conifer forest/grassland steppe and isolated stands of deciduous shrubs/trees. During winter months in warm climates they are assigned to gently sloping areas below 400 m.a.s.l., and associated with either floodplain regions or parkland conifer forest/grassland steppe areas. During both the summer and winter in cold climatic periods, mammoth are assigned to gently sloping areas below 400 m.a.s.l. and floodplain areas (Table 5.3).

Wolf are mostly associated with areas of woodland or tree cover (Jędrzejewski *et al.* 2004, 229; Massolo and Meriggi 1998, 102-103). They make frequent use of river valleys and smaller tributary valleys (Glenz *et al.* 2001, 60), and are probably restricted by steep sloping areas (Paquet *et al.* 2001, 54). For this project, during the summer months in warm climates, wolf are assigned to flat, gentle/medium sloping regions (with no restriction on elevation), and parkland conifer forests with isolated stands of deciduous trees. During the winter months in warm climates, they are also associated with floodplain regions. In summer and winter periods during cold climates, wolf are associated with flat, gentle/medium sloping areas below 1100 m.a.s.l., and steppe/tundra with stands of pine and birch (Table 5.3).

Horses are associated with grassy, shrubby areas including open woodland and woodland margins, heathland and grassland (Clark 1983), and areas of tundra (Boyle 1990). They probably avoid soft marshy ground, areas with deep snow, and large tracks of forest (Linklater *et al.* 2000, 149), and prefer flat or gently sloping areas (Sturdy *et al.* 1997). They inhabit higher elevated regions during the autumn and winter months, and lower regions at the beginning of spring prior to foaling (Linklater *et al.* 2000, 149). They can be found on north-facing aspects, low altitudes (below 400 m.a.s.l.), and on gentle slopes in spring and summer (Butzer 1986, 204; Linklater *et al.* 2000, 139; Sturdy *et al.* 1997). For

Region			Climate/season			
			Warm		Cold	
			Summer	Winter	Summer	Winter
			Vegetation description			
Floodplain			Herbaceous plants and grasses with stands of deciduous trees (F)	Herbaceous plants and grasses with stands of deciduous trees (F)	Steppe/tundra with isolated deciduous stands (F)	Steppe/tundra with isolated deciduous stands (F)
Topographic region						
Altitude	Slope	Aspect				
<400m	Gentle		Parkland conifer forest with grassland steppe (0)	Parkland conifer forest with grassland steppe (0)	Steppe/tundra (0)	Steppe/tundra (0)
	Medium	N	Grassland steppe with isolated stands of conifer trees (-1)	Grassland steppe with isolated stands of conifer trees (-1)	Grassland (-1)	Grassland (-1)
		SWE	Parkland conifer forest with grassland steppe and isolated stands of deciduous shrubs/trees (+1)	Parkland conifer forest with grassland steppe and isolated stands of deciduous shrubs/trees (+1)	Steppe/tundra with pine and birch stands (+1)	Steppe/tundra with pine and birch stands (+1)
	Steep	N	Grassland (-2)	Grassland (-2)	Grassland/bare rock (-2)	Grassland/bare rock (-2)
		SWE	Grassland with conifer woodland and isolated stands of deciduous shrubs/trees (+1)	Grassland with conifer woodland and isolated stands of deciduous shrubs/trees (+1)	Grassland/bare rock (-2)	Grassland/bare rock (-2)
>400m <700m	Gentle		Parkland conifer forest with grassland steppe (0)	Parkland conifer forest with grassland steppe (0)	Steppe/tundra (0)	Steppe/tundra (0)
	Medium	N	Grassland steppe with isolated stands of conifer trees (-1)	Grassland steppe with isolated stands of conifer trees (-1)	Grassland (-1)	Snow (-4)
		SWE	Parkland conifer forest with grassland steppe and isolated stands of deciduous shrubs/trees (+1)	Parkland conifer forest with grassland steppe and isolated stands of deciduous shrubs/trees (+1)	Steppe/tundra with pine and birch stands (+1)	Steppe/tundra with pine and birch stands (+1)
	Steep	N	Grassland (-2)	Grassland steppe (-2)	Grassland/bare rock (-2)	Snow (-4)
		SWE	Parkland conifer forest with grassland steppe and isolated stands of deciduous shrubs/trees (+1)	Parkland conifer forest with grassland steppe and isolated stands of deciduous shrubs/trees (+1)	Grassland/bare rock (-2)	Grassland/bare rock (-2)
>700m < 1100m	Gentle		Parkland conifer forest with grassland steppe (-1)	Snow (-5)	Snow (-4)	Snow (-4)
	Medium	N	Grassland steppe with isolated stands of conifer trees (4)	Snow (-5)	Snow (-4)	Snow (-4)
		SWE	Parkland conifer forest with grassland steppe and isolated stands of deciduous shrubs (-2)	Snow (-5)	Snow (-4)	Snow (-4)
	Steep	N	Grassland (-2)	Snow (-5)	Snow (-4)	Snow (-4)
		SWE	Grassland with stands of conifer trees (-1)	Snow (-5)	Snow (-4)	Snow (-4)
>1100m	Gentle		Grassland (-2)	Snow (-5)	Snow (-4)	Snow (-4)
	Medium	N	Grassland/bare rock (-3)	Snow (-5)	Snow (-4)	Snow (-4)
		SWE	Grassland (-2)	Snow (-5)	Snow (-4)	Snow (-4)
	Steep	N	Bare rock (-4)	Snow (-5)	Snow (-4)	Snow (-4)
		SWE	Grassland/bare rock (-3)	Snow (-5)	Snow (-4)	Snow (-4)

Table 5.2. Palaeovegetation categories assigned to different areas of the the study region, according to hydrology (i.e. floodplains) and topography (i.e. altitude, slope, and aspect) for summer and winter periods during cold and warm climates. F = floodplains, +1 = favourable region, 0 = neutral region, -1, -2, -3, -4, -5 = hostile regions.

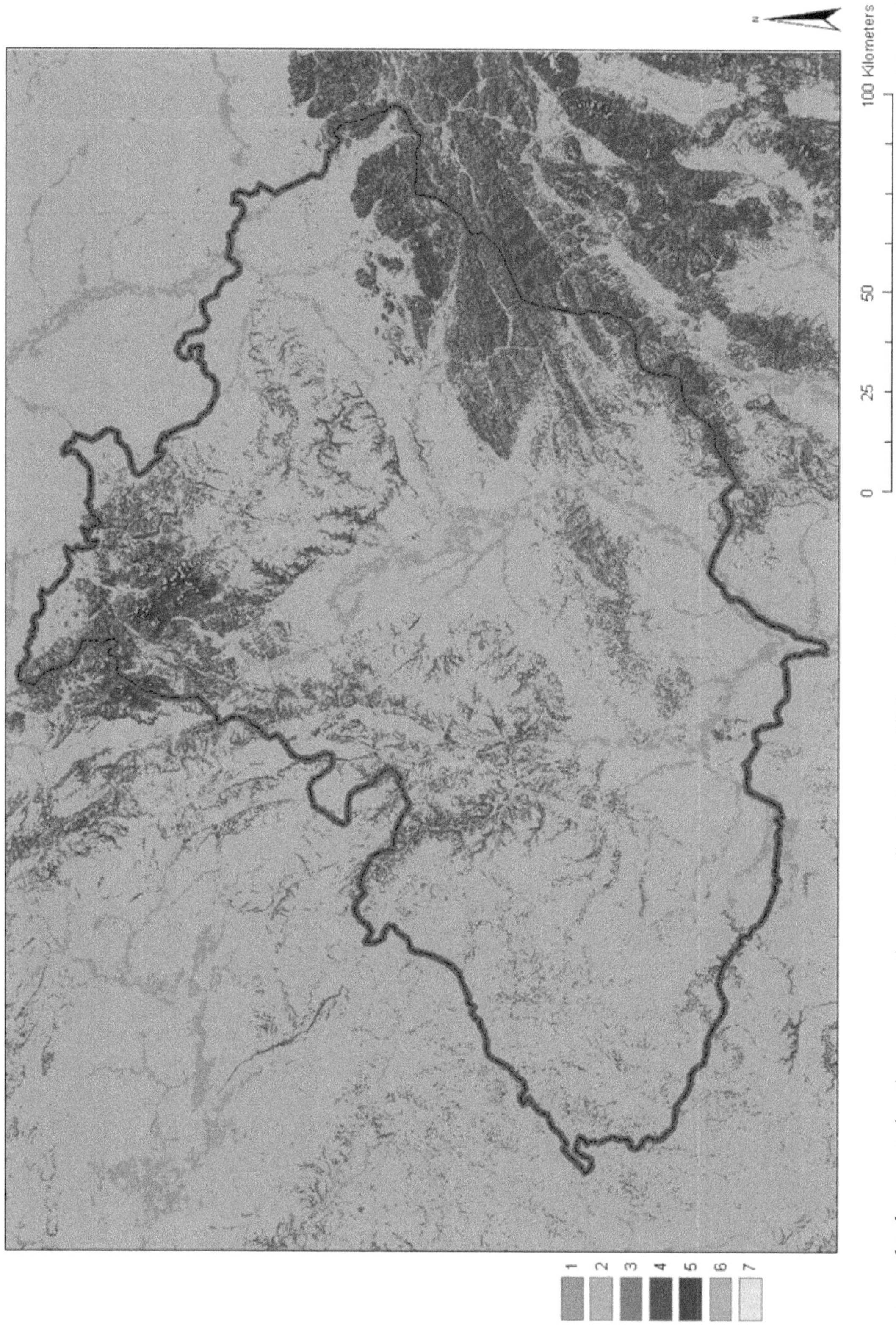

Fig. 5.5. Reconstructed palaeovegetation (summer months, warm climates). *1 = Herbaceous plants and grasses with stands of deciduous trees, 2 = grassland with conifer woodland and isolated stands of deciduous shrubs/trees, 3 = parkland/conifer forest with grassland steppe, 4 = grassland steppe with isolated stands of conifer trees, 5 = grassland, 6 = grassland/bare rock, and 7 = bare rock.*

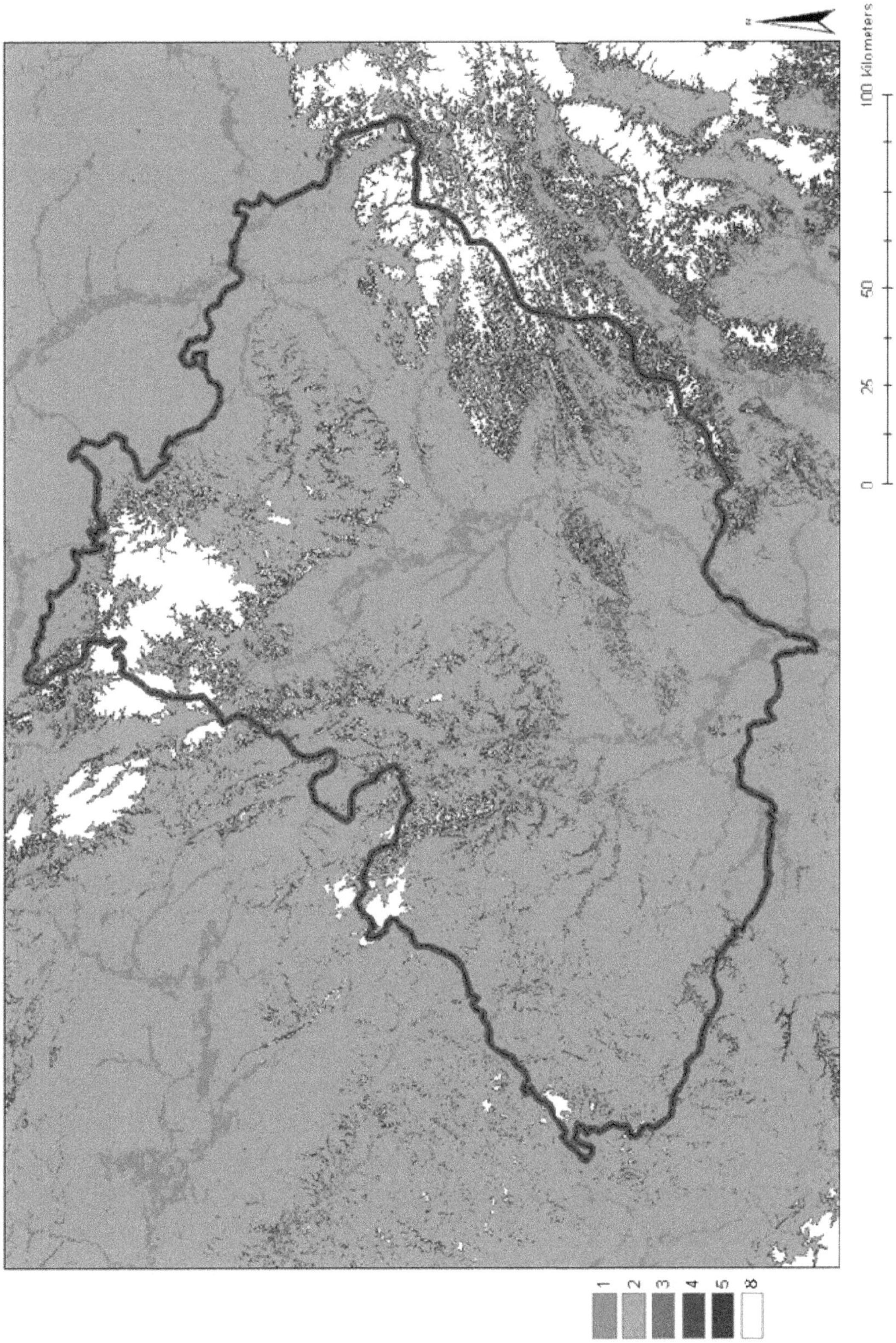

Fig. 5.6. Reconstructed palaeovegetation (winter months, warm climates). *1 = Herbaceous plants and grasses with stands of deciduous trees, 2 = grassland with conifer woodland and isolated stands of deciduous shrubs/trees, 3 = parkland/conifer forest with grassland steppe, 4 = grassland steppe with isolated stands of conifer trees, 5 = grassland, and 8 = snow.*

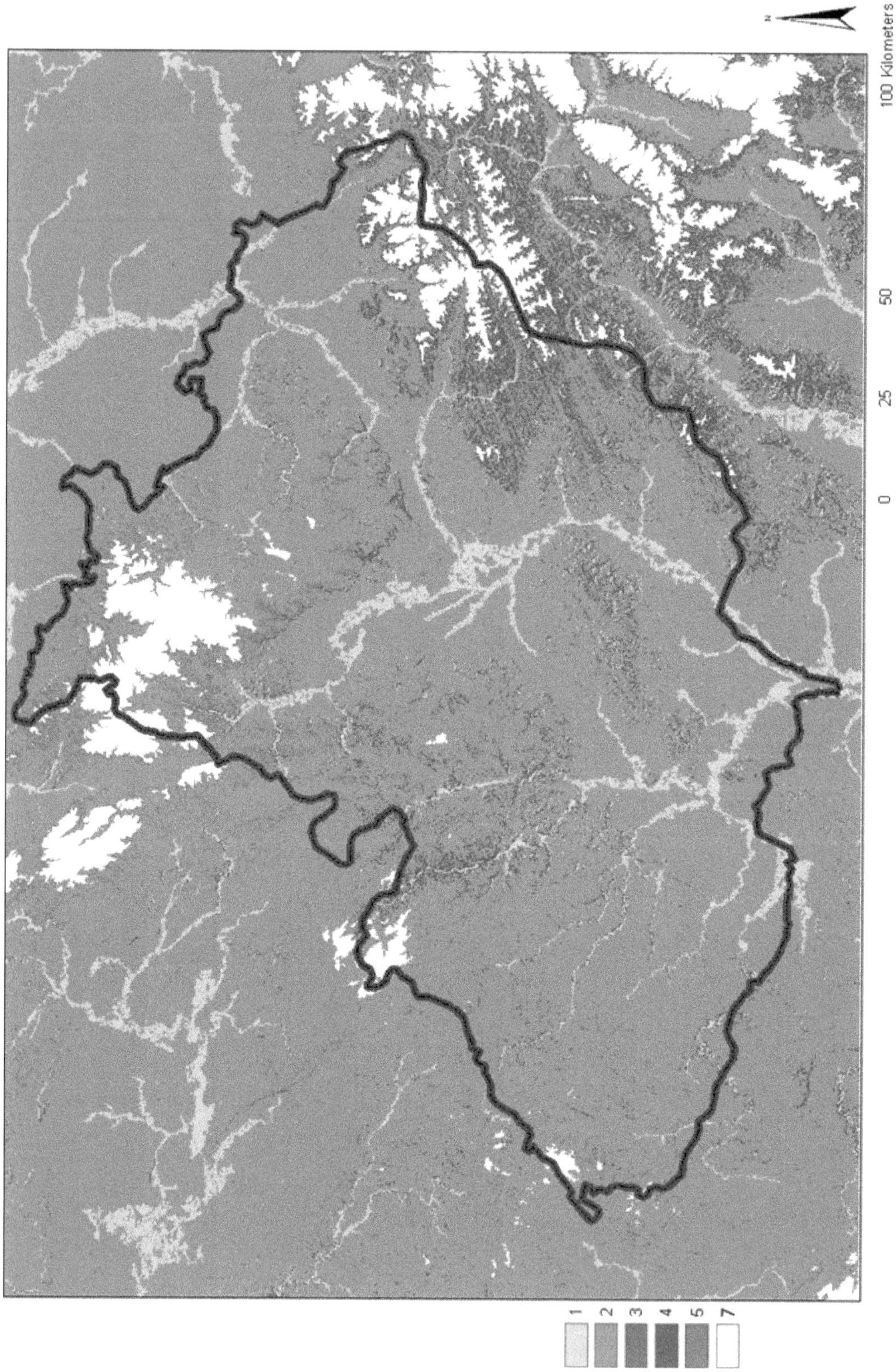

Fig. 5.7. Reconstructed palaeovegetation (summer months, cold climates). *1 = steppe/tundra with herbaceous plants and isolated deciduous stands, 2 = steppe/tundra with pine and birch stands, 3 = steppe/tundra, 4 = grassland, 5 = grassland/bare rock, 7 = snow.*

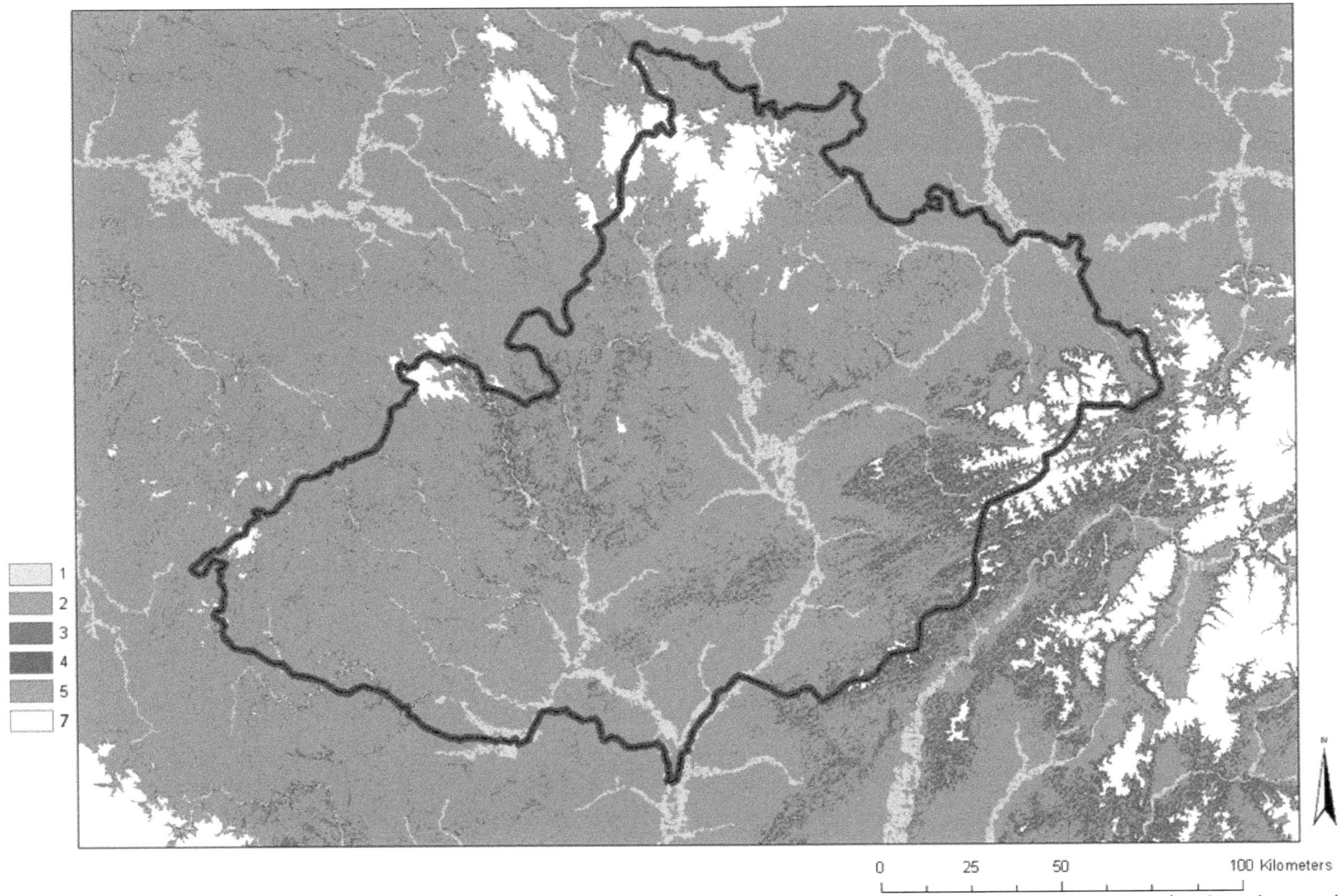

Fig. 5.8. Reconstructed palaeovegetation (winter months, cold climates). *1 = steppe/tundra with herbaceous plants and isolated deciduous stands, 2 = steppe/tundra with pine and birch stands, 3 = steppe/tundra, 4 = grassland, 5 = grassland/bare rock, 7 = snow.*

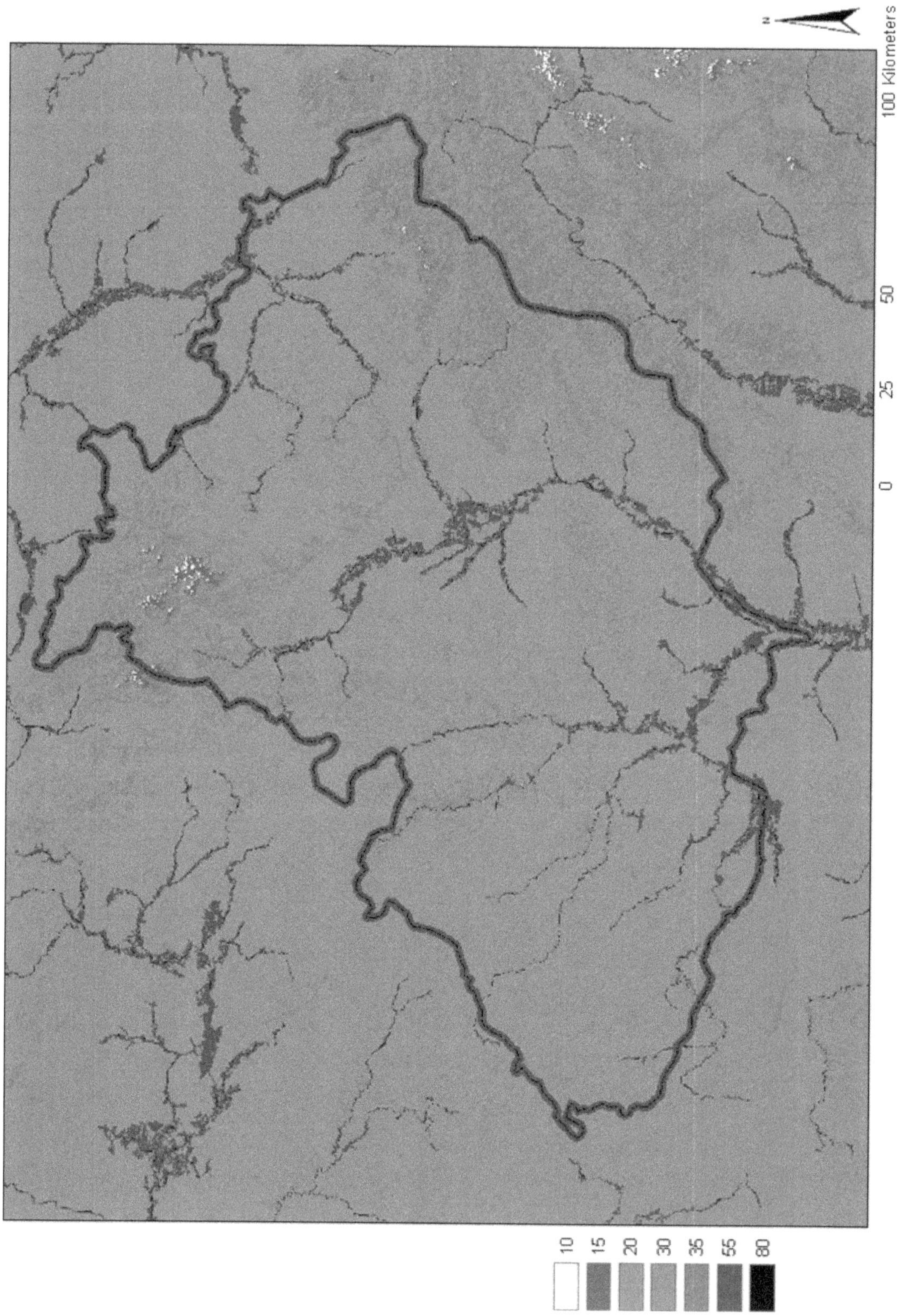

Fig. 5.9. Map of the study region showing derived friction map (summer months, warm climates).Key shows derived friction values (%), with higher values indicating increased friction.

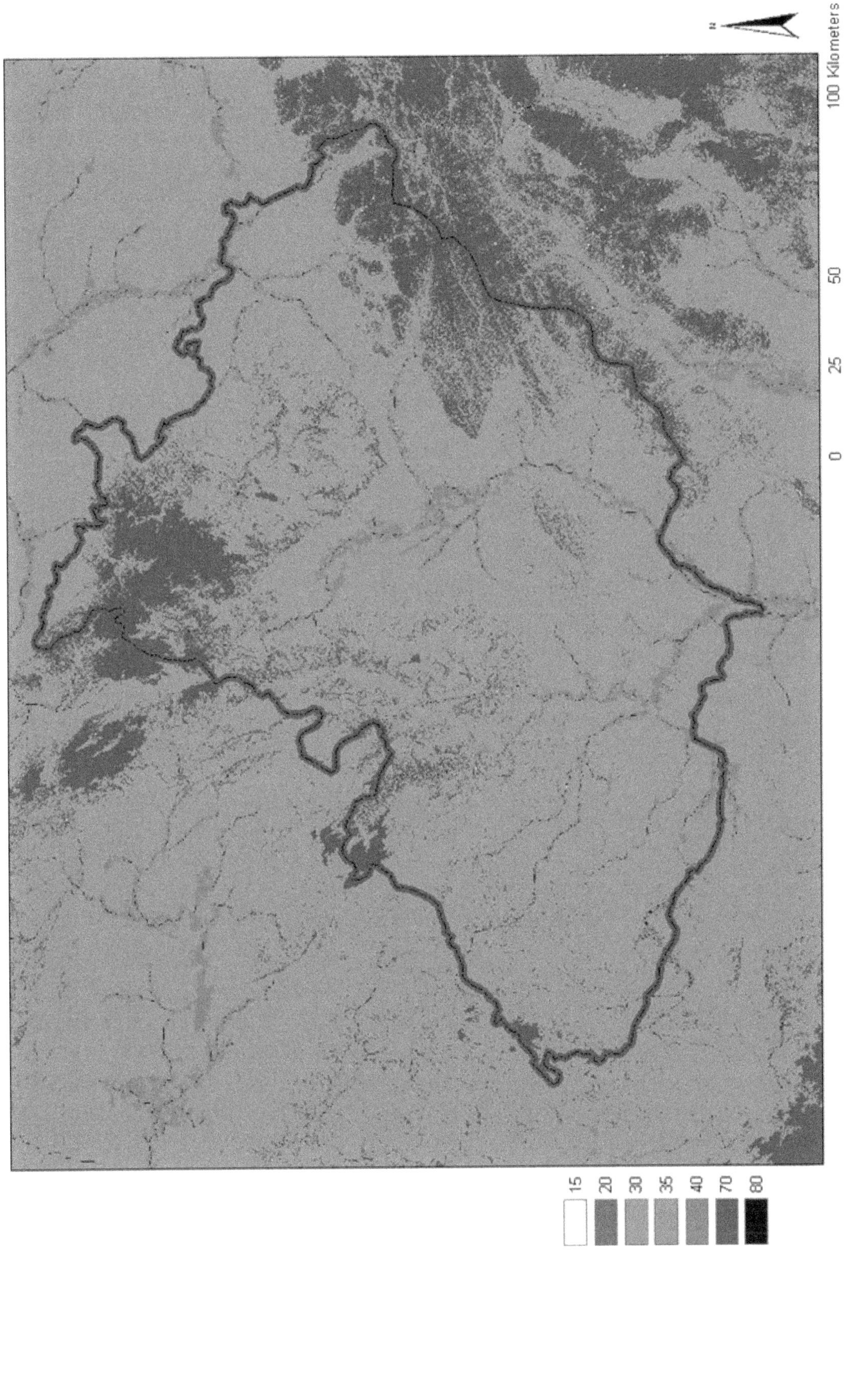

Fig. 5.10. Map of the study region showing derived friction map (winter months, warm climates). Key shows derived friction values (%), with higher values indicating increased friction.

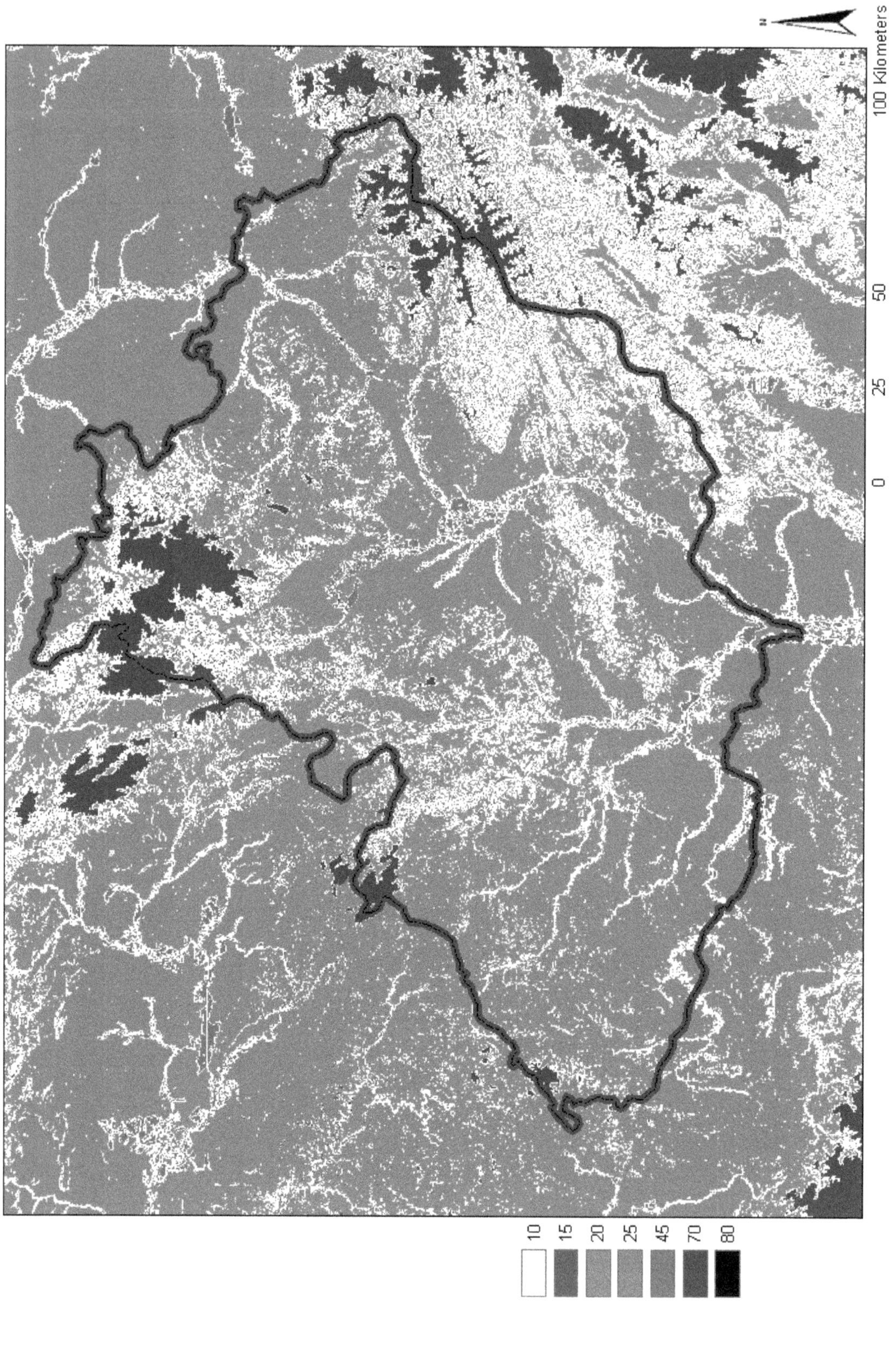

Fig. 5.11 Map of the study region showing derived friction map (summer months, cold climates). Key shows derived friction values (%), with higher values indicating increased friction.

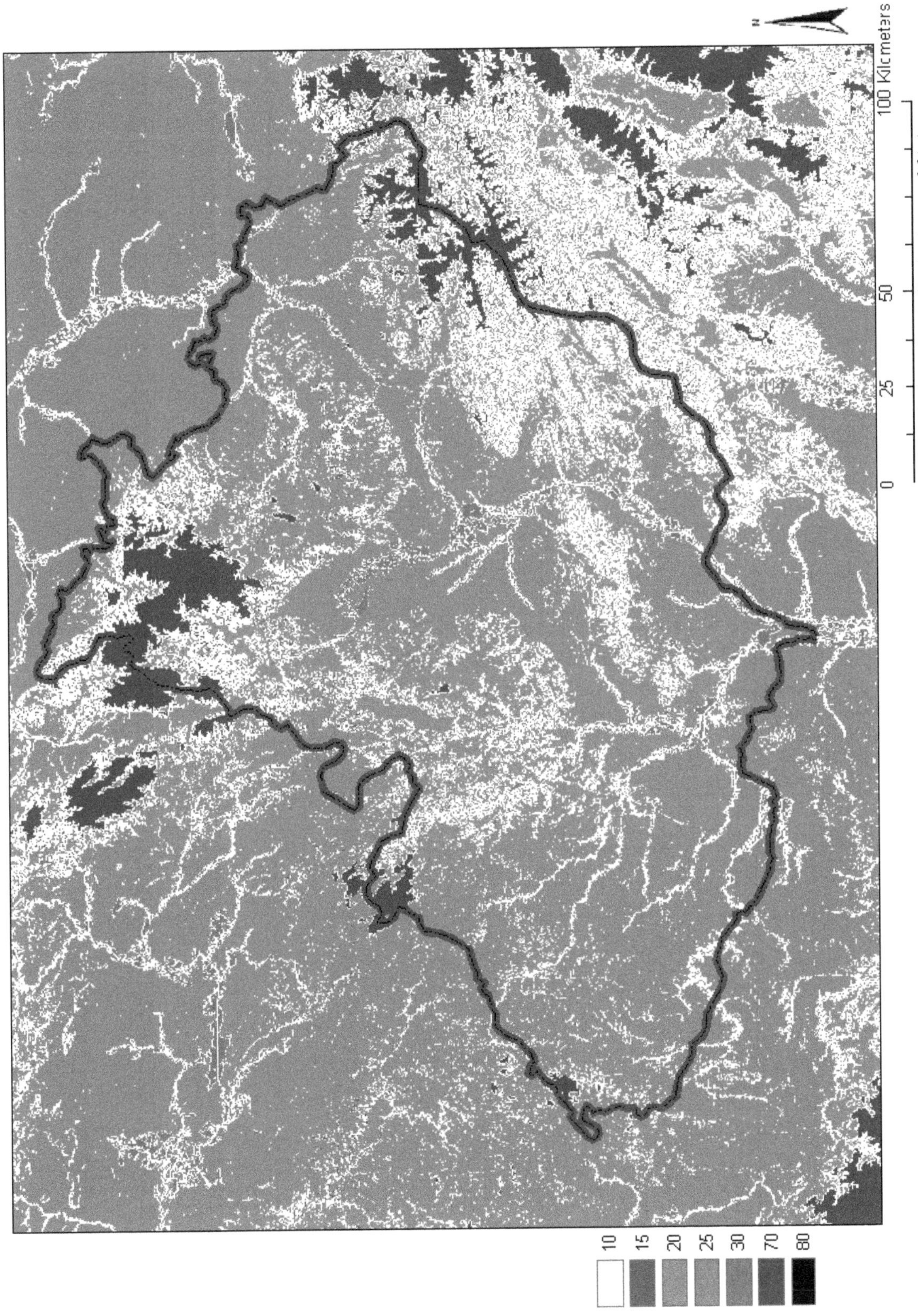

Fig. 5.12. Map of the study region showing derived friction map (winter months, cold climates). Key shows derived friction values (%), with higher values indicating increased friction.

Species	Climate	Season	Topography			Vegetation category						
			Elevation (m.a.s.l.)	Slope	Aspect							
Mammoth												
	Warm					1	2	3	4	5	6	7
		Summer	<700m	None/gentle	-		x	x				
		Winter	<400m	None/gentle	-	x	x					
	Cold											
		Summer	<400m	None/gentle	-	x						
		Winter	<400m	None/gentle	-	x						
Wolf												
	Warm					1	2	3	4	5	6	7
		Summer	-	Non-gentle/medium	-		x					
		Winter	-	Non-gentle/medium	-	x	x					
	Cold											
		Summer	<1100m	Non-gentle/medium	-		x					
		Winter	<1100m	Non-gentle/medium	-		x					
Horse												
	Warm					1	2	3	4	5	6	7
		Summer	<400m	None/gentle	-		x	x	x	x		
		Winter	<700m	None/gentle/medium	NWE			x	x	x		
	Cold											
		Summer	<400m	None/gentle	-			x	x			
		Winter	<700m	None/gentle/medium	NWE			x	x			
Reindeer												
	Warm					1	2	3	4	5	6	7
		Summer	>400m<1100m	None/gentle	-	x	x	x	x			
		Winter	<400m	None/gentle	-	x	x	x	x			x
	Cold											
		Summer	>400m<1100m	None/gentle	-	x	x	x	x			
		Winter	<400m	None/gentle	-	x	x	x	x			x
Bison												
	Warm					1	2	3	4	5	6	7
		Summer	<700m	None/gentle	-	x			x	x		
		Winter	<700m	None/gentle	-	x			x			
	Cold											
		Summer	<700m	None/gentle	-	x		x	x			
		Winter	<700m	None/gentle	-	x		x	x			
Arctic fox												
	Cold					1	2	3	4	5	6	7
		Summer	<1100m	Non-gentle/medium	-		x	x				
		Winter	<1100m	Non-gentle/medium	-	x	x	x				
Hare												
	Cold					1	2	3	4	5	6	7
		Summer	<700m	Non-gentle/medium	-	x	x					
		Winter	>400m	Non-gentle/medium	-	x	x					

Table 5.3. Topographic and palaeovegetation regions to which faunal species discussed in the text are assigned. Vegetation category (1, 2, 3, 4, 5, 6, and 7) correspond to vegetation categories defined in Table 5.1

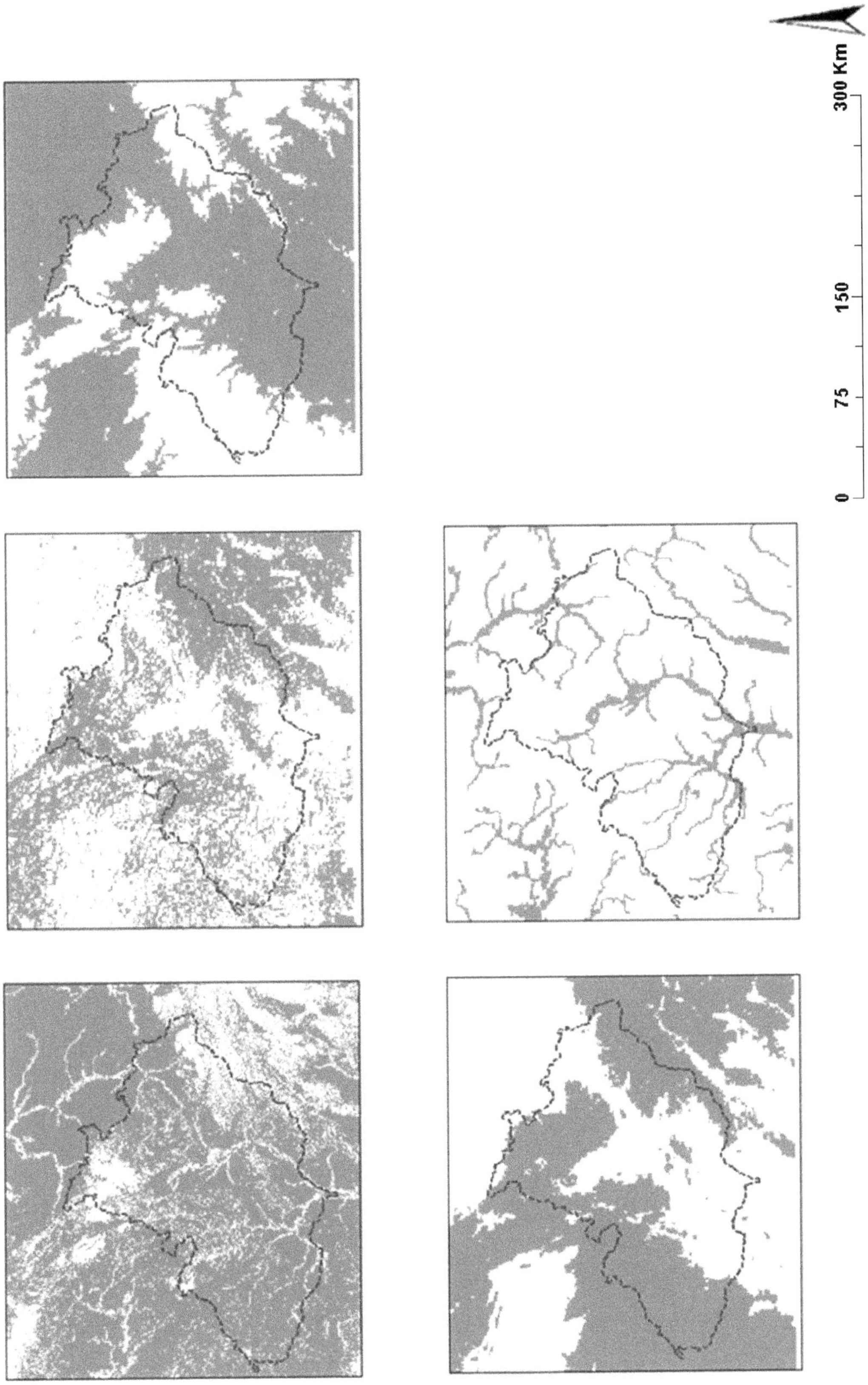

Fig. 5.13. Maps of the study region showing the derived distribution of faunal species discussed in the text (summer months, warm climates). *Top left = mammoth, top middle = wolf, top right = horse, bottom left = reindeer, bottom middle = bison.*

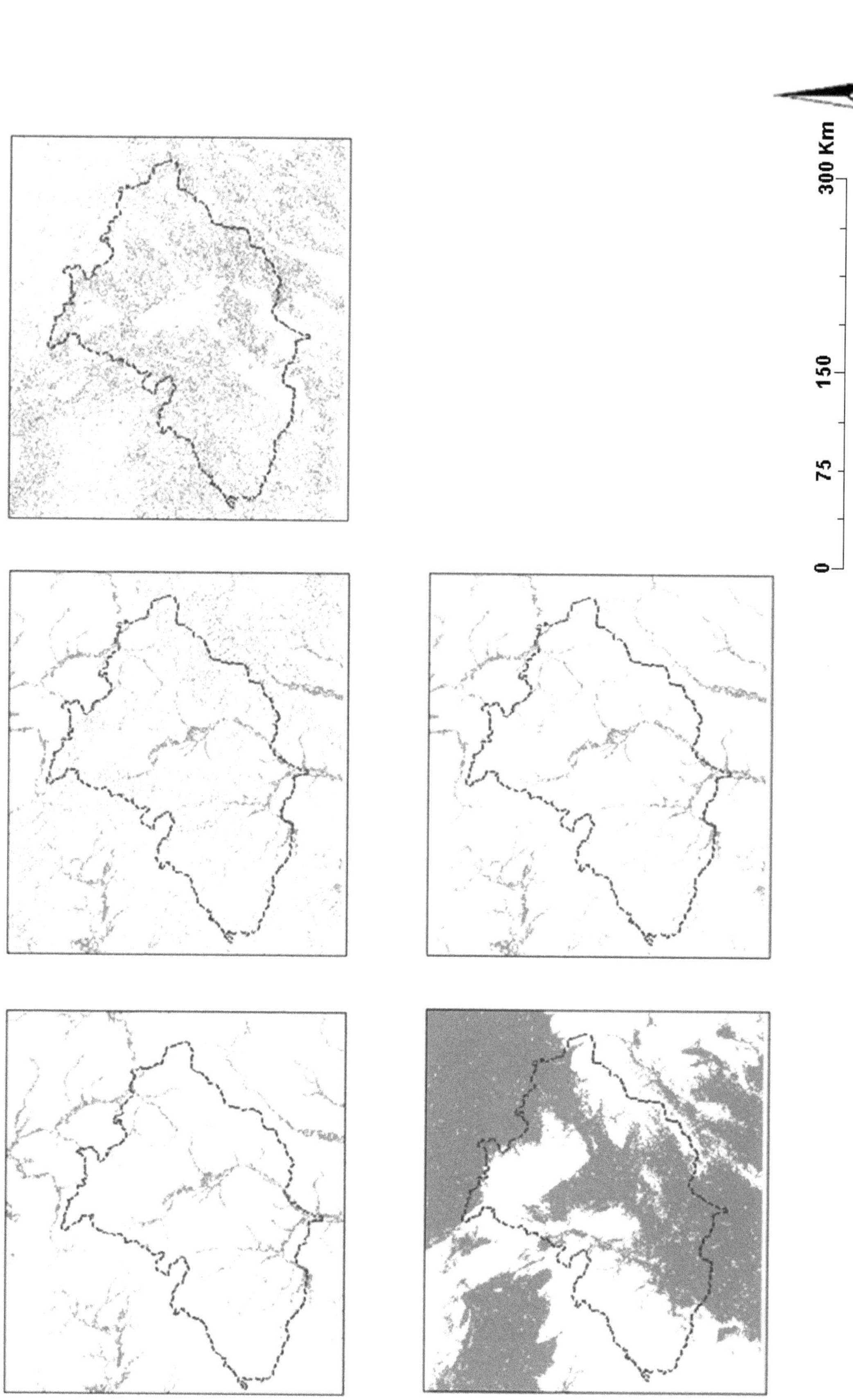

Fig. 5.14. Maps of the study region showing the derived distribution of faunal species discussed in the text (winter months, warm climates). *Top left = mammoth, top middle = wolf, top right = horse, bottom left = reindeer, bottom middle = bison.*

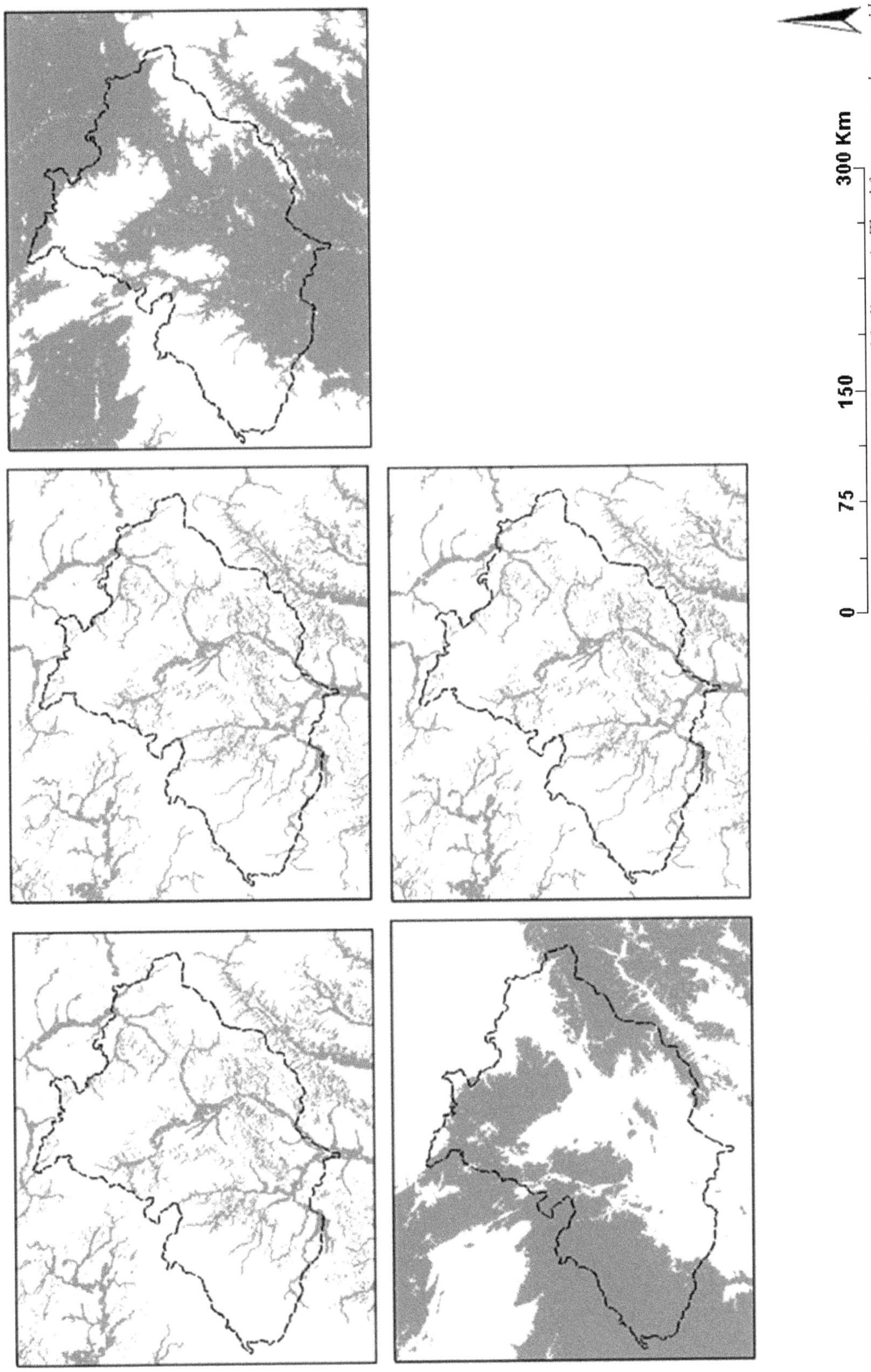

Fig. 5.15. Maps of the study region showing the derived distribution of faunal species discussed in the text (summer months, cold climates). *Top left* = *mammoth, top middle* = *wolf, top right* = *horse, bottom left* = *reindeer, bottom middle* = *bison.*

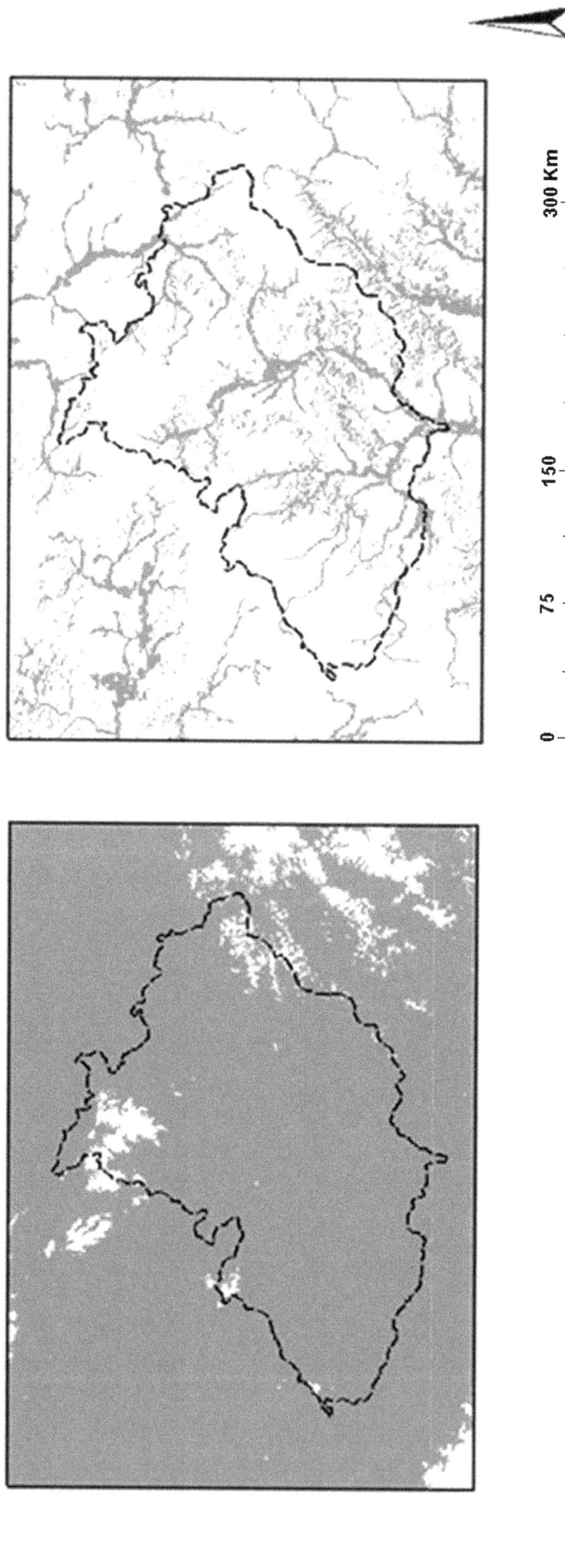

Fig. 5.16. Maps of the study region showing the derived distribution of faunal species discussed in the text (summer months, cold climates (cont.)). *Top left = Arctic fox, top right = hare.*

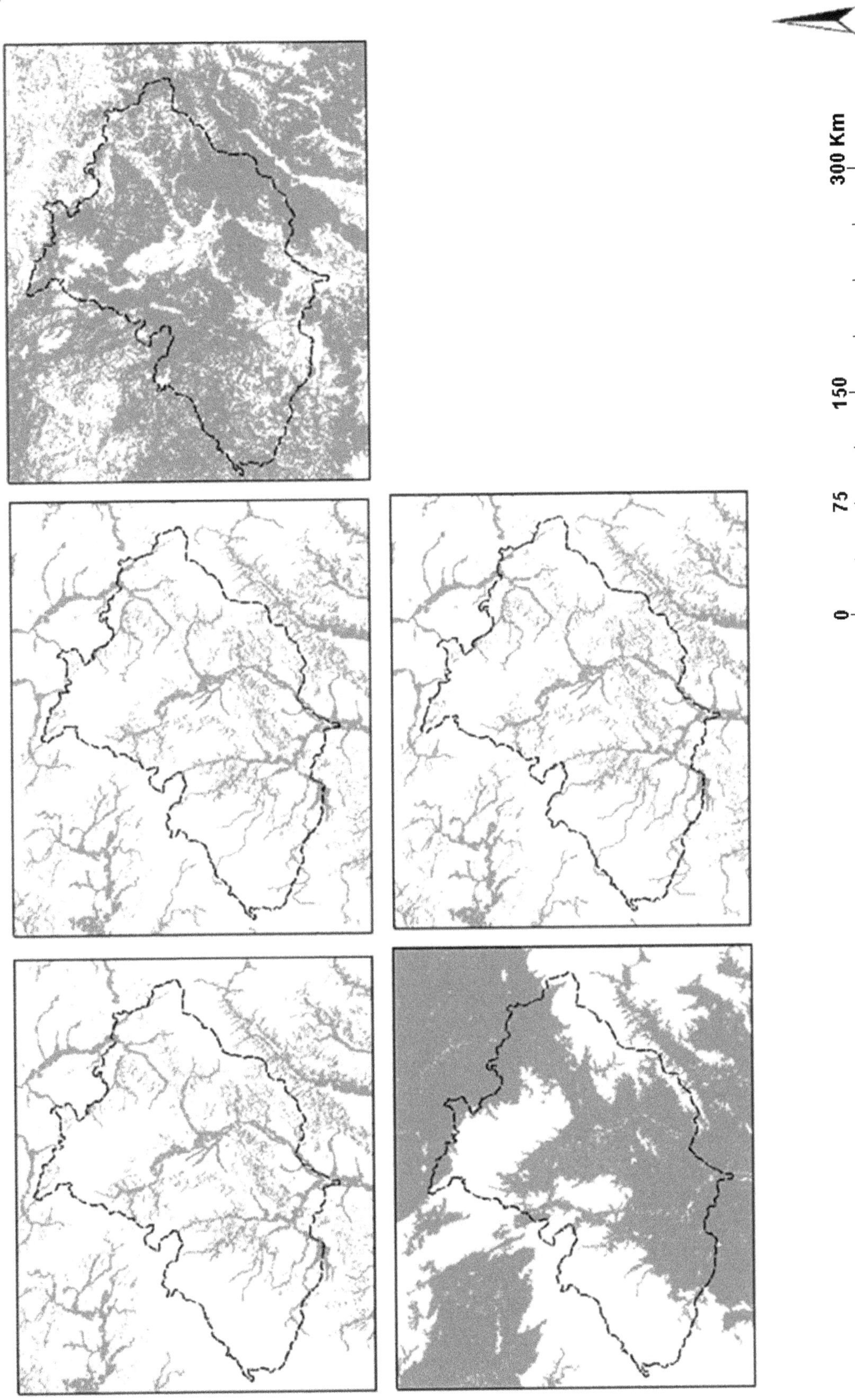

Fig. 5.17. Maps of the study region showing the derived distribution of faunal species discussed in the text (winter months, cold climates). *Top left=mammoth, top middle = wolf, top right = horse, bottom left = reindeer, bottom middle = bison.*

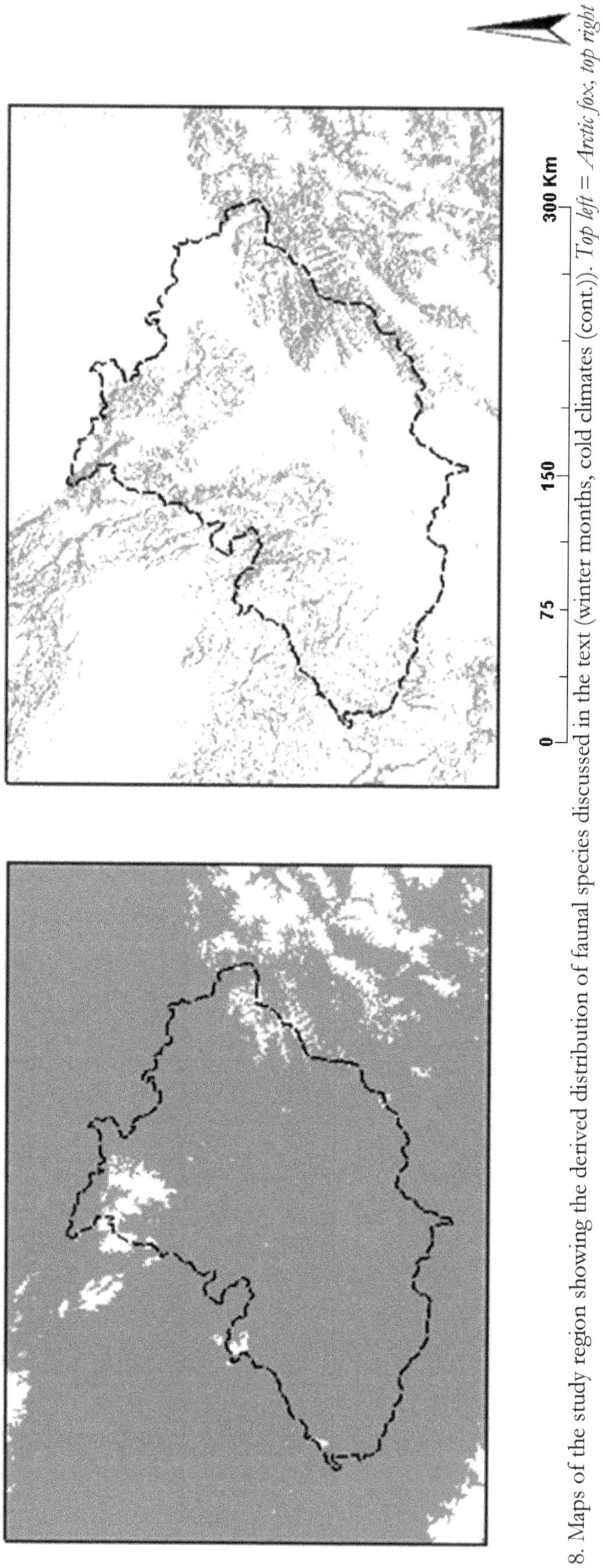

Fig. 5.18. Maps of the study region showing the derived distribution of faunal species discussed in the text (winter months, cold climates (cont.)). *Top left = Arctic fox, top right = hare.*

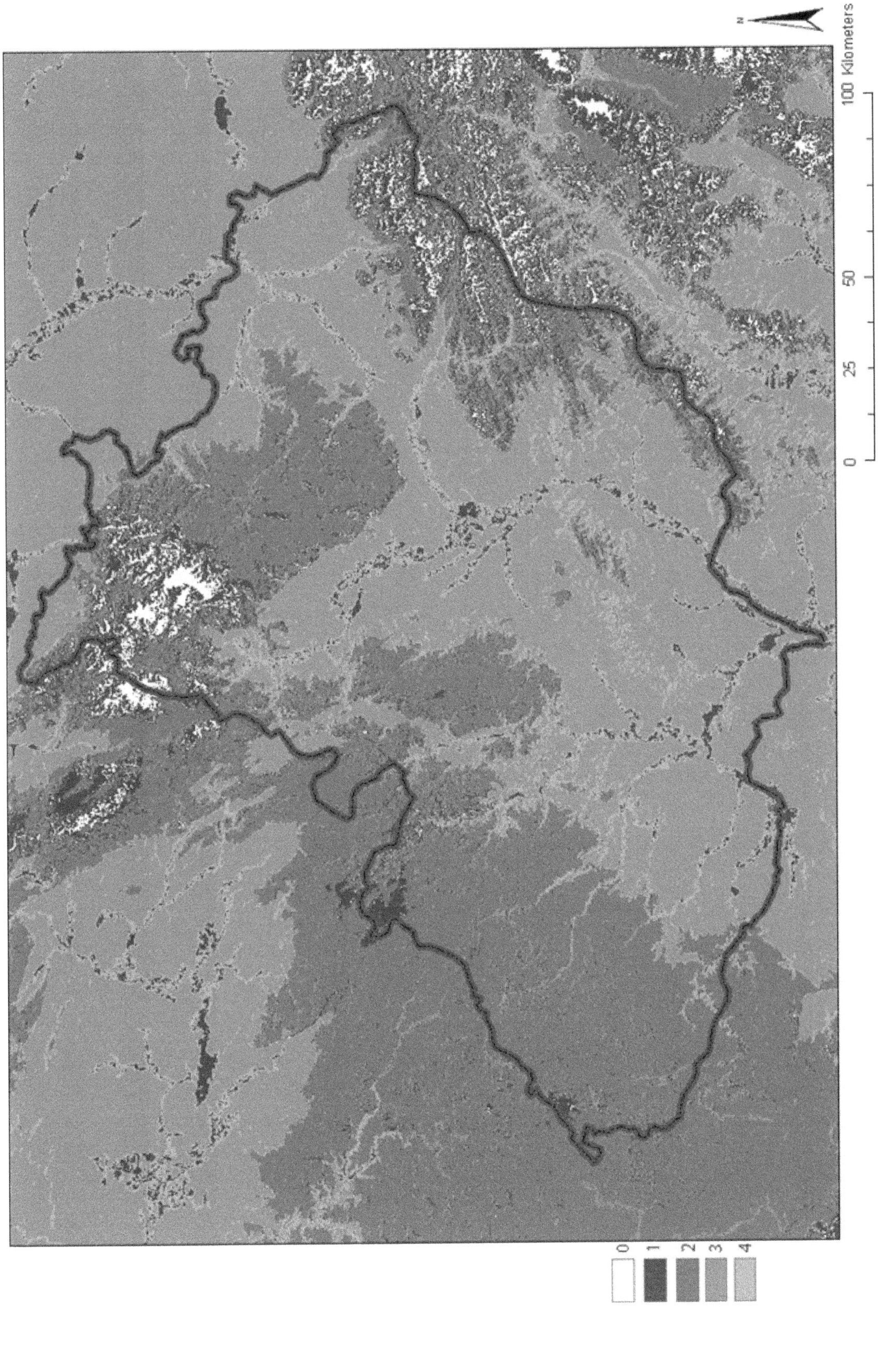

Fig. 5.19. Map of the study region showing derived species diversity (summer months, warm climates). Key shows number of species.

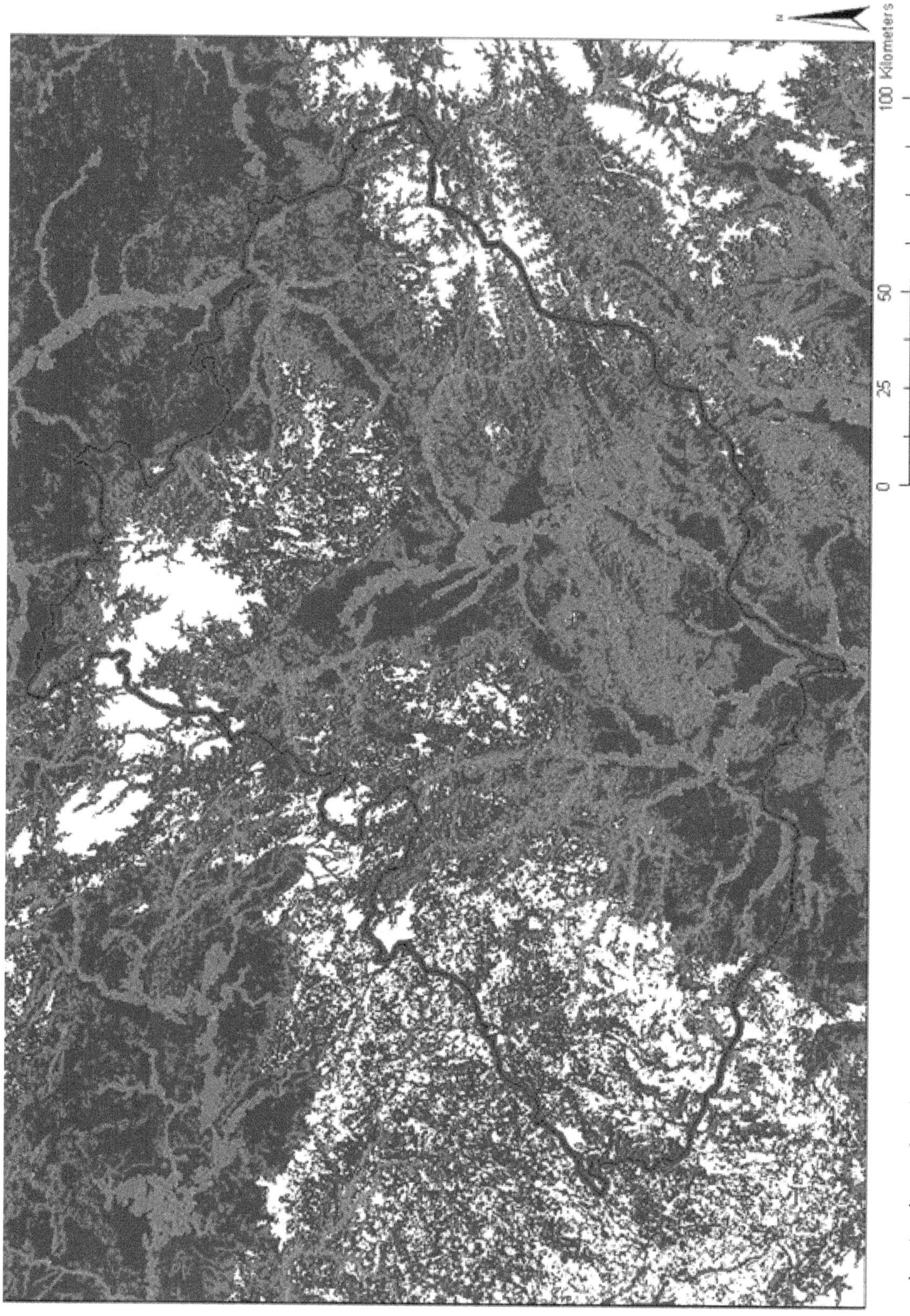

Fig. 5.20. Map of the study region showing derived species diversity (winter months, warm climates). Key shows number of species.

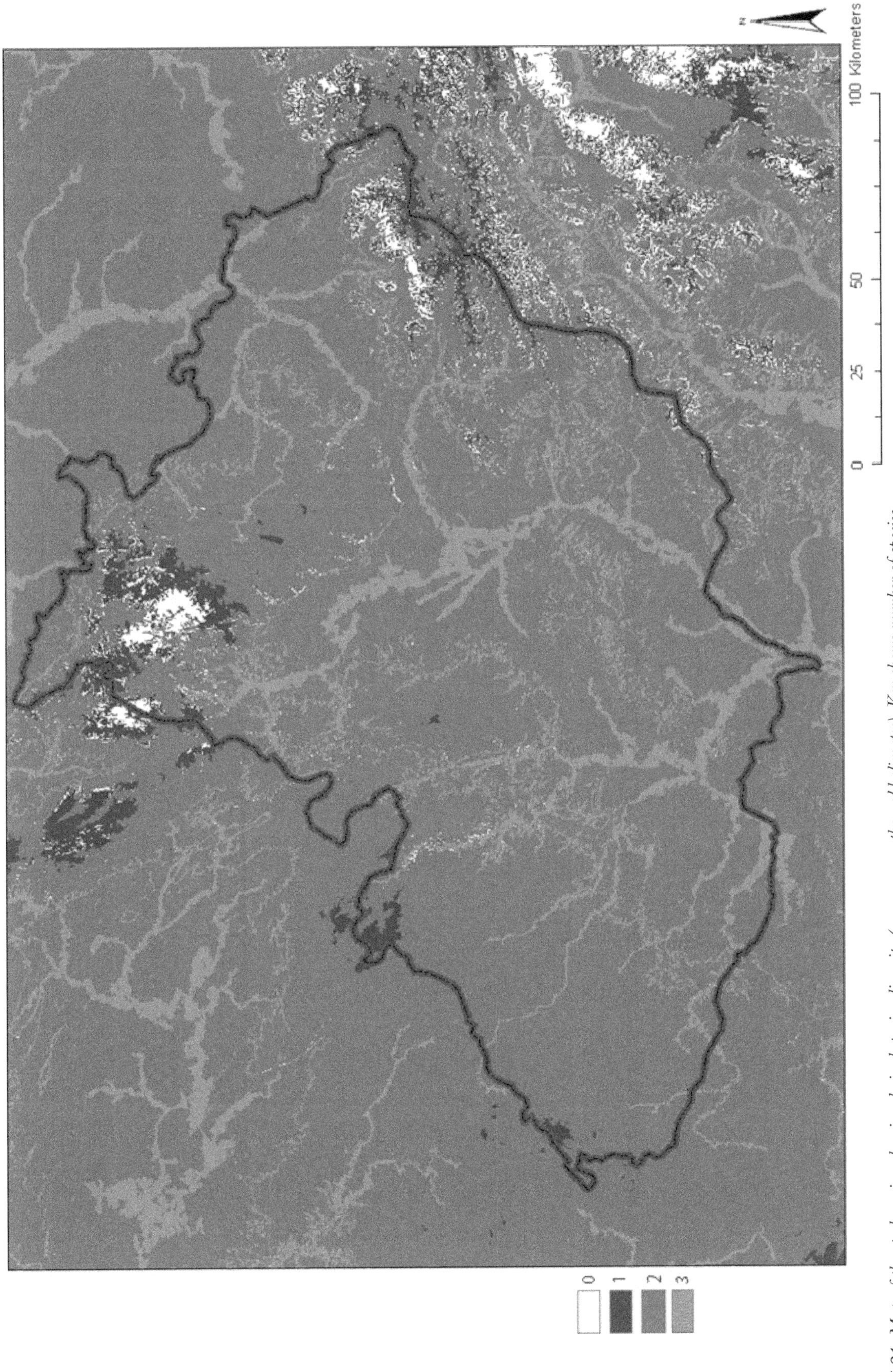

Fig. 5.21. Map of the study region showing derived species diversity (summer months, cold climates). Key shows number of species.

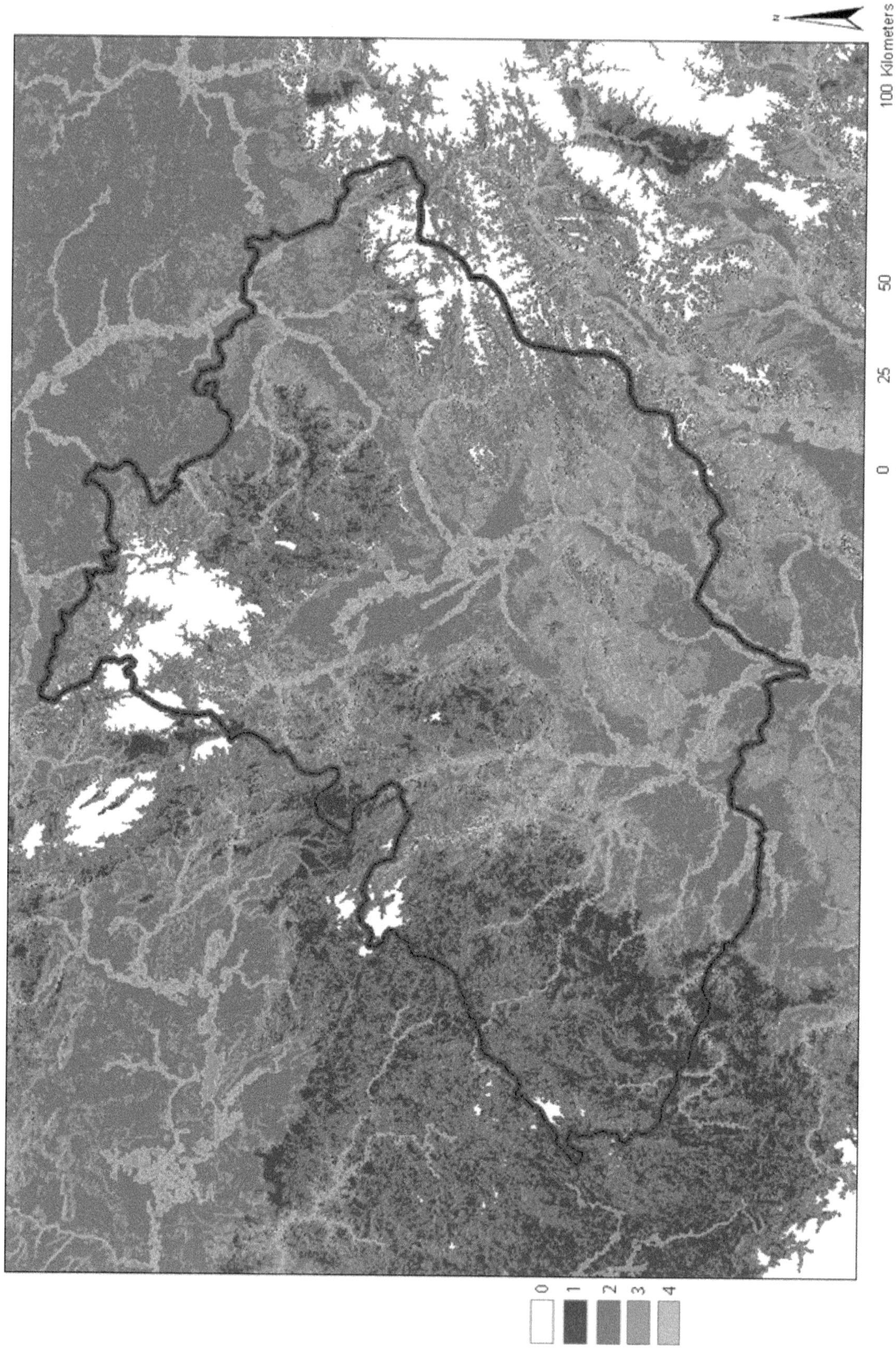

Fig. 5.22. Map of the study region showing derived species diversity (winter months, cold climate). Key shows number of species.

this project, during the summer months in warm climates, horses are confined to flat or gently sloping areas below 400 m.a.s.l., parkland conifer forest with stands of deciduous trees, parkland conifer forest, grassland steppe with isolated stands of conifer trees, and grassland areas. During winter months in warm climates, they are associated with flat areas, north-, west- or east-facing gentle/medium slopes below 700 m.a.s.l., parkland conifer forest areas, areas of grassland steppe with isolated stands of conifer trees, and areas of grassland. During summer months in cold climates, horses are associated with flat or gently sloping areas below 400 m.a.s.l. with steppe/tundra or grassland, and during winter months are confined to flat, gentle/medium-sloping areas below 700 m.a.s.l. with steppe/tundra or grassland (Table 5.3).

Reindeer prefer to eat forbs, lichens, grasses, and deciduous shrubs in areas with high plant diversity such as shallow, wet regions, and plants of high density and in early growth phase (Skogland 1984, 345). They most likely change their preferred habitat seasonally: during the summer they probably eat grasses and sedges, supplemented by leaves of willow and birch, and berries, and during the winter mainly eat lichen (Boyle 1990, 214; Skogland 1984, 349 and 353). Reindeer migrate from higher elevated regions during the summer months to lower sheltered regions in the winter, preferring flat or gently sloping areas, and are able to survive in snowy conditions, preferring shallow-hard snow to deep-soft snow (West 1997, 32). Cliffs and steep screes impede reindeer movement (Sturdy 1972, 164). In this project, during summer periods in warm climates, reindeer are assigned to floodplain areas, parkland conifer forest, and parkland conifer forest with isolated stands of deciduous trees. During winter months in warm climates, reindeer are associated with grassland steppe with isolated stands of conifer trees, and snow-covered areas below 400 m.a.s.l.. During winter months in cold climatic periods, reindeer are associated with flat or gently sloping areas above 400 m.a.s.l. and below 1100 m.a.s.l., and areas below 400 m.a.s.l. during summer months. In both seasons and during cold climates, reindeer are associated with floodplain areas, steppe/tundra regions, steppe/tundra areas with pine and birch stands, and areas with grassland. During winter months reindeer are also assigned to snow covered areas (Table 5.3).

Bison follow the routes of valley bottoms, gullies and small meadows, and are primarily associated with grasses, sedges and reedgrass (Bruggeman *et al.* 2007; Reynolds *et al.* 1978, 581, 588; Velichko and Zelikson 2005, 137). Although they prefer wet habitats, they can also feed on meadows and ruminate in forested areas (Reynolds *et al.* 1978, 581, 587, 588), and are confined to flat or gently sloping areas (Butzer 1986, 204; Kurten 1968; Sturdy *et al.* 1997). For this project, bison are confined to flat or gently sloping areas below 700 m.a.s.l. for all periods. During both summer and winter periods in warm climates they are associated with floodplain areas, and areas with parkland-conifer forest. During summer months in cold climates, they are largely confined to flat or gently sloping areas below 700 m.a.s.l., floodplain areas, and areas with grassland steppe and isolated stands of conifer trees. During winter months in cold climates they are associated with grassland type vegetation (Table 5.3).

Arctic foxes prefer areas of alpine tundra, open woodland and ice covered water areas (Hall 1981). Here, Arctic fox are associated with flat, gentle/medium-sloping areas below 1100 m.a.s.l. with areas of steppe/tundra and pine and birch stands during summer periods, and during winter months are also associated with floodplain regions (Table 5.3).

Hares are found in woodland environments, and areas of grassland and deciduous trees, deep, damp valley regions, and areas with shrubs along water courses (Boyle 1990, 96), and feed on grasses, herbs and plants such as dwarf birch and willow twigs. During the winter, they prefer higher elevated regions, and in the spring move to lower elevations (West 1997, 37). Here, during summer months, hare are associated with flat, gentle/medium-sloping areas below 700 m.a.s.l., and during the winter are assigned to flat, gentle/medium-sloping areas below 400 m.a.s.l.. In these regions they occupy floodplain areas and places with steppe/tundra and pine and birch stands (Table 5.3).

The results of the mapped prey species preferred habitats associated with summer months in warm climatic periods indicate that mammoth are potentially associated with most geographic regions, except for floodplain regions, the northern areas of the Jeseníký Mountains and areas of the Carpathians. Horses are associated with regions corresponding mostly to the Brno Unit and much of the southern region of the study area. Wolf are largely inversely related to the distribution of the Brno Unit, but this is to some extent sporadic in nature. Reindeer are distributed in a similar fashion to wolf, but more consistently associated with the spatial extents of the Bohemian Massif and the Carpathian Mountains (Fig. 5.13). During winter months in warm climates, mammoth and bison are associated mostly with floodplain regions, wolf are associated with floodplain regions and areas of the Carpathian Mountains and the Bohemian Massif, horses are sporadically associated with most regions except for areas coinciding with the Brno Unit, and reindeer are distributed mostly in accordance with the Brno Unit (Fig. 5.14). During summer months in cold climates, mammoth are affiliated with floodplain regions, but also sporadically associated with other areas. Wolves are largely associated with floodplain regions and less often with regions of the Bohemian Massif and the Carpathian Mountains, and some areas of the Brno Unit. The distribution patterns of horses are sporadically associated with regions of the Bohemian Massif and the Carpathian Mountains, and the distribution of reindeer largely follows the geographical extents of the Brno Unit. Bison are associated mainly with the floodplain regions, Arctic fox are potentially associated with most areas of the study area, and hare are mostly associated with floodplain regions and sporadically with some other areas (Fig. 5.15 and Fig. 5.16). During winter periods in cold climates, mammoth are largely associated with floodplain regions, wolves are largely associated with floodplain regions and more sporadically with regions of the Bohemian Massif and the Carpathian Mountains, and less commonly with some areas of the Brno Unit. Horses are sporadically associated with regions of the Bohemian Massif and the Carpathian Mountains, and the distribution of reindeer largely follows the geographical extents of the Brno Unit. Bison are associated mainly with floodplain regions. Arctic foxes are associated with most regions to some extent, and hares are sporadically associated with areas of the Bohemian Massif and the Carpathian Mountains (Fig. 5.17 and Fig. 5.18).

Species diversity maps are created by assigning a value of '1' to the geographical regions associated with the already mapped prey species preferred habitats. Contemporaneous faunal distribution areas (i.e. distribution areas associated with the same season and climate) are geographically summed in ArcGIS9.2 to provide prey species diversity maps for summer and winter periods in warm and cold climates. Results associated with the summer months in warm climates (Fig. 5.19) indicate that areas with the highest prey species diversity value (4) are located on the periphery of floodplains, and foothill regions. Areas with a species diversity of 3 are mainly associated with low-lying regions. Regions with a prey species diversity value of 2 are associated with the Bohemian Massif, some floodplain areas, and some areas of the Carpathian Mountains. Areas with a prey species diversity value of 1 are rare and located in some areas of the Bohemian Massif. Areas with a prey species diversity value of 0 are associated with the Jeseníký Mountains and some areas of the Carpathians. Results associated with the winter months in warm climates (Fig. 5.20) indicate that areas with the highest species diversity values (3) are mostly associated with floodplain regions. Areas with a species diversity value of 2 are mainly associated with the lower regions of the Carpathians and the foothills of the Bohemian Massif. Areas with a species diversity value of 1 are associated with low-lying regions adjacent to many of the floodplain regions, and areas of the Bohemian Massif and the eastern regions of the Carpathian Mountains. Areas not associated with prey species are generally located in regions of the Bohemian Massif, in particular the Jeseníký Mountains and the highest regions of the Carpathians in the southeast of the study region. Results associated with summer periods in cold climates (Fig. 5.21) indicate that regions with the highest diversity values (3) are associated with floodplain regions, and sporadically with other areas of the study region. Regions with a species diversity value of 2 are associated with most areas of the study region apart from floodplain regions and the higher regions of the Bohemian Massif and the Carpathian Mountains. Areas with species diversity values of 1 and 0 are mainly associated with the highest regions of the Bohemian Massif (i.e. Jeseníký Mountains) and the Carpathians towards the south-east regions of the study area. Results associated with winter periods in cold climates (Fig. 5.22) indicate that areas with the highest species diversity value (4) are associated with floodplain regions and sporadically with other regions. Areas with a prey species diversity of 3 are generally associated with the low-lying regions of the Carpathians and the foothills of the Bohemian Massif. Areas with a species diversity value 2 are associated with the low-lying flat regions in areas surrounding the floodplain regions and more sporadically with some other regions. Areas with a species diversity value of 1 are mainly associated with the mid-high elevated regions of the Bohemian Massif, and areas with no species affiliation are associated with the highest regions of the Bohemian Massif (Jeseníký Mountains) and the highest regions of the Carpathians.

Summary and conclusions

In this chapter, I have attempted to construct a digital GIS framework that can be used for the remaining part of the study. I mapped site locations, lithic raw material outcrops, topography (elevation, slope, and aspect), palaeohydrology, palaeovegetation, friction maps, prey species preferred habitat and diversity. The next three chapters will make extensive use of the digital maps produced in this chapter to reconstruct hominin and cave-bear distribution patterns and associated interactions between the two.

Chapter 6: Reconstructing hominin distribution patterns

Introduction

The objective of this chapter is to reconstruct hominin distribution patterns. I use already mapped hominin site locations, prey species diversity maps, and friction maps, together with Least-cost Surfaces and Least-cost Corridors. I first describe the methodology used and then present and summarise the results.

Mapping hominin distribution patterns

Hominin distribution patterns are constructed using the already mapped hominin site locations and prey species diversity maps, together with Least-cost Surfaces to derive home ranges, and Least-cost Corridors to derive pathways. Home ranges are constructed using the Least-cost Surfaces algorithm available in ArcGIS9.2, with the already mapped hominin sites used as central location points. The maximum radius of the derived least-cost surface is set to 5 km, and offset using the seasonally and climatically relevant friction map. Least-cost corridors are derived using the Least-cost Corridor algorithm available in ArcGIS9.2, again offset using relevant friction maps, and with the maximum extents of the corridor set to a width equivalent to 105% of the minimum least-cost corridor energy expenditure value. Where the location of the lithic raw material outcrop is less than 5 km from the settlement site, only home ranges are mapped, and this is done for three of the seven hominin case study sites (Šipka Cave, Vedrovice V, and Stránská skála IIA and IIIA). In cases where lithic raw material outcrops are located more than 5 km from the settlement site, only least-cost corridors are derived. Least-cost corridors are calculated for Kůlna Cave, Bohunice-kejbalý (I and II), Dolní Věstonice (I, II, and III), and Pavlov I. Prey species diversity maps are combined with the derived home ranges and least-cost corridors to produce overall hominin distribution patterns (Table 6.1). The results are presented below.

The derived home range associated with Šipka Cave during the summer in warm climates (Fig. 6.1) is irregular in shape, with a maximum diameter of approximately 4 km from the most extreme south-westerly point to the furthest north-easterly location, and approximately 3 km from the most extreme north-westerly point to the south-eastern perimeter. The narrowest regions are located towards the northeast, northwest, and southwest of the home range area, with central and eastern areas being the most expansive. Most of the derived region is associated with prey species, but this is less so in regions within about 1 km of Šipka Cave. Areas associated with prey species mostly have a species diversity of 1 and are associated with reindeer. Areas with a species diversity of 2 are less common, mostly located about 0.5 km to the southeast and northwest of Šipka Cave, towards the far southeast, and are associated with reindeer and mammoth. Results associated with Šipka Cave during the winter months in warm climates (Fig. 6.2) are similar to those derived for summer periods in terms of the relative location of Šipka Cave and the shape and size of the derived home range. Prey species are associated with nearly all areas of the region for this period, apart from some areas within about 0.5 - 1 km radius of Šipka Cave. All areas associated with prey species have a species diversity value of 2, and are associated with horse and reindeer.

Results associated with Kůlna Cave during the summer in warm climates (Fig. 6.3) indicate that the hominin site is located towards the east region of the derived least-cost corridor, which is roughly 9 km in an east-west direction and between about 5 km and 1 km in a north-south direction. The area is divided into two main areas that run parallel to each other in an east-west direction, and conjoin about 1.5 km east of the raw material (Cretaceous

Case study sites	Climate	Hunted fauna	Raw material	Home range/least-cost corridors
Šipka Cave	Warm	Reindeer, mammoth and horse	Flysch sandstone (< 5 km)	Home range
Kůlna Cave	Warm	Reindeer, mammoth and horse	Cretaceous chert (> 5 km)	Least-cost corridors
Vedrovice V	Warm	Reindeer, mammoth, horse, and bison	Krumlovský les chert (< 5 km)	Home range
Bohunice-kejbalý (I and II)	Warm	Reindeer, mammoth and horse	Stránská skála chert (> 5 km)	Least-cost corridors
Stránská skála (IIA and IIIA)	Cold	Reindeer, mammoth and horse	Stránská skála chert (< 5 km)	Home range
Dolní Věstonice (I, II, and III)	Cold	Mammoth, reindeer, wolf, hare, and Arctic fox	Erratic flint (> 5 km)	Least-cost corridors
Pavlov I	Cold	Mammoth, reindeer, horse, wolf, hare, and Arctic fox	Erratic flint and radiolarite (> 5 km)	Least-cost corridors

Table 6.1. Overview of the chosen case study sites, affiliated climate, associated hunted fauna, and most commonly procured lithic raw material (with approximate distance from hominin site), together with the method used to map hominin distribution patterns.

chert) procurement area, and about 1.25 km and 4 km west of Kůlna Cave. Nearly all areas of the least-cost corridor are associated with prey species, with mammoth and reindeer associated with regions within about 6 km radius of Kůlna Cave, and only mammoth associated with regions further west of this area. Results associated with Kůlna Cave during the winter in warm climates (Fig. 6.4) are largely similar to those during the summer months in terms of the spatial parameters of the least-cost corridor and the relative locations of the hominin site and raw material outcrop. However, during the winter months only areas further than about 6 km to the west of Kůlna Cave are associated with prey species. These have a uniform diversity value of 2, and are associated with mammoth and reindeer.

Reconstructed home ranges associated with Vedrovice V during summer months in warm climates (Fig. 6.5) have a maximum north-south diameter of about 4 km, and a maximum east-west diameter of about 2.5 km. The site of Vedrovice V is located towards the north-central region of the derived home range, close to the raw material outcrop (Krumlovský les chert). Most areas are associated with prey species apart from a small area about 1.25 km in length that runs in a north-south direction about 0.5 km east of Vedrovice V. All areas associated with prey species have diversity values of 2, and correspond with mammoth and horse. The reconstructed home ranges associated with Vedrovice V during winter months in warm climates (Fig. 6.6) are similar to those during the summer months, both in terms of the spatial extents of the home range and the relative location of Vedrovice V and the raw material outcrop. Most areas within the home range are associated with prey species, apart from a small region about 0.5 km to the east of Vedrovice V and some isolated areas to the north. Most areas that are associated with prey species have a diversity value of 1, and are associated with reindeer. Areas with a diversity value of 2 are less common, and are associated with reindeer and horse, and these regions are located generally to the south-east, north-east, north-west, and sporadically in areas to the south-west of Vedrovice V.

The derived least-cost corridor associated with Bohunice-kejbalý (I and II) during summer months in warm climates (Fig. 6.7) is diamond-shaped, with the narrowest regions occurring close to the settlement site and the raw material procurement location, and the widest regions located towards the central areas. The corridor measures approximately 9 km in an east-west direction, and about 7 km in a north-south direction. Prey species are associated mostly towards the west, within about 2.5 km of the hominin settlement site, and towards the eastern areas, within about 4 km radius of the raw material procurement place. All areas associated with prey species have a diversity value of 2 and are associated with mammoth and horse. Results associated with Bohunice-kejbalý (I and II) during the winter period in warm climates (Fig. 6.8) are largely similar in spatial extents and relative location of the hominin and raw material procurement sites. However, during this period only about 1/3 of the overall region is associated with prey species, and these are mostly located in central regions in a north-south direction through the centre of the corridor and measuring about 2 km in an east-west direction. Although other areas are associated with prey species, they are sporadic in nature and less common. All regions associated with prey species have a diversity value of 2 and are associated with mammoth and horse.

The reconstructed home range associated with Stránská skála (IIA and IIIA) during the summer in cold climates (Fig. 6.9) is about 5 km north-south, and about 4 km in an east-west direction. The hominin site and the associated raw material outcrop (Stránská skála chert) are located towards the central-northern area of the region. Most areas are associated with prey species to some extent, apart from a very small region immediately to the northwest of the raw material source. All areas with prey species have a diversity value of 2, and are associated with horse and mammoth. The derived home range for winter months in cold climates (Fig. 6.10) is largely similar with regard to spatial extents and the relative location of Stránská skála (IIA and IIIA) and the raw material outcrop. Again, all areas are associated with prey species, except for a small region immediately northwest of the hominin site, and this has a diversity value of 1 (reindeer) and 3 (reindeer, horse, and mammoth).

The derived least-cost corridor associated with Dolní Věstonice (I, II, and III) during the summer in cold climates (Fig. 6.11) is about 125 km in length, and between about 3 and 17 km in width. The hominin site is located towards the far southwest of the region, and the raw material outcrop (erratic flint) is situated towards the far northeast. The derived least-cost corridor runs south-west from Dolní Věstonice for about 35 km, north-east, north, and north-east again towards the raw material procurement zone. All areas are associated with prey species to some degree, except for some isolated regions. By and large most areas associated with prey species have a diversity value of 2 (Arctic fox and horse) and 4 (Arctic fox, hare, wolf, and mammoth). The higher prey species diversity values generally coincide with central regions of the corridor, and the highest concentration of high diversity values occur in the north-east region, about 40 km southeast of the raw material outcrop. Least-cost corridors for winter months in cold climates (Fig. 6.12) are largely similar in spatial extents and relative locations of Dolní Věstonice and the raw material source. Prey species diversity values are also distributed in a similar fashion, i.e. higher values are located towards the centre of the corridor and lower values are located towards the outer regions. During this period, however, low prey species diversity areas have a value of 2 and are associated with Arctic fox and reindeer, and high prey species diversity areas have a value of 3 and correspond to mammoth, wolf, and Arctic fox.

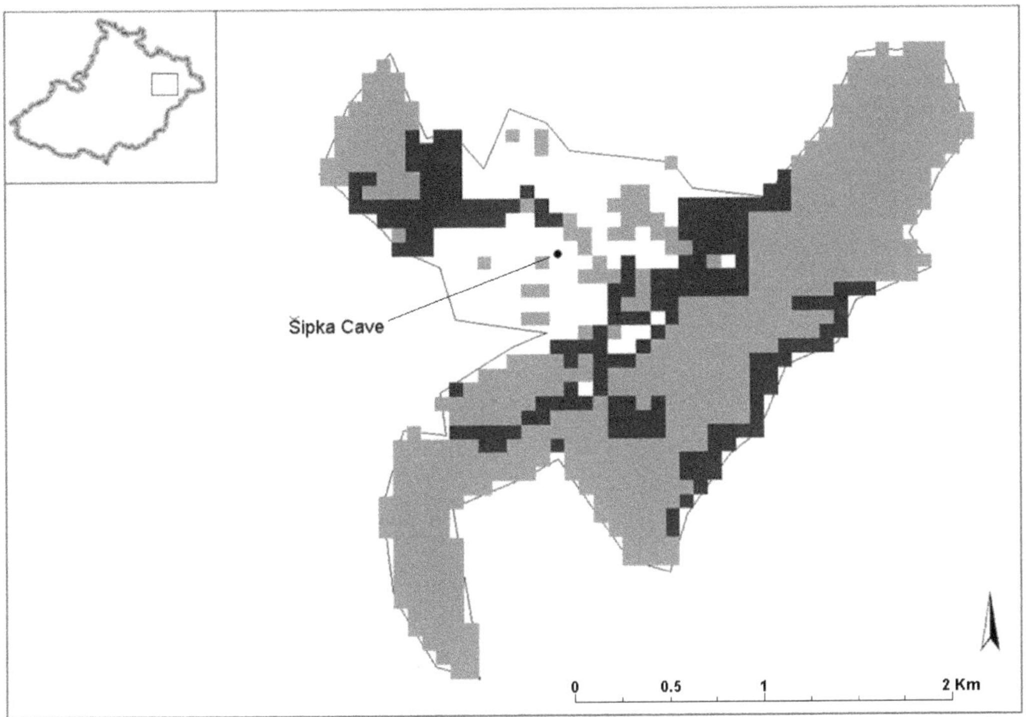

Fig. 6.1. Reconstructed home range and prey species diversity associated with Šipka Cave (summer months, warm climates). Prey species diversity – light grey = 1 (reindeer), and dark grey = 2 (reindeer and horse).

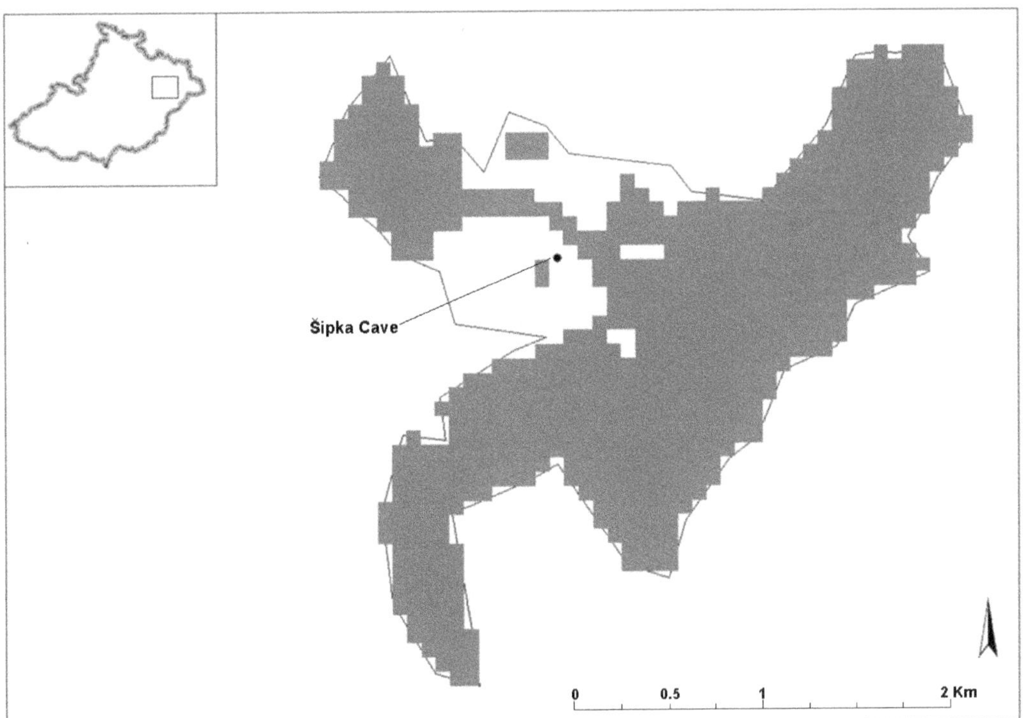

Fig. 6.2. Reconstructed home range and prey species diversity associated with Šipka Cave (winter months, warm climates). Prey species diversity – grey = 2 (horse and reindeer).

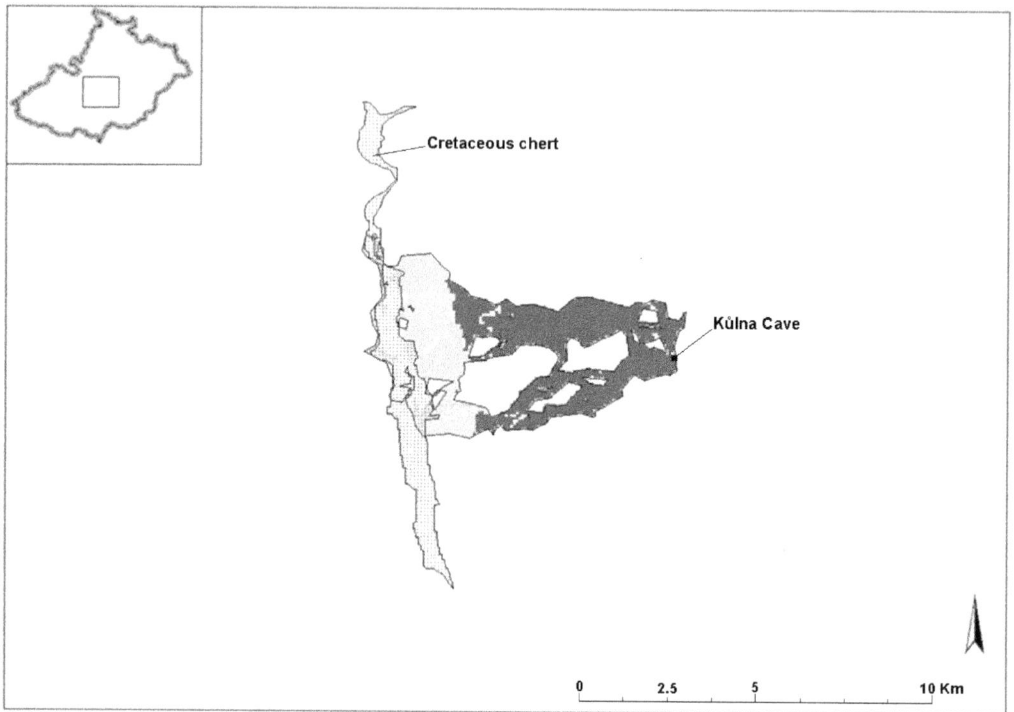

Fig. 6.3. Reconstructed least-cost corridor and prey species diversity associated with Kůlna Cave (summer months, warm climates). Prey species diversity – light grey = 1 (mammoth), and dark grey = 2 (reindeer and mammoth).

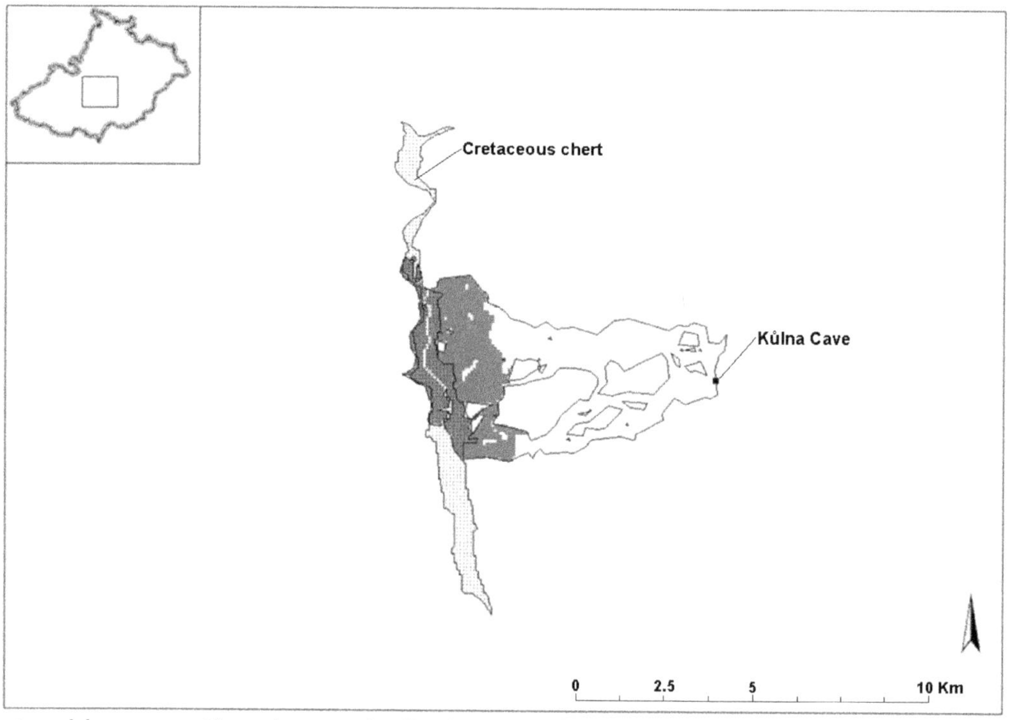

Fig. 6.4. Reconstructed least-cost corridor and prey species diversity associated with Kůlna Cave (winter months, warm climates). Prey species diversity – grey = 2 (reindeer and mammoth).

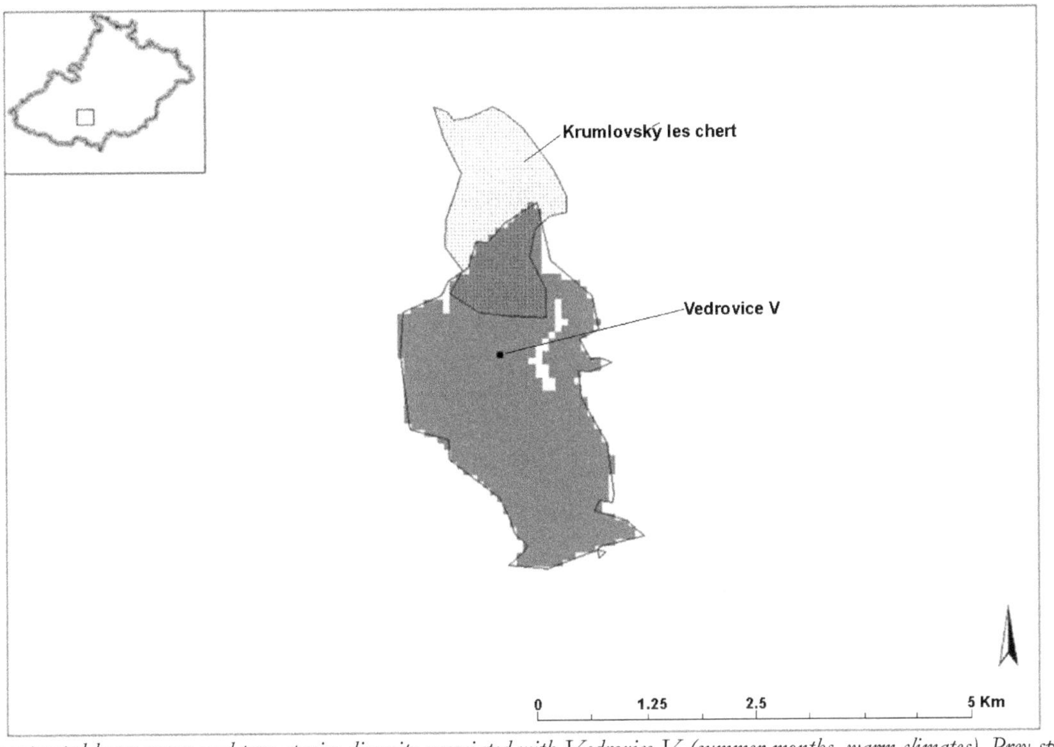

Fig. 6.5. Reconstructed home range and prey species diversity associated with Vedrovice V (summer months, warm climates). Prey species diversity – grey = 2 (horse and mammoth).

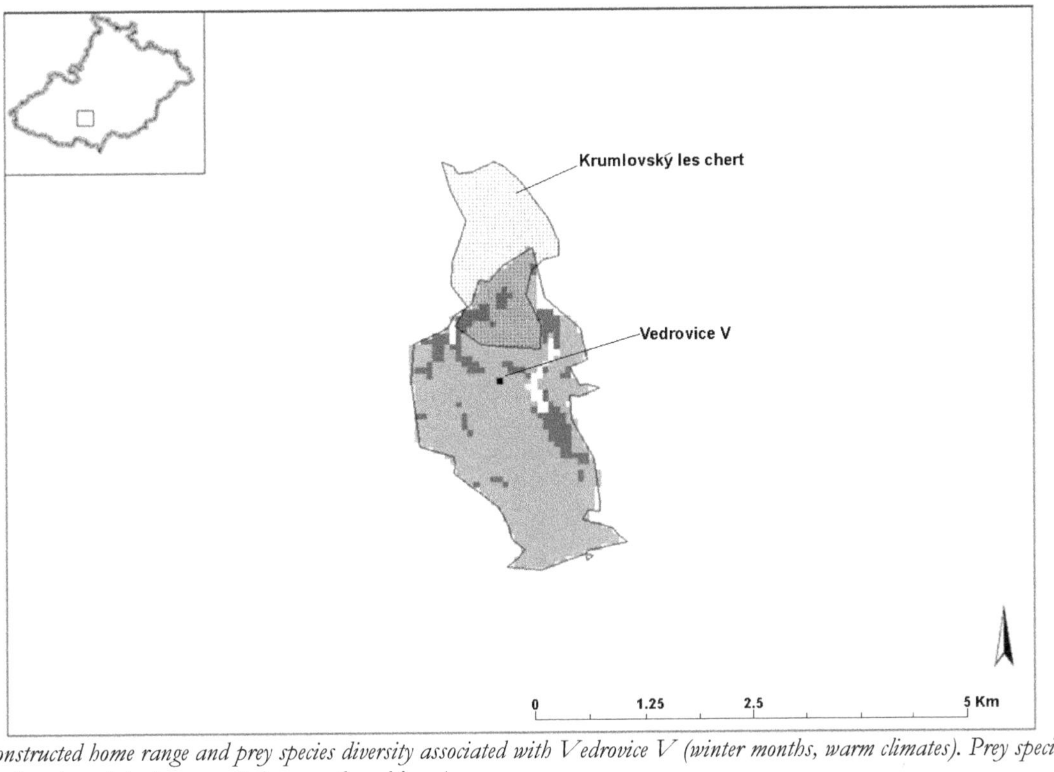

Fig. 6.6. Reconstructed home range and prey species diversity associated with Vedrovice V (winter months, warm climates). Prey species diversity – light grey = 1 (horse), and dark grey = 2 (mammoth and horse).

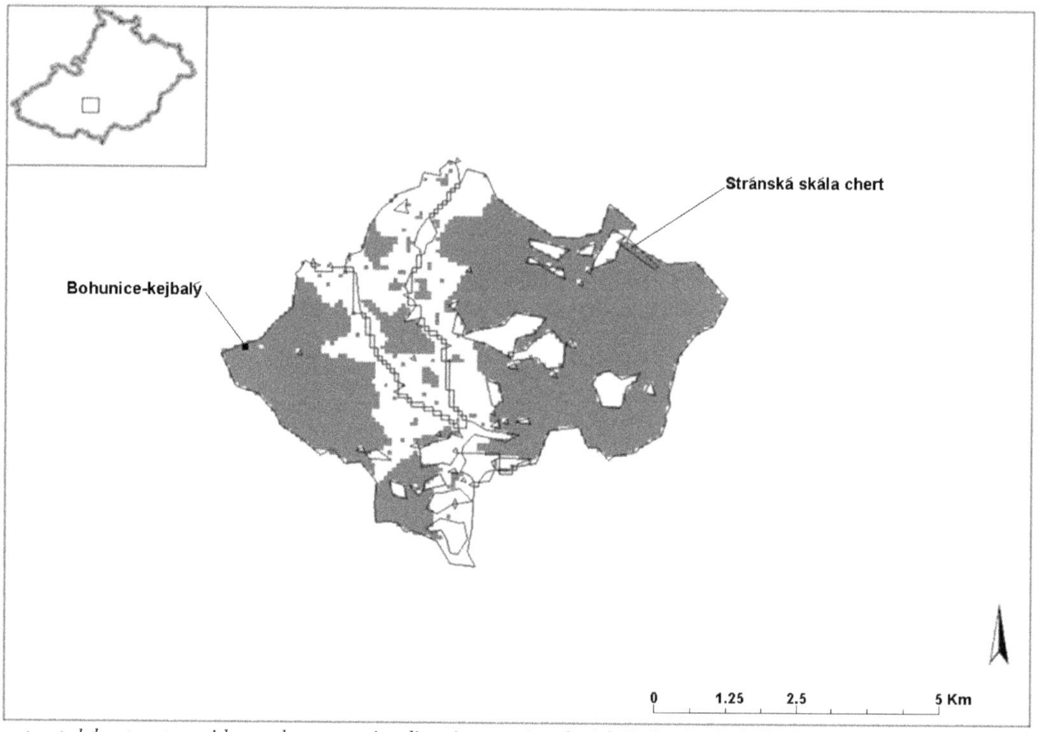

Fig. 6.7. Reconstructed least-cost corridor and prey species diversity associated with Bohunice-kejbalý (I and II) (summer, warm climates). Prey species diversity – grey = 2 (horse and mammoth).

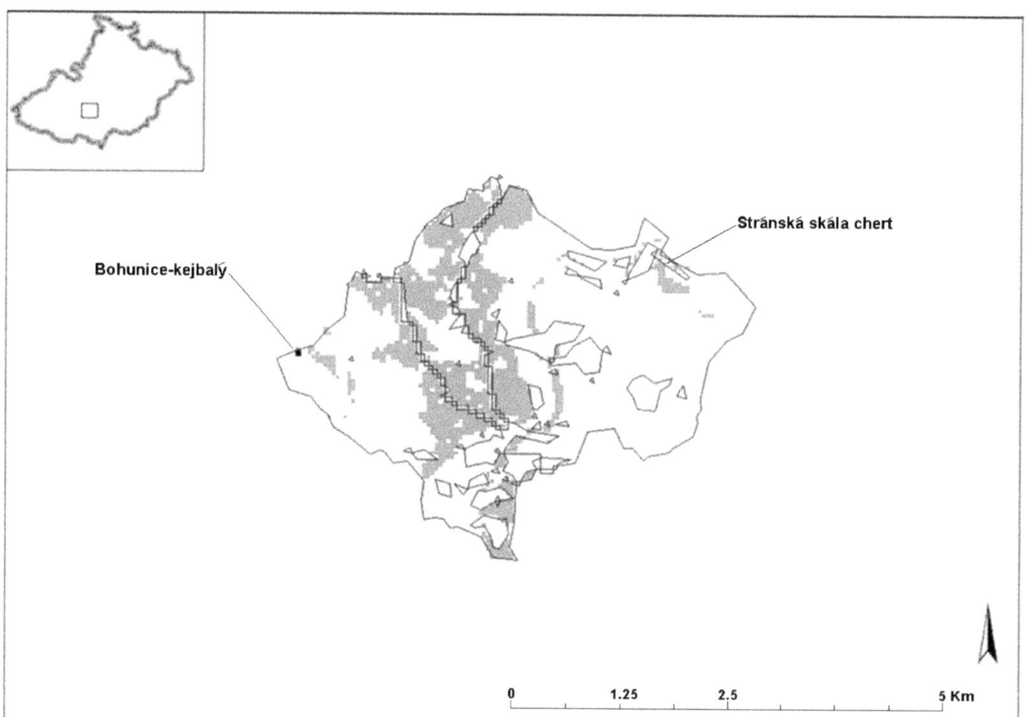

Fig. 6.8. Reconstructed least-cost corridor and prey species diversity associated with Bohunice-kejbalý (I and II) (winter, warm climates). Prey species diversity – grey = 2 (horse and mammoth).

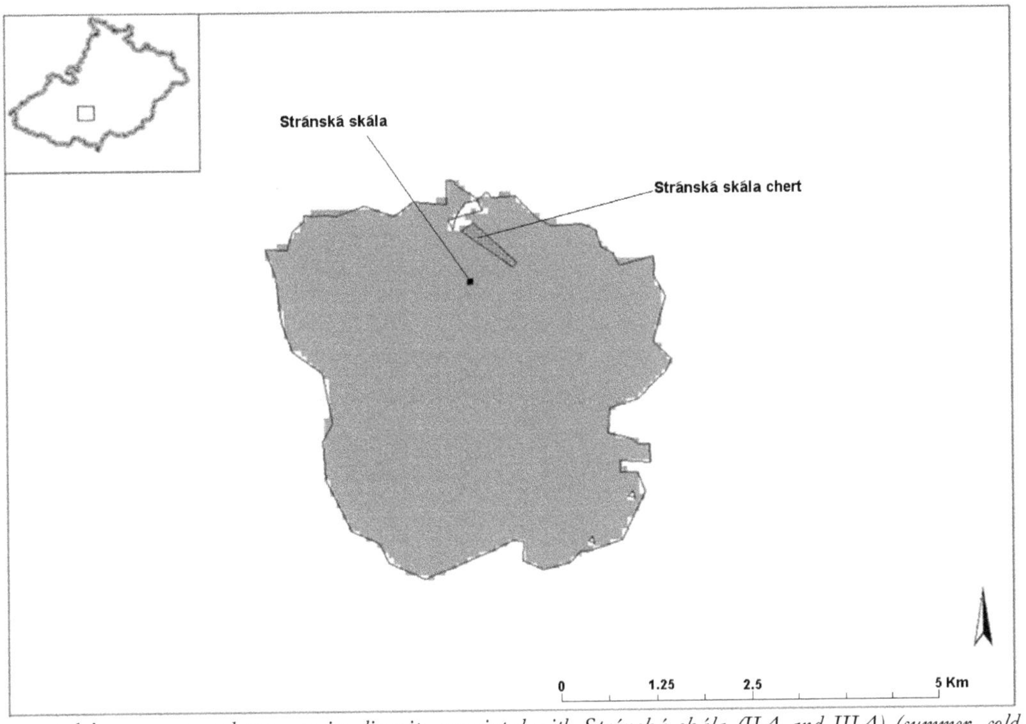

Fig. 6.9. Reconstructed home range and prey species diversity associated with Stránská skála (IIA and IIIA) (summer, cold climates). Prey species diversity – grey = 2 (horse and mammoth).

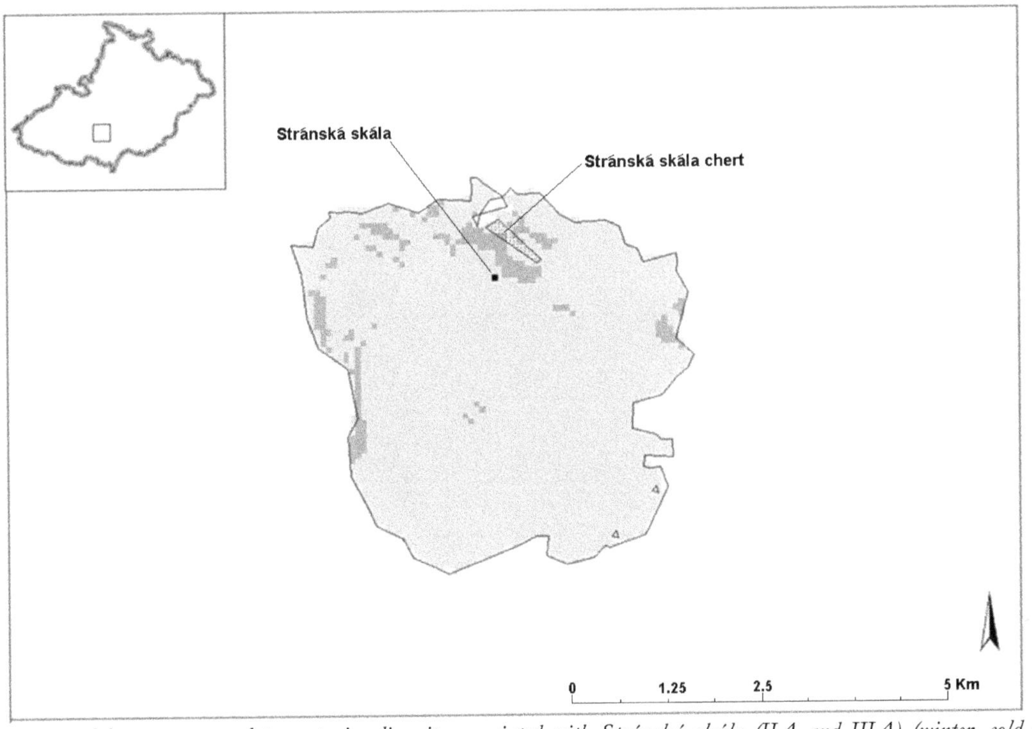

Fig. 6.10. Reconstructed home range and prey species diversity associated with Stránská skála (IIA and IIIA) (winter, cold climates). Prey species diversity – light grey = 1 (reindeer), and dark grey = 3 (reindeer, horse, and mammoth).

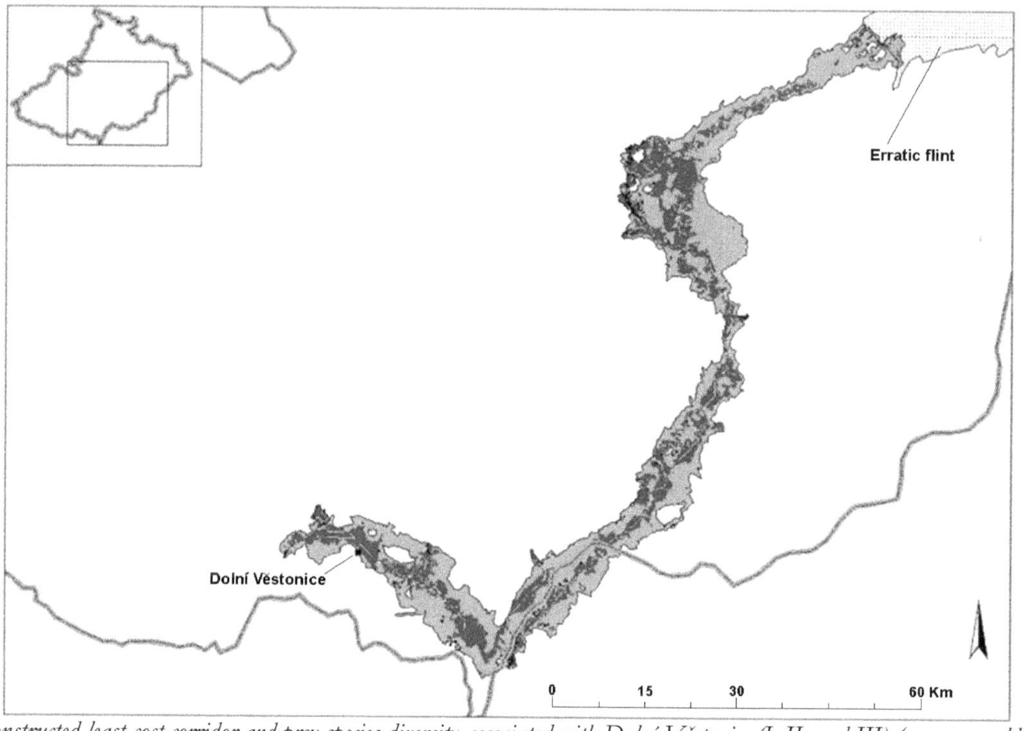

Fig. 6.11. Reconstructed least-cost corridor and prey species diversity associated with Dolní Věstonice (I, II, and III) (summer, cold climates). Prey species diversity – light grey = 1 (Arctic fox and horse), and dark grey = 4 (mammoth, Arctic fox, hare, wolf).

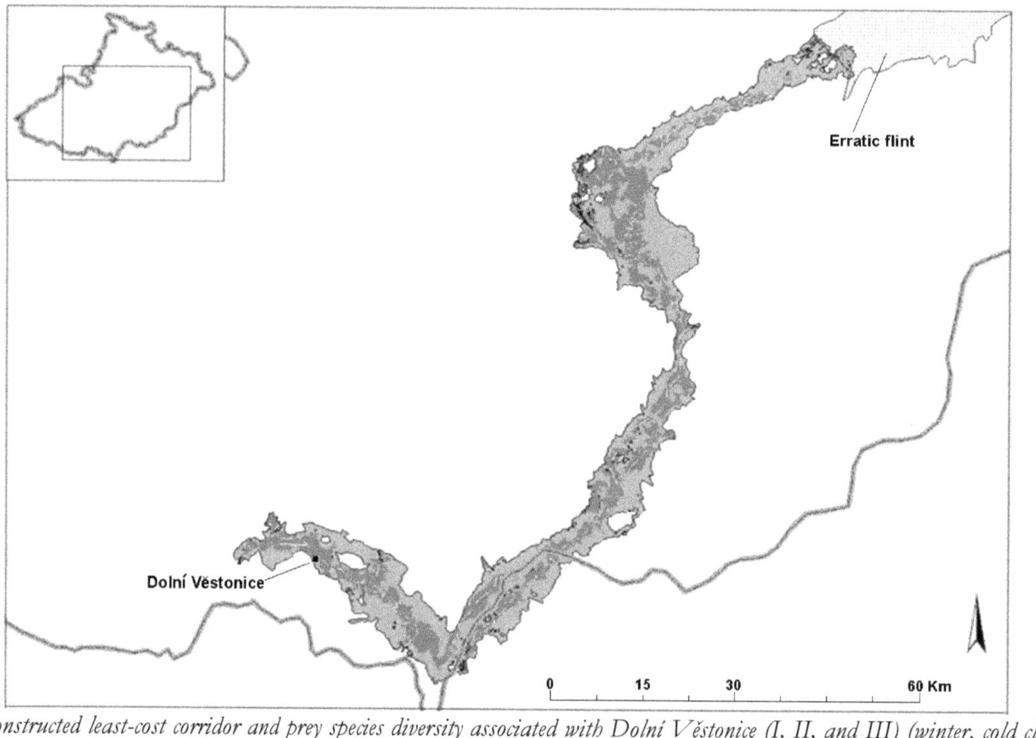

Fig. 6.12. Reconstructed least-cost corridor and prey species diversity associated with Dolní Věstonice (I, II, and III) (winter, cold climates). Prey species diversity – light grey = 2 (Arctic fox and reindeer), and dark grey = 3 (mammoth, wolf, and Arctic fox).

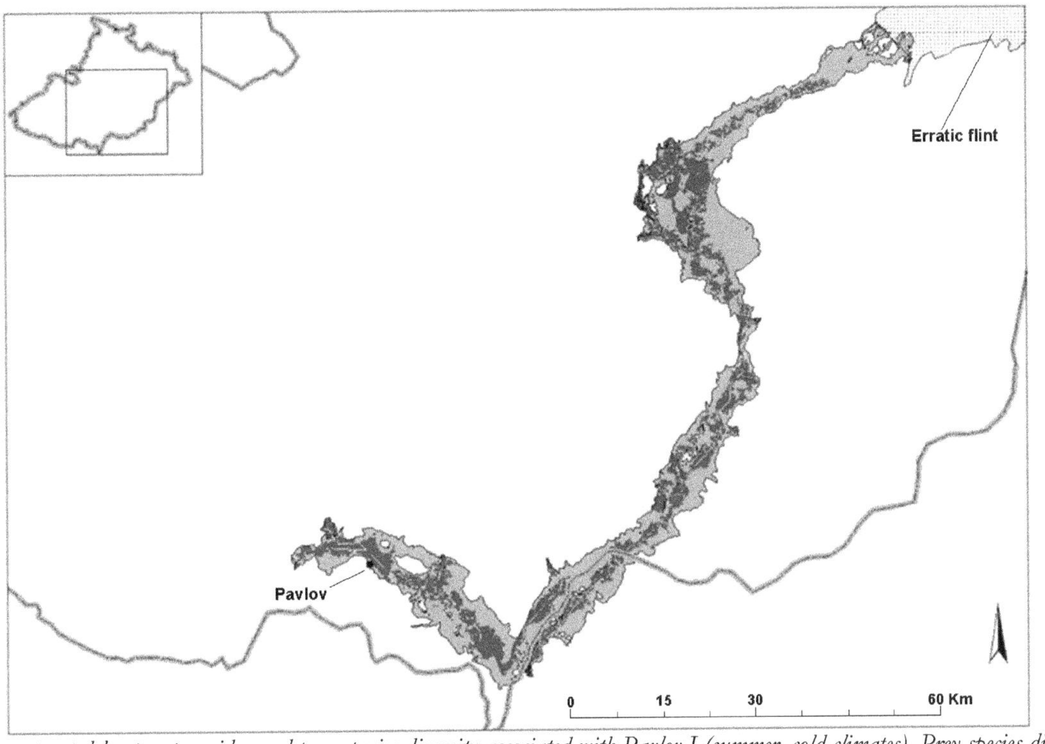

Fig. 6.13. Reconstructed least-cost corridor and prey species diversity associated with Pavlov I (summer, cold climates). Prey species diversity – light grey = 2 (Arctic fox and horse), and dark grey = 4 (wolf, mammoth, Arctic fox, and hare).

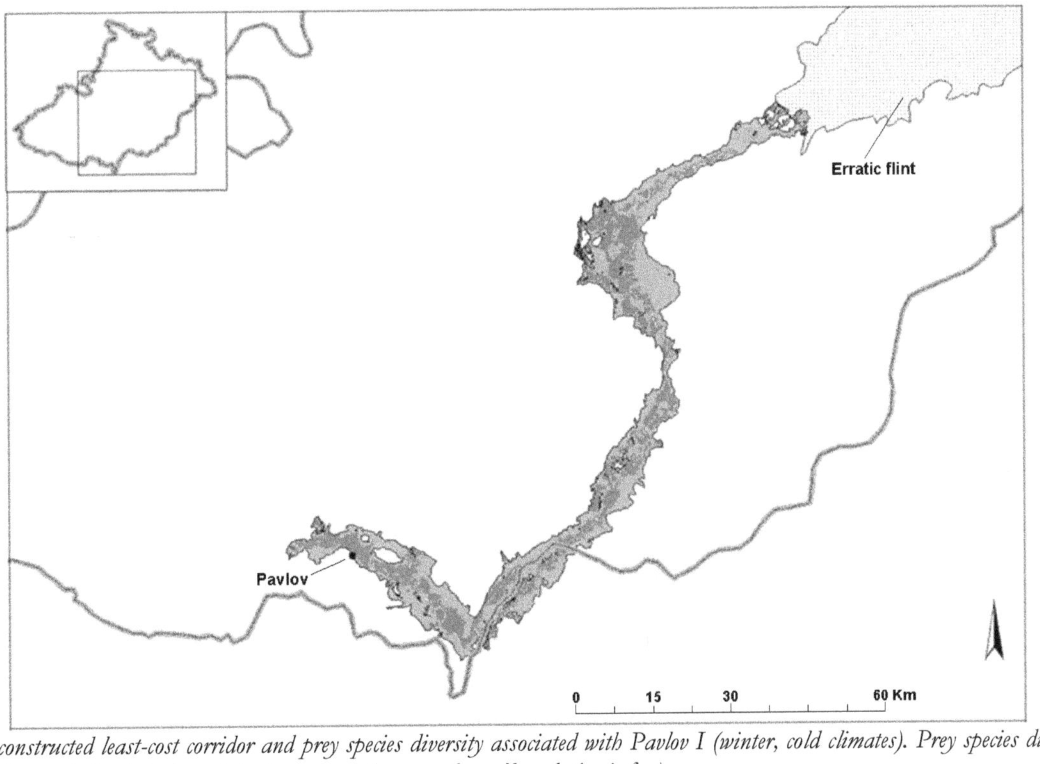

Fig. 6.14. Reconstructed least-cost corridor and prey species diversity associated with Pavlov I (winter, cold climates). Prey species diversity – light grey = 2 (reindeer and Arctic fox), and dark grey = 3 (mammoth, wolf, and Arctic fox).

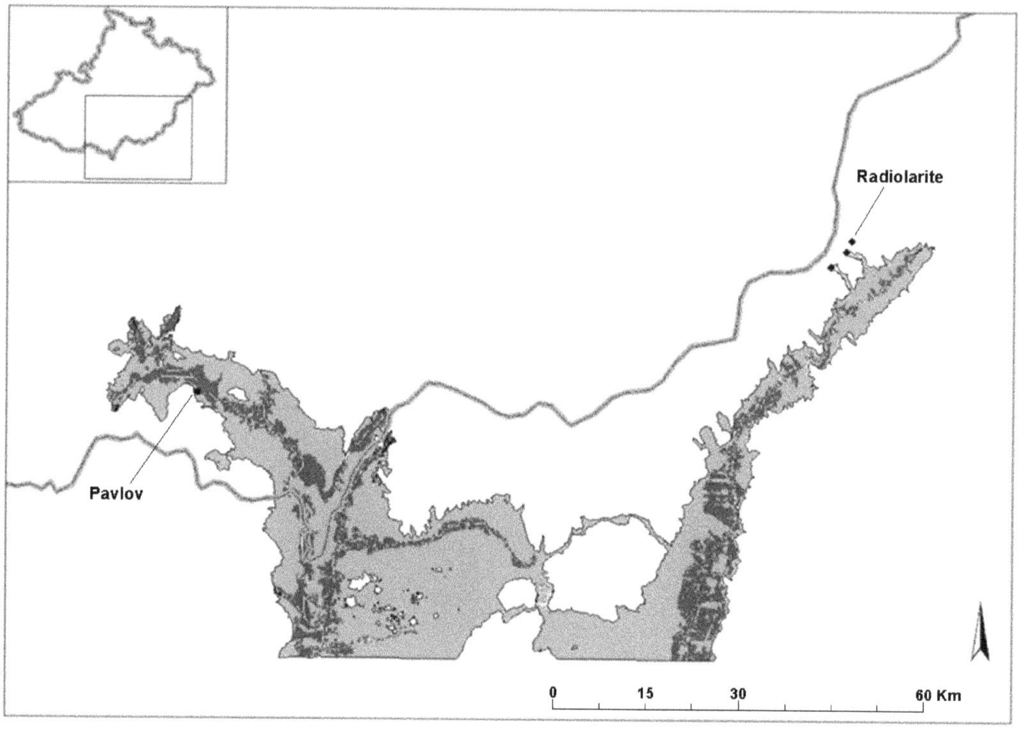

Fig. 6.15. Reconstructed least-cost corridor and prey species diversity associated with Pavlov I (summer, cold climates). Prey species diversity – light grey = 2 (Arctic fox and horse), and dark grey = 4 (wolf, mammoth, Arctic fox, and hare).

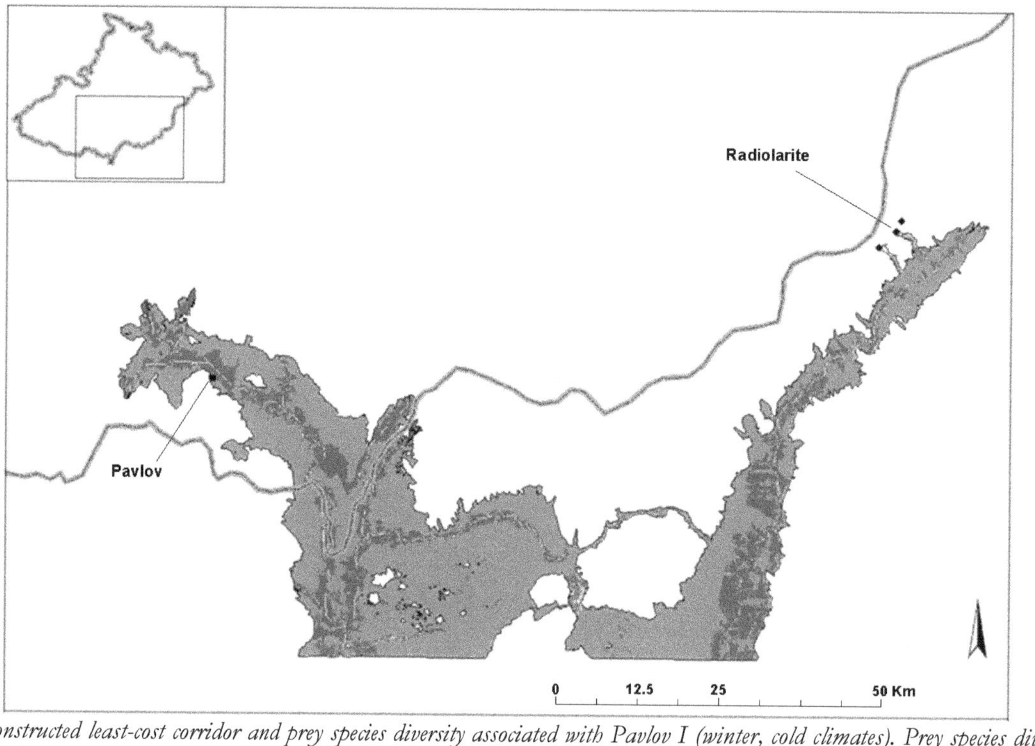

Fig. 6.16. Reconstructed least-cost corridor and prey species diversity associated with Pavlov I (winter, cold climates). Prey species diversity – light grey = 2 (Arctic fox and horse), and dark grey = 3 (wolf, mammoth, and Arctic fox).

Derived least-cost corridors associated with Pavlov I (erratic flint) during the summer in cold climates (Fig. 6.13) are largely similar to those derived for Dolní Věstonice (I, II, and III). Areas with the lowest prey species diversity values are generally located towards the outer regions of the corridor, have a value of 2, and are associated with horse and Arctic fox. Regions with the highest diversity values correspond with the central regions of the corridor. These areas mainly have a prey species diversity value of 4 and correspond with mammoth, wolf, hare, and Arctic fox. Least-cost corridors associated with winter months in cold climates (Fig. 6.14) are largely the same as they are during the summer periods in terms of the extents of the derived corridor, the relative locations of Pavlov, the raw material source, and the relative distribution of high and low prey species diversity values. During this period, however, areas with the lowest prey species diversity value have a value of 2 and correspond with Arctic fox and reindeer, and areas with the highest prey species diversity have a value of 3 and are associated with mammoth, wolf, and Arctic fox.

Reconstructed least-cost corridors associated with Pavlov I (radiolarite raw material source) during the summer in cold climates (Fig. 6.15) are located almost universally to the east and south-east of the hominin settlement site, with the raw material procurement places located at the far east of the corridor. The overall length of the corridor is about 130 km – 150 km. The area is split into two main regions; an area to the west, located immediately to the east of Pavlov, and an area to the east, and these regions are divided by very narrow areas towards the centre of the corridor. Prey species are associated with nearly all regions of the corridor, apart from some isolated areas close to Pavlov and some regions about 50 km south-east of Pavlov. Of the areas that are associated with prey species, most have a species diversity value of 2 (Arctic fox and horse) or 4 (wolf, mammoth, Arctic fox, and hare). Most high species diversity areas are located either in the interior regions of the western half of the corridor, or the eastern and north-eastern areas of the eastern half of the corridor. Areas with low prey species diversity values are located in most other regions. The derived least-cost corridors associated with winter months in cold climates (Fig. 6.16) are mostly the same in terms of the size and extents of the derived corridor and relative location of Pavlov, the raw material outcrop, and the distribution of areas with high and low prey species diversity values. During the winter, however, areas associated with prey species have diversity values of either 2 or 3. Those areas with a diversity value of 2 are associated with reindeer and Arctic fox, and areas with a diversity value of 3 are associated with mammoth, wolf, and Arctic fox.

Overall, results associated with hominin distribution patterns during warm climatic periods (Šipka Cave, Kůlna Cave, Vedrovice V, and Bohunice-kejbalý I and II) are both similar and different in nature. The size of the distribution patterns associated with each site is largely similar, ranging between about 4 km and 10 km in their maximum extents. In all cases, hominin distribution patterns have largely the same spatial extents during summer and winter months, and the relative location of the hominin settlement site and raw material procurement sites remain mostly unchanged with seasonal change. However, in most cases derived hominin distribution patterns and association with prey species does change: in all four cases the distribution of prey species and associated diversity values differ between summer and winter months, and this is most evident in the case of Kůlna Cave and Bohunice-kejbalý (I and II) where the spatial distribution associated with prey species diversity are inversely related in summer and winter periods. As is the case with hominin distribution patterns associated with warm climates, the spatial extents of the derived distribution patterns for cold climates remain largely the same during both summer and winter periods. However, the derived distribution patterns associated with these climatic periods, with the exception of Stránská skála (IIA and IIIA), are much larger. Distribution patterns associated with Dolní Věstonice (I, II, III) and Pavlov I (erratic flint and radiolarite) are more than 100 km in length. The geographical distribution of high and low prey species diversity values associated with Stránská skála (IIA and IIIA) change seasonally and are largely unpredictable, but in the case of Dolní Věstonice (I, II, and III) and Pavlov I (erratic flint and radiolarite), the relative location of high and low species diversity values largely remain the same during both summer and winter months.

The results of this chapter demonstrate that the derived hominin distribution areas associated with warm and cold climatic periods are different. The main difference is the overall extent of the derived regions, and prey species diversity. During warm climates, derived distribution patterns are all local in nature, never reaching more than about 10 km in diameter, and there is little or no obvious pattern in the distribution of high and low prey species diversity areas, with the possible exception of Kůlna Cave, and Bohunice-kejbalý (I and II). This is in contrast to the derived distribution patterns associated with cold climatic periods: apart from Stránská skála (IIA and IIIA), the spatial extents of the distribution areas are considerably larger than those associated with warm periods, and again with the exception of Stránska skála (IIA and IIIA), there seems to be a high degree of predictability in the spatial distribution patterns of prey species diversity.

Summary and conclusions

This chapter set out to reconstruct hominin distribution patterns. I used already mapped site locations, prey species diversity maps, lithic raw material outcrops, together with Least-cost Surfaces and Least-Cost Corridors. In conclusion, it is clear that the most significant factor to come out of this chapter is the evident difference in the relative size and extents of the derived distribution areas associated with warm and cold periods. In the next chapter, I reconstruct the distribution patterns of cave bears in preparation for revealing potential interactions between hominins and cave bears in Chapter 8.

Chapter 7: Reconstructing cave-bear distribution patterns

Introduction

The objective of this chapter is to reconstruct the distribution patterns of cave bears. I first undertake odontometric and tooth-wear analyses, and then map cave-bear distribution areas (home range, preferred habitat, and relative cave-bear densities – RCD). In each case, I discuss the methodology used and associated results.

Odontometric and tooth-wear analyses

The cave-bear teeth used for this part of the project come from museum collections stored in the Budišov museum and the Anthropos Institute in Brno, both located in Moravia, Czech Republic. The teeth were examined during the summer of 2007 when I visited the study region, and the data were initially recorded by hand and subsequently digitally collated into a relational database using a laptop PC and Microsoft Access database software (Fig. 7.1). In total, more than 1700 teeth were examined, with photographs taken of all teeth either individually or in groups. In all cases, general information about the tooth was recorded, such as the site from which the tooth originated, the museum in which the tooth was stored, the box or container in which the tooth was located, the type of tooth, and the articulation of the tooth. In most cases, it was also possible to record the depositional layer from which the tooth originated, although this was not possible with some teeth as they had not been allocated to any particular layer. Canine teeth were measured using digital measuring pincers, and by measuring the breadth and length of all measurable canine teeth, across the crown of the tooth; although only the breadth is shown here. All measurements were taken three times and recorded in millimetres to an accuracy of two decimal places, and the result entered into the relational database. In some cases teeth were so badly damaged that it was impossible to take any measurements, and only the presence of these teeth was recorded. Cave-bear molars (Lower Molar 1, LM1; Lower Molar 2, LM2; Lower Molar 3, LM3; Upper Molar 1, UM1; Upper Molar 2, UM2; see Fig. 7.2 and Fig. 7.3) were assessed by visually comparing the degree of wear apparent on the teeth with the wear chart provided by Stiner *et al.* (1996, 312) (Fig. 7.4 and Fig. 7.5); the results were then entered into the relational database. In most cases, it was possible to assess the degree of wear, and only occasionally was a tooth so badly damaged that this was not possible. In these cases, only the presence of the tooth was recorded (Appendix 1, Table A1.1).

Only teeth from deposits that could be confidently associated with either warm or cold climatic periods are included here, and these are grouped accordingly. The results of the odontometric analysis are presented as histograms with the number of teeth in each measurement category presented as percentages, and with the total number of teeth measured displayed in the top right-hand corner. Tooth breadth is displayed in millimetre divisions, with the 'breadth (mm)' measurement representing the maximum possible size of the tooth. For example, a tooth falling into the 13 mm category equates to a tooth measuring between 12 mm and 13 mm. Tooth-wear results are presented collectively rather than individually. Results are again presented in histogram format, with the number of teeth for each wear stage presented as a percentage, and the total number of teeth sampled

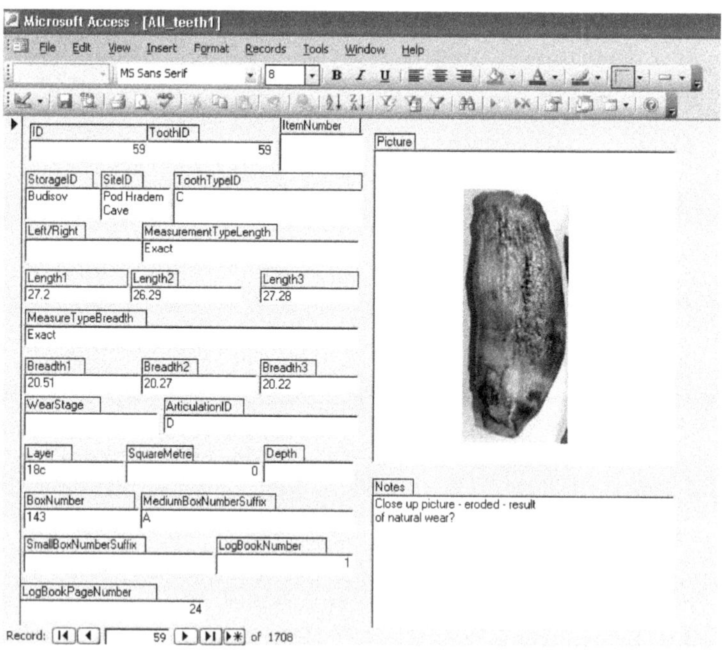

Fig. 7.1. Entry form of the relational database used for entering cave-bear tooth data.

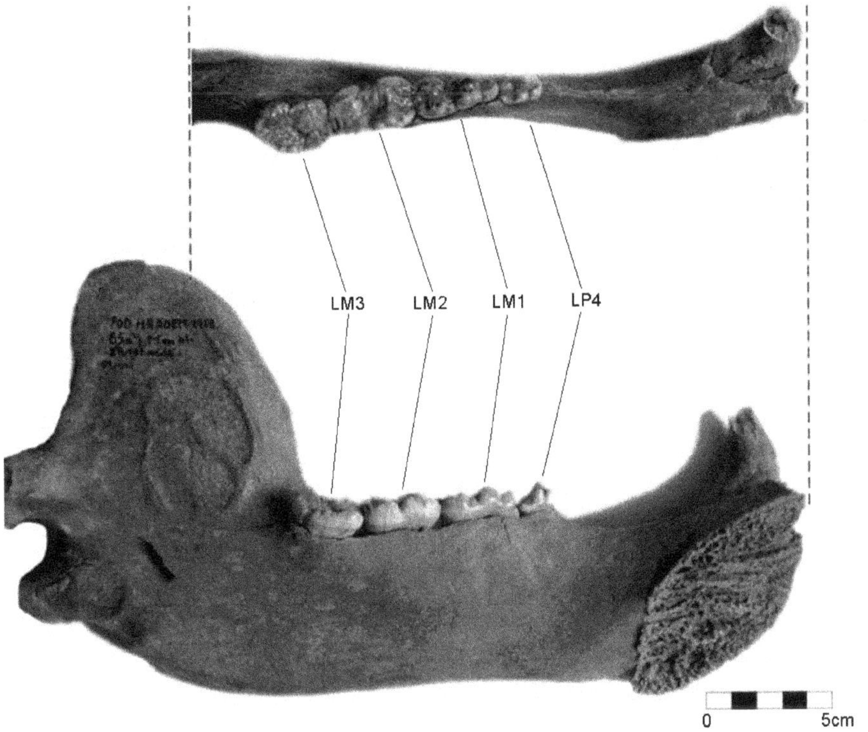

Fig. 7.2. Cave-bear mandible showing lower molar (LM3, LM2, and LM1) and premolar (LP4) teeth. (Photograph taken by author.)

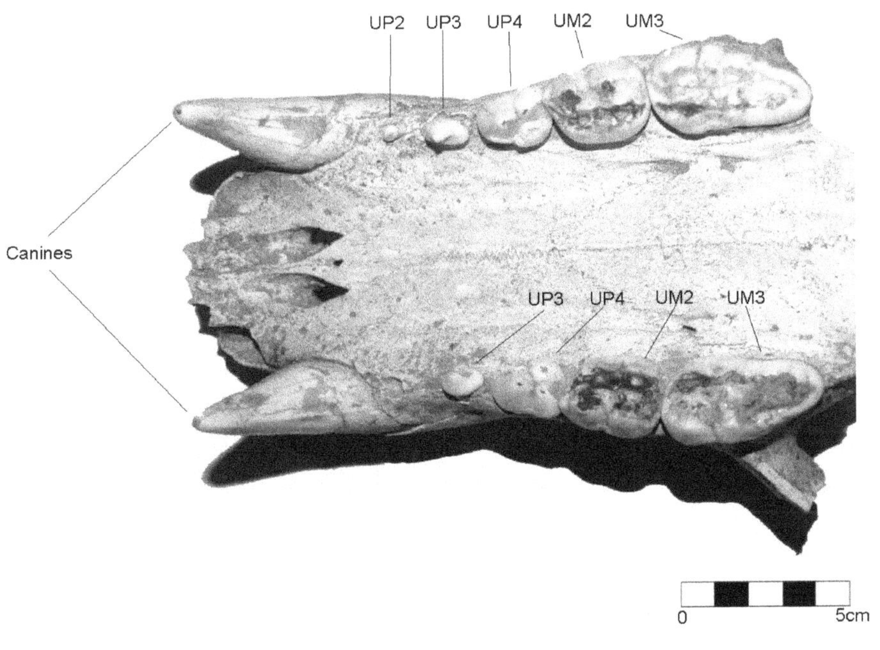

Fig. 7.3. Cave-bear maxilla, showing upper molar (UM2 and UM3), premolar (UP2, UP3, UP4) and canine teeth. (Photograph taken by author.)

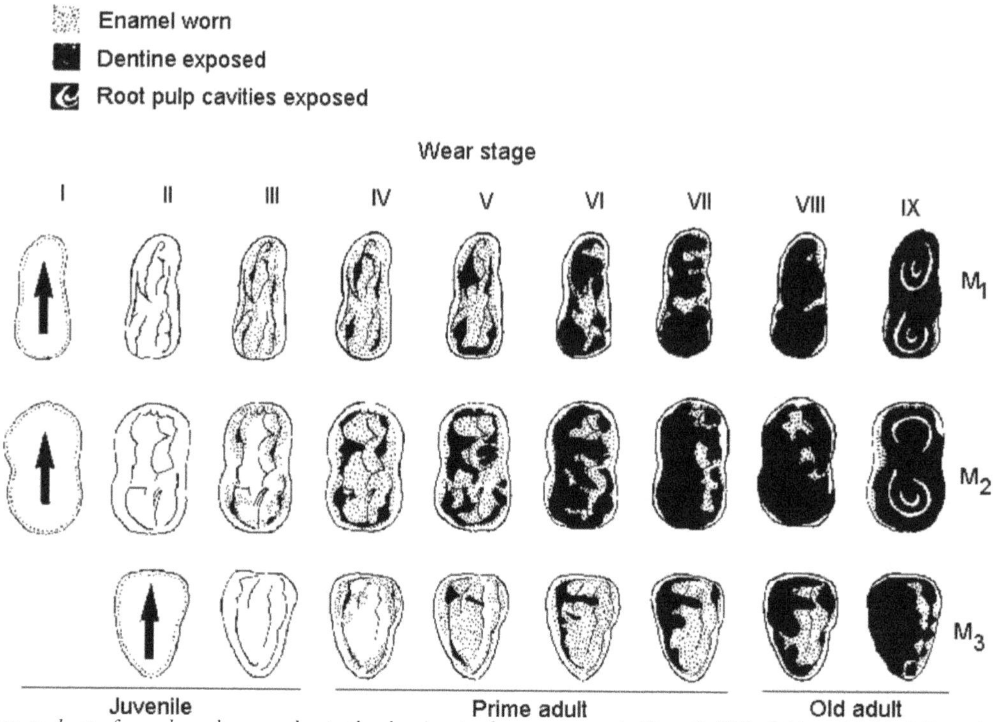

Fig. 7.4. Tooth-wear chart of cave-bear lower molar teeth, showing tooth-wear stages 1 (I) to 9 (IX). M1 = LM1, M2 = LM2, and M3 = LM3. (After Stiner et al. 1996, 312.)

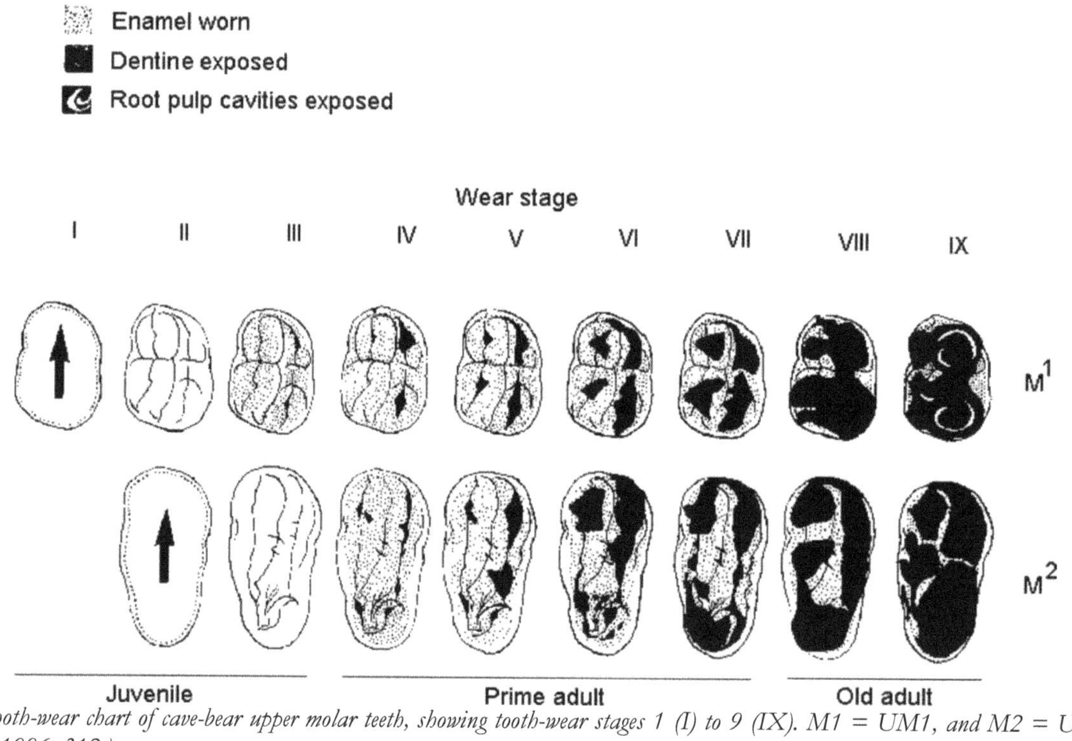

Fig. 7.5. Tooth-wear chart of cave-bear upper molar teeth, showing tooth-wear stages 1 (I) to 9 (IX). M1 = UM1, and M2 = UM2. (After Stiner et al. 1996, 312.)

displayed in the top right-hand corner of the histogram chart. I discuss the results of the odontometric and tooth-wear analyses below. In each case, results are grouped according to climatic affiliation.

Ninety canine teeth were measured from Šipka Cave (Layers III and IV). The tooth breadth ranges from 13 mm to 24 mm, with the highest proportion of teeth occurring at 16 mm, and the lowest at 19 mm, 21 mm and 24 mm. Overall, most teeth measure between 14 mm and 18 mm, with proportionally fewer occurring between 18 mm and 24 mm. The overall distribution would seem to be bi-modal and pivotal around 19 mm (Fig. 7.6). In total, 41 canine teeth from Pod hradem Cave (warm climates, Layers 17 and 18) were measured. The breadth of the teeth ranges between 12 mm and 23 mm, with proportionally more occurring at 16-18 mm and 19-22 mm, and proportionally fewer occurring at 12-15 mm, 18-19 mm and 22-23 mm. There is an overall bi-modal distribution with the centre point occurring at 19 mm (Fig. 7.7). In total, 100 canine teeth were measured from Barová Cave. Tooth breadth ranges from 13 mm to 25 mm, with the highest proportion of teeth occurring between 16 mm and 17 mm, and the lowest proportion of teeth occurring at 24-25 mm. Overall, the majority of teeth fall between 15 mm and 18 mm, with proportionally fewer occurring at 13-15 mm and 18-25 mm. The distribution seems to be bi-modal with the centre point occurring at 20 mm (Fig. 7.8). In total, 14 canine teeth associated with Pod hradem Cave (cold climates, 5, 7, 8, 9 and 10) were measured. The breadth of these teeth falls between 14 mm and 24 mm, with the highest proportion of teeth occurring at 22-23 mm, and the lowest proportion of teeth occurring at 14-15 mm, 16-19 mm and 23-24 mm. There seems to be an overall bi-modal distribution, pivotal around 20 mm (Fig. 7.9). Overall, these results indicate that the most apparent difference seems to be with regard to the proportion of male and female cave bears associated with Pod hradem Cave in comparison with Šipka Cave and Barová Cave: in the case of Šipka Cave and Barová Cave the female:male ratio is about 70:30, whereas in the case of Pod hradem Cave the female:male ratio is approximately 42:58. As Pod hradem Cave is the only site which contains cave-bear remains confidently associated with cold climates, similar comparisons cannot be made for these periods; however, it is interesting to note that the female:male ratio is almost identical to that for Pod hradem Cave during warm climates (Fig. 7.10 and Fig. 7.11).

In total, 263 teeth were assessed for their wear from Šipka Cave (Layers III and IV), and the teeth range from wear stage I to wear stage IX. The most common is wear stage III and the least common is wear stage I. The overall wear stage distribution of teeth from this site is weighted mostly towards the earlier stages, wear stages II, III and IV, and somewhat towards the latter stages, wear stages VIII and IX, with proportionally fewer occurring in the middle stages of wear (V, VI, and VII) (Fig. 7.12). In total, 128 teeth were assessed for their wear from Pod harden Cave (warm climates, Layers 17 and 18). Teeth range from wear stage I to IX, with the highest proportion of teeth occurring in wear stage II, and the fewest in wear stage I. Overall, the distribution of teeth is heavily weighted towards the lower stages, with the exception of wear stage I, and the higher stages of wear, with far fewer teeth occurring in the middle wear stages, III, IV, V, VI, VII (Fig. 7.13). In total, 261 teeth were assessed for their wear from Barová Cave. The teeth range from wear stages II to IX. The most common is wear stage III and the least common is wear stage IX. No teeth fall into the category of wear stage I. Overall, the distribution of tooth wear is weighted mostly towards the early and middle stages of wear, wear stages III and IV, falling off exponentially to wear stage IX (Fig. 7.14). In total, 124 teeth were assessed for their wear from Pod hradem Cave (cold climates, Layers 5, 7, 8, 9, and 10). The wear stages of the teeth range from I to IX. The most common is wear stage II, the least common is wear stage VI and no teeth fall into wear stage V. Overall, the distribution is weighted mostly towards the earlier stages of wear, particularly wear stages II and III, and tooth-wear stage IX, with fewer teeth occurring between wear stages IV and VIII (Fig. 7.15). Overall, results of the tooth-wear analysis demonstrate that the remains of cave bears of all ages are found at the three cave-bear case study sites in association with warm and cold climates. With respect to the results associated with warm climates, the most significant difference seems to be the apparent increase in the number of remains associated with prime-aged adults at Barová Cave (Fig. 7.16). This is in stark contrast to the results associated with Pod hradem Cave, which demonstrate a relatively 'normal' age profile with more juvenile and old adult remains than prime-aged adult remains, and Šipka Cave which demonstrates considerably more juvenile remains than prime-aged and old adult remains, and only slightly fewer old remains compared with prime-aged adult remains (Fig. 7.16). It is not possible to make direct comparisons between results associated with cold climates: remains from Pod hradem Cave (Layers 5, 7, 8, 9, 10) are the only remains confidently associated with cold climates; however, it is interesting to observe that the age profile is similar to the results associated with Pod hradem Cave for warm climates, with more juvenile remains than old adult remains, and more old adult remains than prime-aged adult (Fig. 7.17). A summary of the results is presented in Table 7.1.

Although it is not possible to make accurate statements about the living cave-bear populations associated with the three cave-bear case study sites on the basis of these results (Chapter 2), these data do at least suggest that the general demographic character of cave bears using caves within the study region for hibernation purposes was not uniform. Although some caves were used by cave bears of all ages and both sexes relatively uniformly (e.g. Pod hradem Cave), others were used more often by particular demographic groups, such as young adults (e.g. Barová Cave), or female cave bears (e.g. Barová Cave and Šipka

Cave). However, the fact that such differences are apparent in cave-bear remains associated with caves within close proximity to each other (Pod hradem Cave and Barová Cave) indicates that such differences in cave-bear demography were not indicative of a regional phenomenon: they were specific to the cave site and the area immediately surrounding the cave. As a result, it is unlikely that any differences in the demographic character of cave-bears using caves for hibernation purposes would have significantly affected the overall demographic make-up of living cave-bear populations.

Comparison with other similar studies indicate that such differences in the way that male and female, and different aged cave bears used caves was not restricted to the case study area. For instance, Stiner *et al.* (1998, 85) have demonstrated that male cave bears used Yarimburgaz Cave in Turkey less often than female cave bears, with a male:female ratio of 73:100. Conversely, according to Debeljak (2007, 481) the number of male and female adults that used Mokrica Cave in northern Slovenia was almost equal. Also, although Debeljak (2007, 478) commented that the use of Mokrica Cave by different aged cave bears was relatively equal, Stiner *et al.*'s (1998, 88) analyses of cave bear teeth from Yarimburgaz Cave, indicate a predominance of young adult cave bears. Thus, it would seem that the results shown in this study are not idiosyncratic to the study region, but are manifest of a larger pattern of complex cave-bear behaviour where caves are not used equally by cave bears of both sexes and all age groups, but instead differ significantly from cave to cave.

Although the results from this study compare well with those from other similar studies, some caution must be exercised. The validity of any zooarchaeological data is dependent on the sample used to obtain the results. Davis (1987, 29) made the point that obtaining a faunal sample that is representative of the animal population is important in ensuring that the results obtained from a zooarchaeological study are reliable. One aspect of sampling that is particularly pertinent to the study carried out here is sample size. Reitz and Wing (2008, 113) made the point that a sample should be large enough to reflect the composition or character of the problem being studied. Also, Stiner *et al.* (1998, 84) commented that a sufficiently large sample of cave bear remains is necessary when trying to understand their population characteristics. This point is especially relevant with respect to canine teeth recovered from Layers 5, 7, 8, 9, and 10 of Pod hradem Cave (i.e. canine teeth associated with cold climatic periods), and teeth associated with newly born cave bears (i.e. tooth-wear stage I). Indeed, sex-ratio determinations for cave bears associated with cold climates are based on only 14 canine teeth, and teeth associated with newly born cave bears are curiously almost completely absent, considering that their expected mortality rate is likely to be the highest of all cave bear age-groups (Debeljak 2007, 478). In both cases, the lack of remains is probably a result of taphonomic circumstances and changes in the overall survival rate of cave-bear teeth in different climatic conditions, and the increased fragility of teeth belonging to newly born cave bears. Whatever the reason for the lack of such data, however, their relative absence means that it is clearly necessary to be cautious about making associated inferences. In saying that, the relatively large number of teeth associated with cave bears living in warm climatic periods (Pod hradem Cave – Layers 17 and 18, Barová Cave, and Šipka Cave – Layers III and IV), suggests that the general conclusions regarding the apparent age/sex, cave-specific idiosyncrasies demonstrated in this study are largely real, at least in terms of cave bears living in warm climatic conditions. Moreover, the apparent similarity, albeit tentative, between the general age and sex mortality rates of cave bears associated with Pod hradem Cave during both cold and warm climates, suggests that similar patterns of cave use by cave bears were occurring during both warm and cold climatic conditions. However, as cave-bear remains associated with cold climatic periods can only be confirmed as being associated with Pod hradem Cave, it is not possible to make any confident statements about how idiosyncratic their use of caves was during cold climates.

Mapping cave-bear distribution patterns

Home ranges associated with different demographic groups (cave-bear females with cubs, cave-bear females with yearlings, lone females, and lone male cave bears) for the three cave-bear case study sites are reconstructed using brown bear home range surface areas provided by Dahle and Swensen (2003a) (Chapter 3). Brown bears are a good substitute for cave bears: brown bears have a similar diet to that of cave bears, as they are generally considered to be omnivorous with a large part of their diet made up of vegetation. Also, apart from polar bears who are carnivores and therefore not a suitable analogy for cave bears, brown bears are the largest of all extant bear species and the closest in size to cave bears, and therefore probably have similar metabolic rates, associated energy expenditure and home ranges. The radii of the home ranges associated with each brown bear demographic group are derived using the equation $A=\pi r^2$, and the derived radii are multiplied by two to take into account the possibility that cave-bear hibernation sites are on the periphery of the home range. Home ranges are digitally mapped using the Least-cost Surfaces algorithm available in ArcGIS9.2, with the maximum extents set to the newly derived cave-bear home range radius values. Preferred cave-bear habitat areas are identified using the already digitally mapped palaeovegetation regions (Chapter 4), and the expected dietary habits of cave bears (i.e. mostly vegetarian and with preferences for areas rich in deciduous vegetation) (Chapter 3). A 50:50 female:male population distribution is assumed, with 33% of the females assumed to be lone females, 33% of the females with cubs, and 33% with yearlings (Chapter 3). Relative cave-bear densities (RCDs) are calculated for each cave-bear demographic group by dividing the

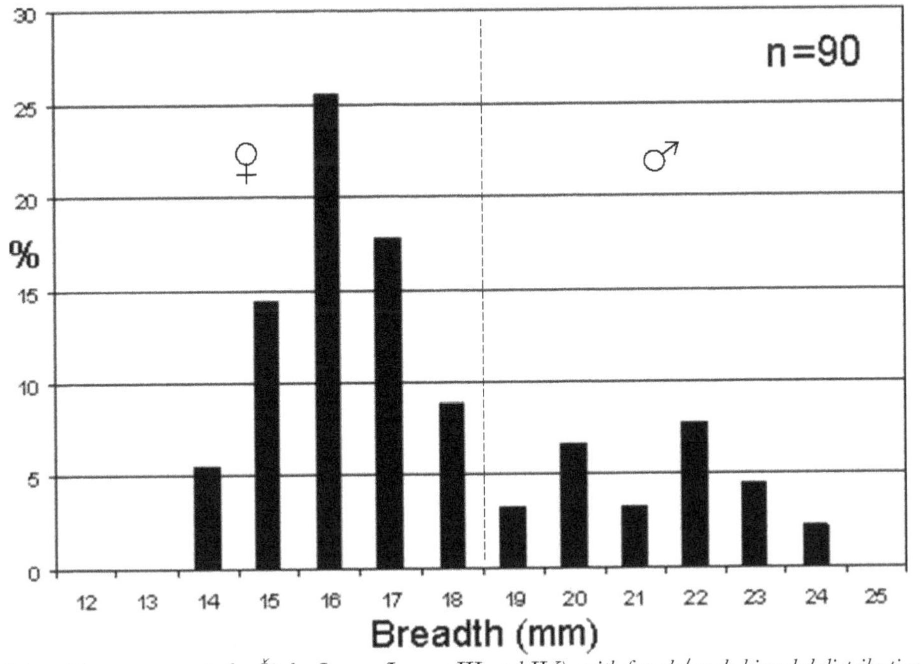

Fig. 7.6. Canine teeth breadth measurements for Šipka Layer (Layers III and IV), with female/male bimodal distribution point shown (dashed line).

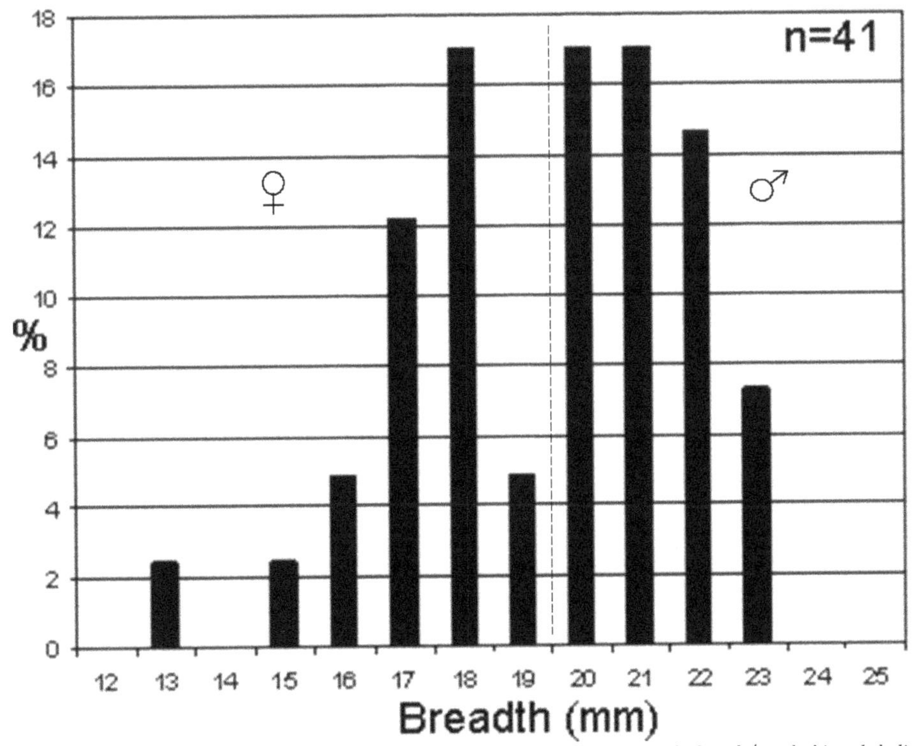

Fig. 7.7. Canine teeth breadth measurements for Pod hradem Cave, Layers 17 and 18, with female/male bimodal distribution point shown (dashed line).

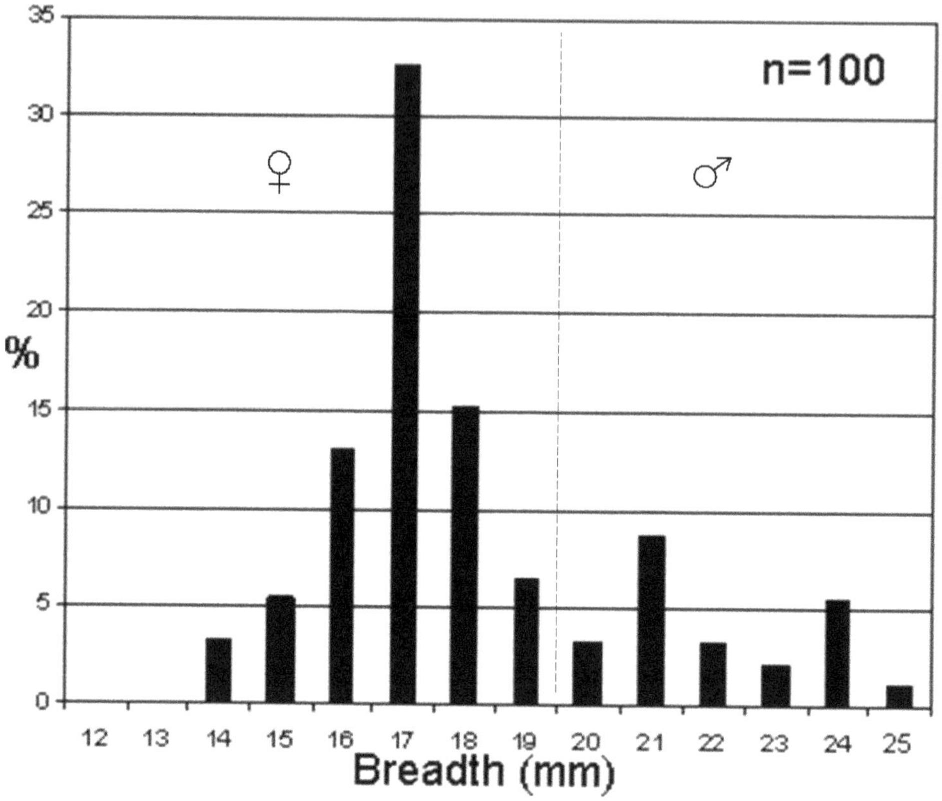

Fig. 7.8. Canine teeth breadth measurements for Barová Cave, with female/male bimodal distribution point shown (dashed line).

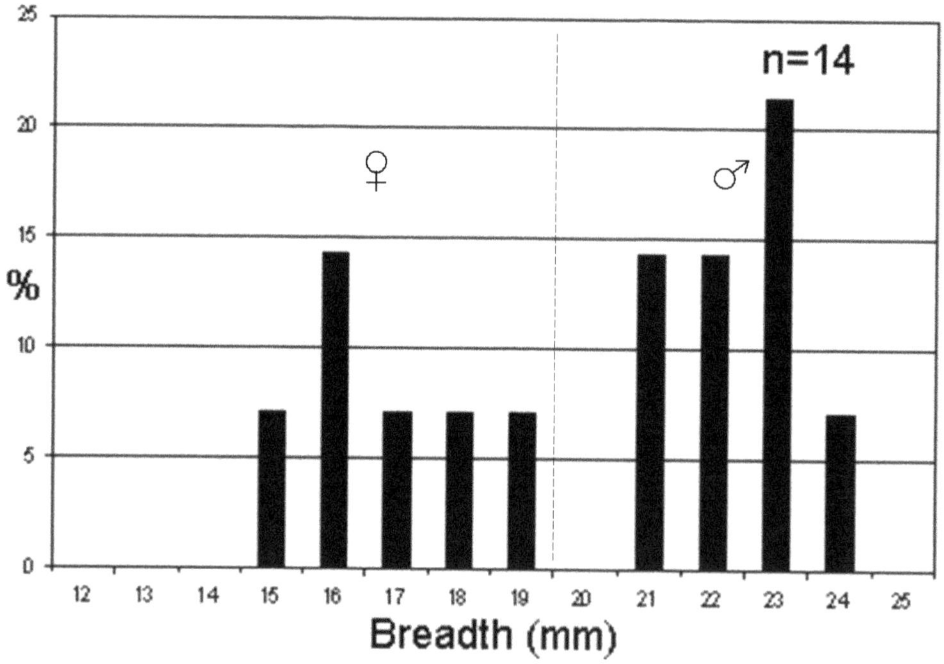

Fig. 7.9. Canine teeth breadth measurements for Pod hradem Cave, Layers 5, 7, 8, 9, and 10, with female/male bimodal distribution point shown (dashed line).

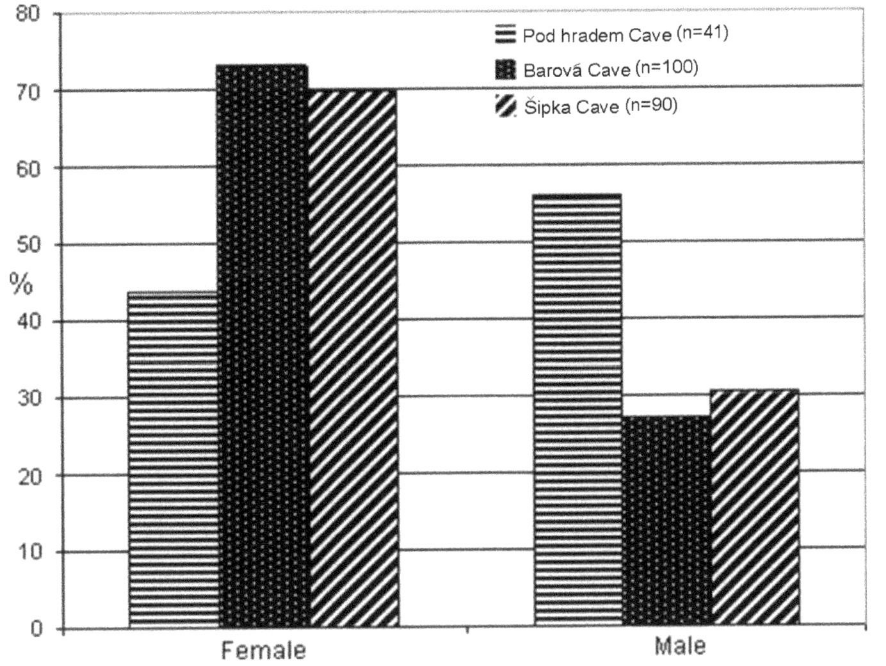

Fig. 7.10. Proportion of female and male cave bears from warm associated deposits (i.e. Pod hradem Cave – Layers 17 and 18, Barová Cave, and Šipka Cave – Layer III and IV).

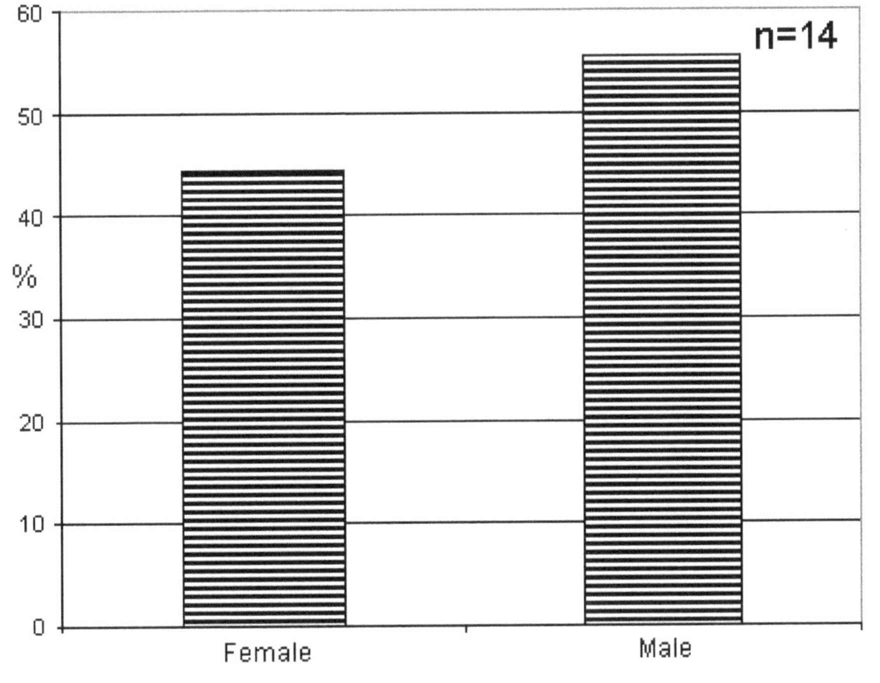

Fig. 7.11. Proportion of female and males cave bears from cold associated deposits (i.e. Pod hradem Cave – Layers 5, 7, 8, 9, and 10).

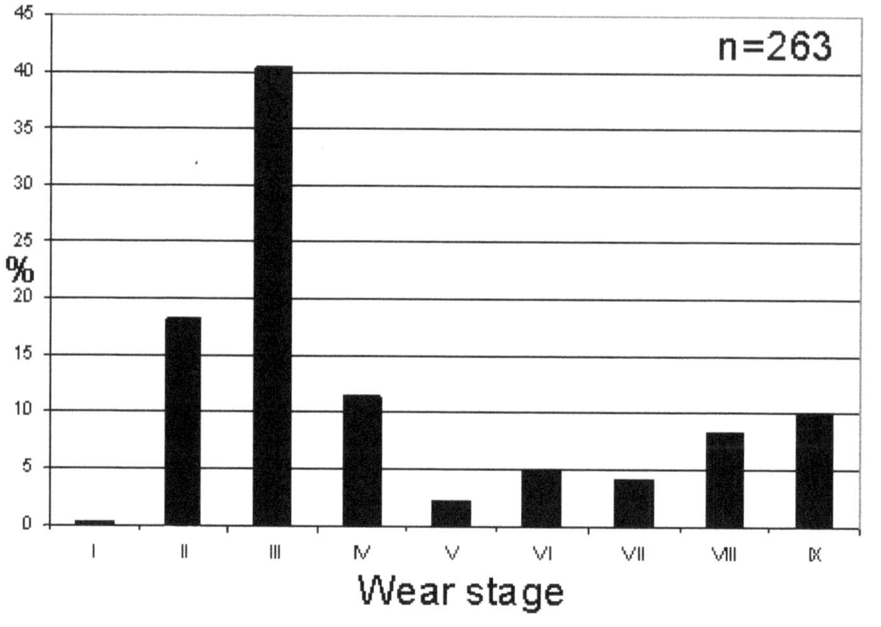

Fig. 7.12. Tooth-wear results from Šipka Cave, Layers III and IV.

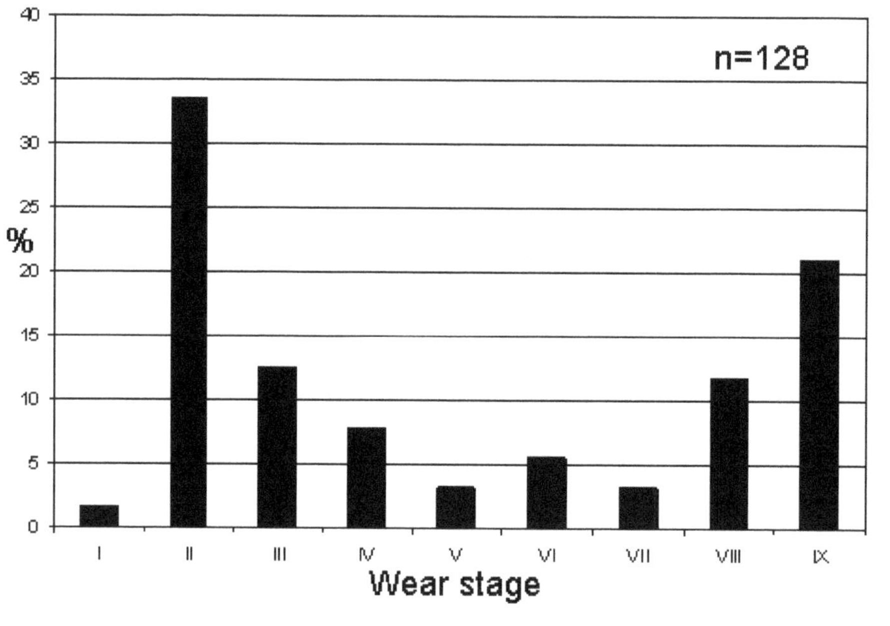

Fig. 7.13. Tooth-wear results from Pod hradem Cave, Layers 17 and 18.

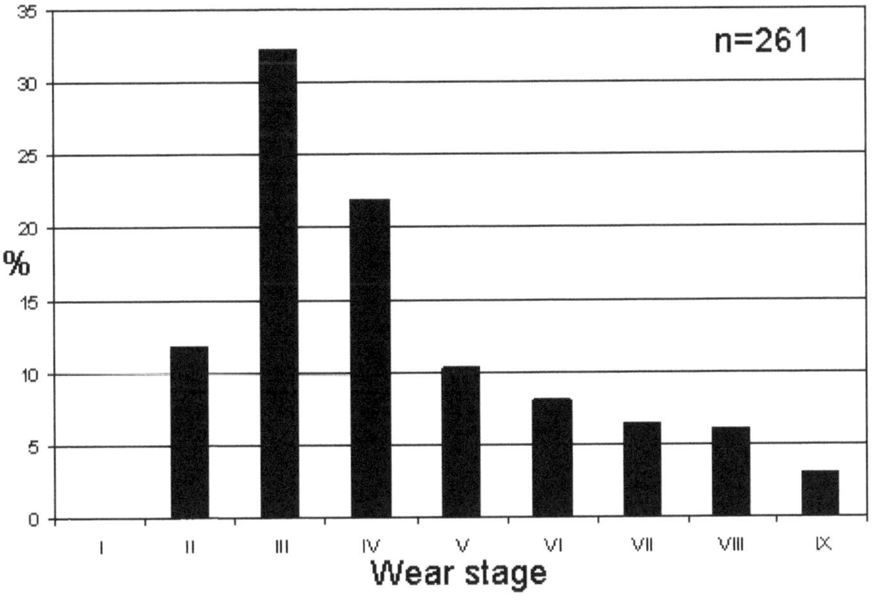

Fig. 7.14. Tooth-wear results from Barová Cave.

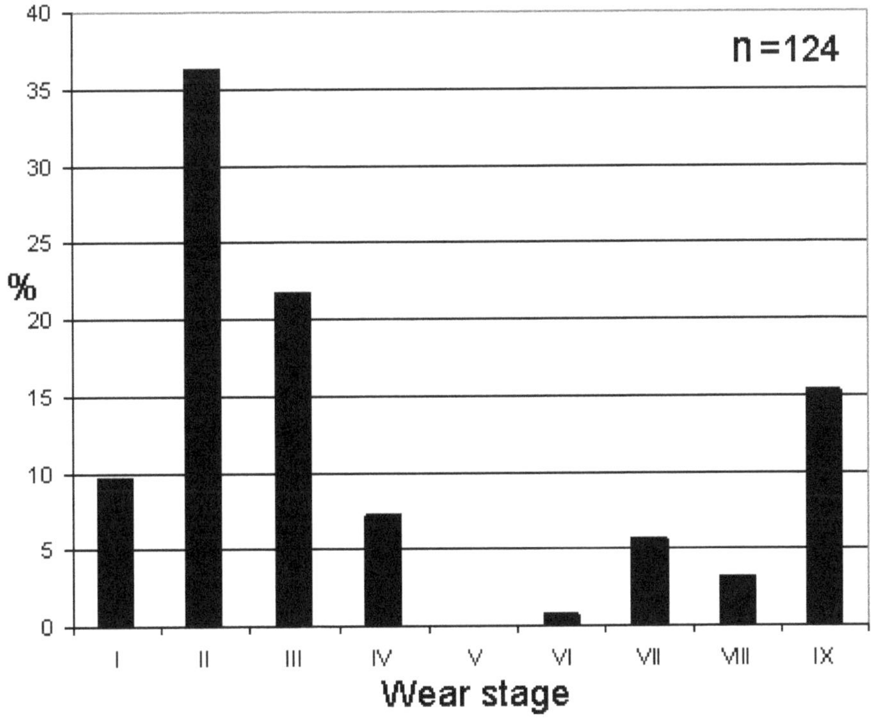

Fig. 7.15. Tooth-wear results for Pod hradem Cave, Layers 5, 7, 8, 9, and 10.

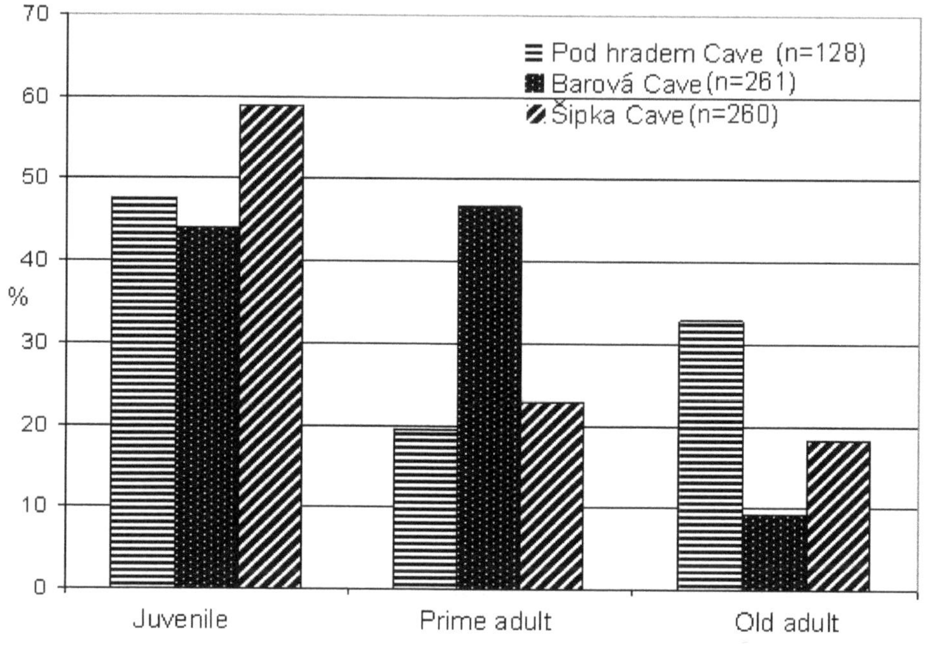

Fig. 7.16. Proportion of juvenile, prime-aged, and old cave bears associated with warm climatic periods (i.e. Pod hradem Cave – Layers 17 and 18, Barová Cave, and Šipka Cave – Layer III and IV).

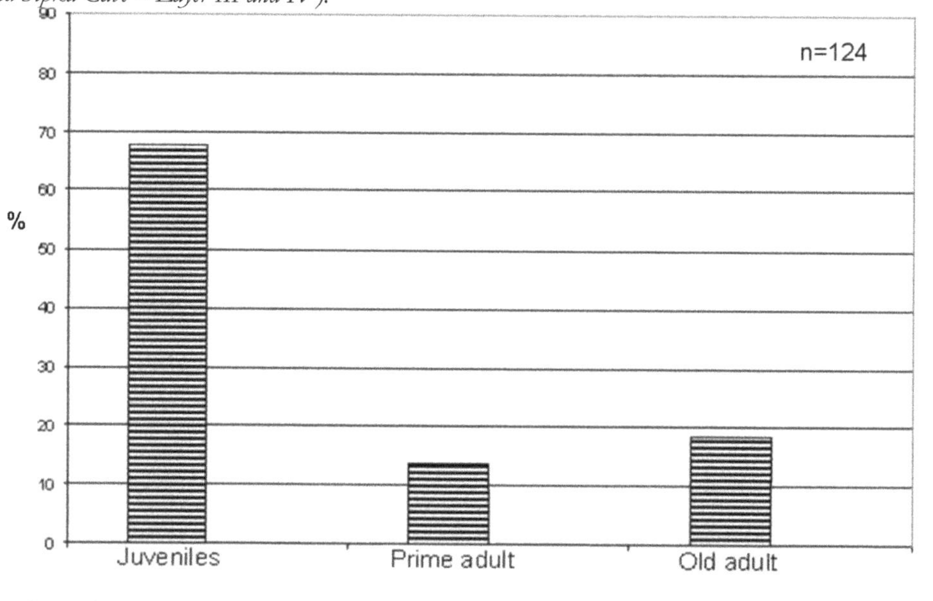

Fig. 7.17. Proportion of juvenile, prime adult, and old adult cave bears associated with cold climatic periods (i.e. Layers 5, 7, 8, 9, and 10 of Pod hradem Cave).

		Sex %		Age %		
		Male	**Female**	**Juvenile (wear stages I, II, III)**	**Prime-aged (wear stages IV, V, VI, VII)**	**Old (wear stages VIII and IX)**
Pod hradem Cave	Warm	Assuming bi-modal point of 20 mm (< 20 mm = female)				
		58	**42**	**47**	**19**	**34**
	Cold	Assuming bi-modal point of 20 mm (< 20 mm = female)				
		57	**43**	**67**	**14**	**19**
Šipka Cave	Warm	Assuming bi-modal point of 19 mm (< 19 mm = female)				
		29	**71**	**60**	**23**	**17**
Barová Cave	Warm	Assuming bi-modal point of 20 mm (< 20 mm = female)				
		24	**76**	**44**	**47**	**9**

Table 7.1. Overall summary of odontometrioc and tooth-wear analyses, showing the percentage of male and female, and juvenile, prime-aged, and old cave bears for Pod hradem Cave, Barová Cave, and Šipka Cave during warm and cold climates.

expected relative population associated with each demographic group by the associated surface area. Where home ranges overlap, RCD values are added to take into account the cumulative effect of cave bears from different demographic groups occupying the home ranges of other cave bears. RCDs are divided by the highest overall common denominator, i.e. the highest RCD, to produce a series of figures corresponding to estimated RCDs for each cave-bear demographic group during mating and post-mating periods in warm and cold climates, with the highest overall RCD being 1.00.

Results associated with Šipka Cave during the mating period in warm climates (Fig. 7.18) indicate that home range extents associated with each demographic group (i.e. females with cubs, females with young, lone females, and lone males) are concentric. Home ranges for females with cubs have the smallest overall extents, measuring less than about 5 km in both an east-west and a north-south direction, and have a RCD value of 1.00 (Table 7.2). Home ranges associated with female cave bears with yearlings are larger than those associated with females with cubs, measuring about 12 km in a north-south direction, and about 10 km in an east-west direction; although the southeast regions of this home range are not much larger than the home range associated with females cave bears with cubs. The RCD value associated with these cave bears is 0.34 (Table 7.2). Home ranges for lone females are larger than those for female cave bears with yearlings, and measure about 20 km in a north-south direction, and about 15 km in an east-west direction. In places, however, home ranges associated lone female cave bears are only slightly larger than those for females with yearlings, and this is particularly the case in the eastern region of the home range. The associated RCD value is 0.14 (Table 7.2). Home ranges for lone male cave bears are considerably larger than home ranges associated with all other cave-bear groups, and measure about 50 km in a north-south and east-west direction. However, in certain areas, particularly towards the south, home ranges associated with lone male cave bears are only about 10-15 km larger than those for lone females cave bears. In the north, lone male home ranges are about 30 km larger than those associated with lone females. The RCD value associated with lone male cave bears during this period is 0.05 (Table 7.2). Preferred cave-bear habitat areas associated with Šipka Cave during this period are mostly sporadic in nature, but areas to the north, north-west, and south-west of Šipka Cave are particularly densely occupied.

Results associated with Pod hradem Cave during the cave-bear mating period in warm climates (Fig. 7.18) are also concentric. Home ranges associated with female cave bears with cubs are the smallest, and measure about 2-3 km in both a north-south and east-west direction. The RCD value associated with cave-bear females with cubs is 1.00 (Table 7.2). Home ranges for female cave bears with yearlings are larger than those associated with female cave bears with cubs, and measure between about 7-10 km in both a north-south and east-west direction; although towards the south of this home range, the extents are only marginally larger than those associated with female cave bears with cubs. The RCD value associated with cave-bear females with yearlings is 0.34 (Table 7.2). Home ranges for lone female cave bears are larger than those for females with yearlings, measuring about 12 km in a north-south and east-west direction, but the southern regions are only marginally larger than home ranges associated with female cave bears with yearlings. The RCD value for lone female cave bears is 0.14 (Table 7.2). Home ranges for lone male cave bears are larger than those for lone female cave bears, and measure about 35 km in a north-south direction and about 30 km in an east-west direction. The RCD value for lone males during this period is 0.05 (Table 7.2). Preferred cave-bear habitat areas associated with Pod hradem Cave during this period are mostly sporadic in nature, although areas to the west of Pod hradem Cave are more densely occupied than other regions.

Results associated with Barová Cave during the mating period in warm climates (Fig. 7.18) indicate that the extents of the home ranges are also concentric in nature. Home ranges for female cave bears with cubs are the smallest, and measure no more than about 2-3 km diameter in all directions. The RCD value for female cave bears with cubs is 1.00 (Table 7.2). Home ranges associated with female cave bears with yearlings are larger than those for female cave bears with cubs, and measure about 7-10 km in an east-west and north-south direction. The RCD value for female cave bears with yearlings is 0.34 (Table 7.2). Home ranges for lone female cave bears are larger than those associated with female cave bears with yearlings, and are approximately 10 km in an east-west direction and about 12 km in a north-south direction. The RCD value for lone female cave bears is 0.14 (Table 7.2). Home ranges for lone male cave bears are considerably larger than those for lone female cave bears, and measure about 25 km in a north-south direction, and approximately 20-25 km in an east-west direction. The RCD for lone male cave bears is 0.05 (Table 7.2). Preferred cave-bear habitat areas are mostly sporadic in nature, but regions located immediately to the south and northwest of the cave are most densely occupied.

Results associated with Šipka Cave during the post-mating period in warm climates (Fig. 7.19) indicate that home ranges for female cave bears with cubs are almost the same size and shape as home ranges associated with female cave bears with yearlings. Both home ranges are about 10 km in a north-south direction and about 10 km in an east-west direction. The RCD value associated with female cave bears with cubs is 0.85, and the RCD value associated with females with yearlings is 0.55 (Table 7.2). Home ranges for lone female cave bears are larger than those associated with female cave bears with cubs and female cave bears with yearlings, and these measure about 15 km in a north-south direction, and about 12-15

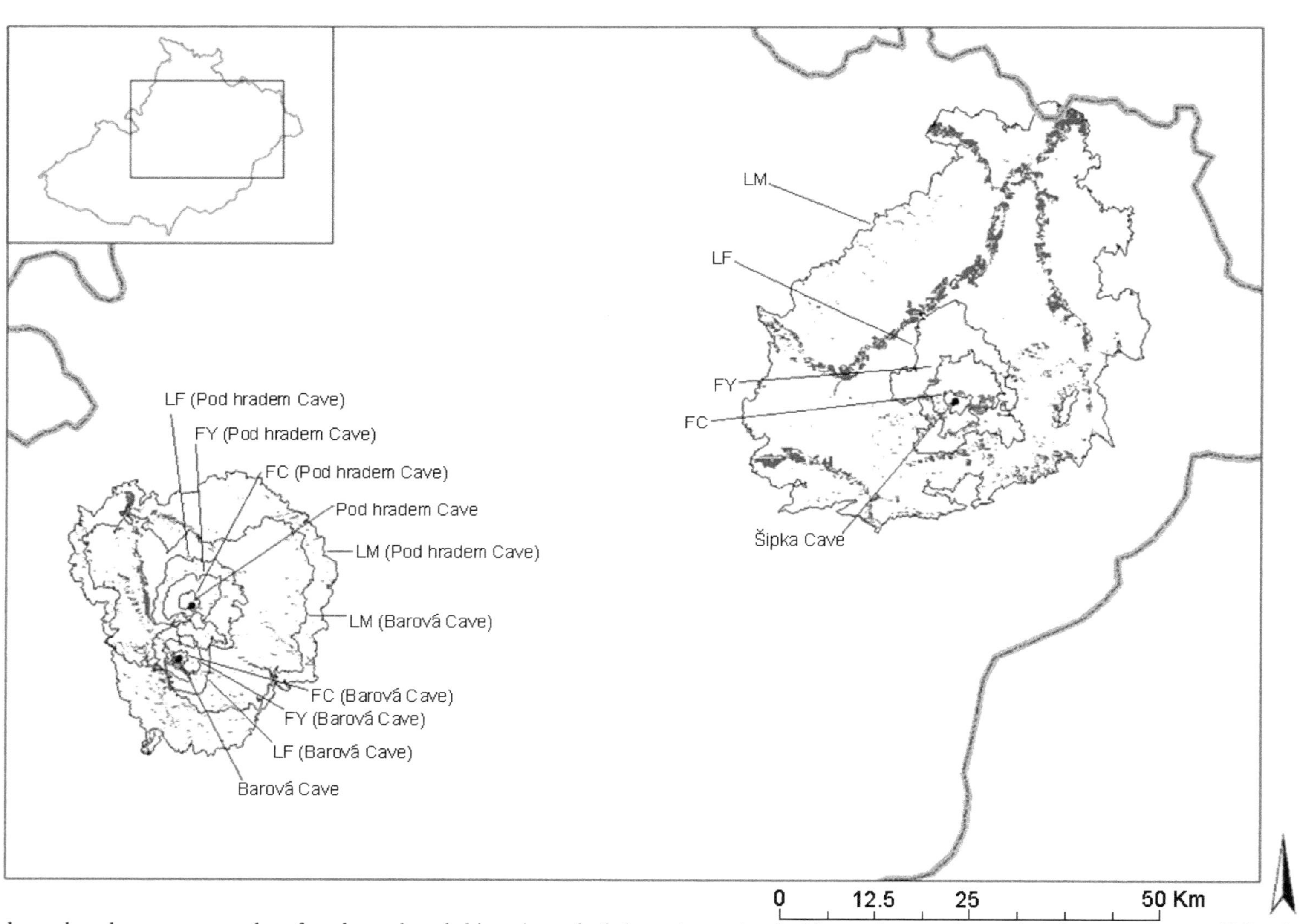

Fig. 7.18. Derived cave-bear home ranges and preferred cave-bear habitats (grey shaded areas) associated with cave-bear mating periods in warm climates. *FC = female cave bears with cubs, FY = female cave bears with yearlings, LF = lone female cave bears, and LM = lone male cave bears.*

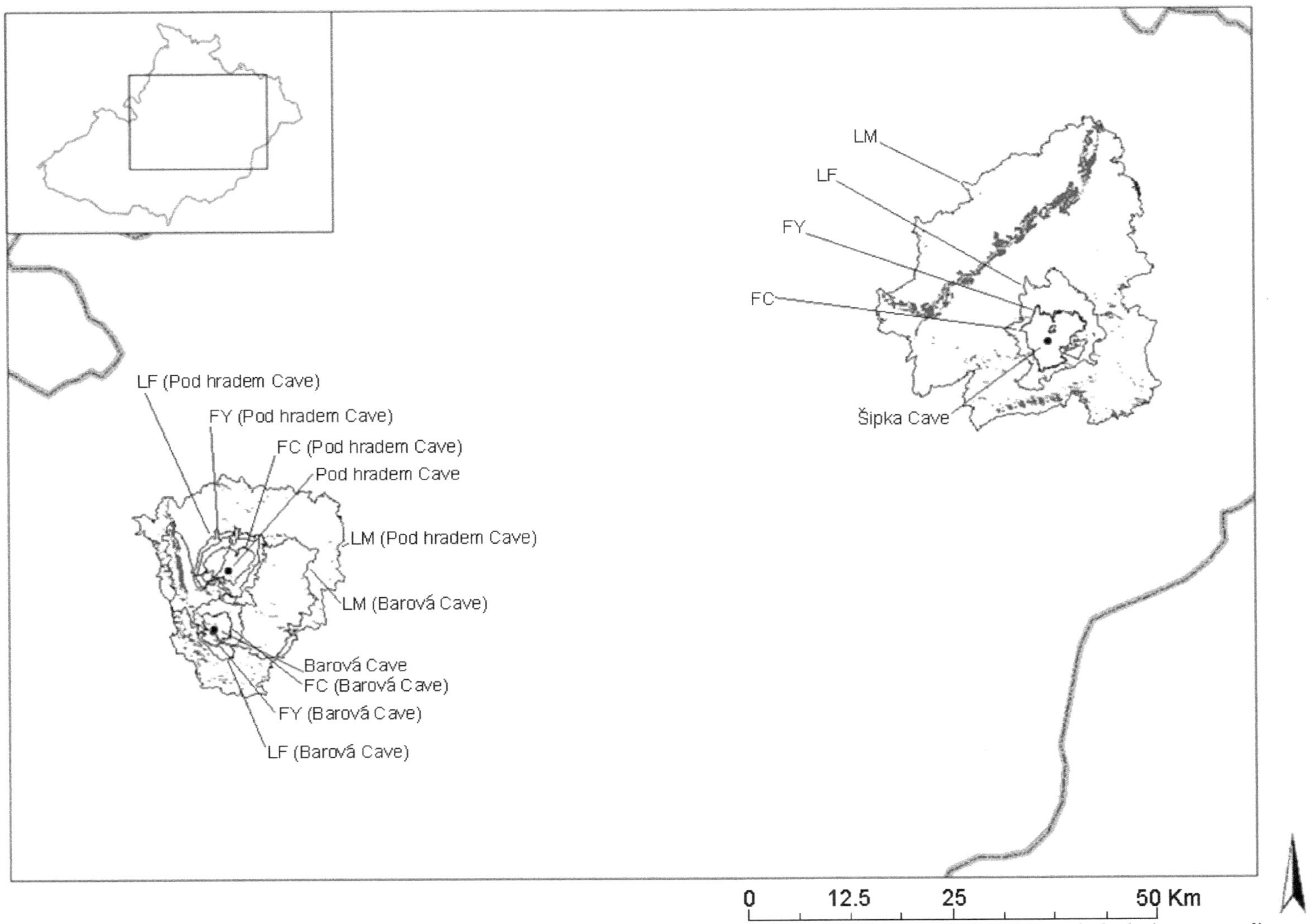

Fig. 7.19. Derived cave-bear home ranges and preferred cave-bear habitats (grey shaded areas) associated with cave-bear post-mating periods during warm climates. *FC = female cave bears with cubs, FY = female cave bears with yearlings, LF = lone female cave bears, and LM = lone male cave bears.*

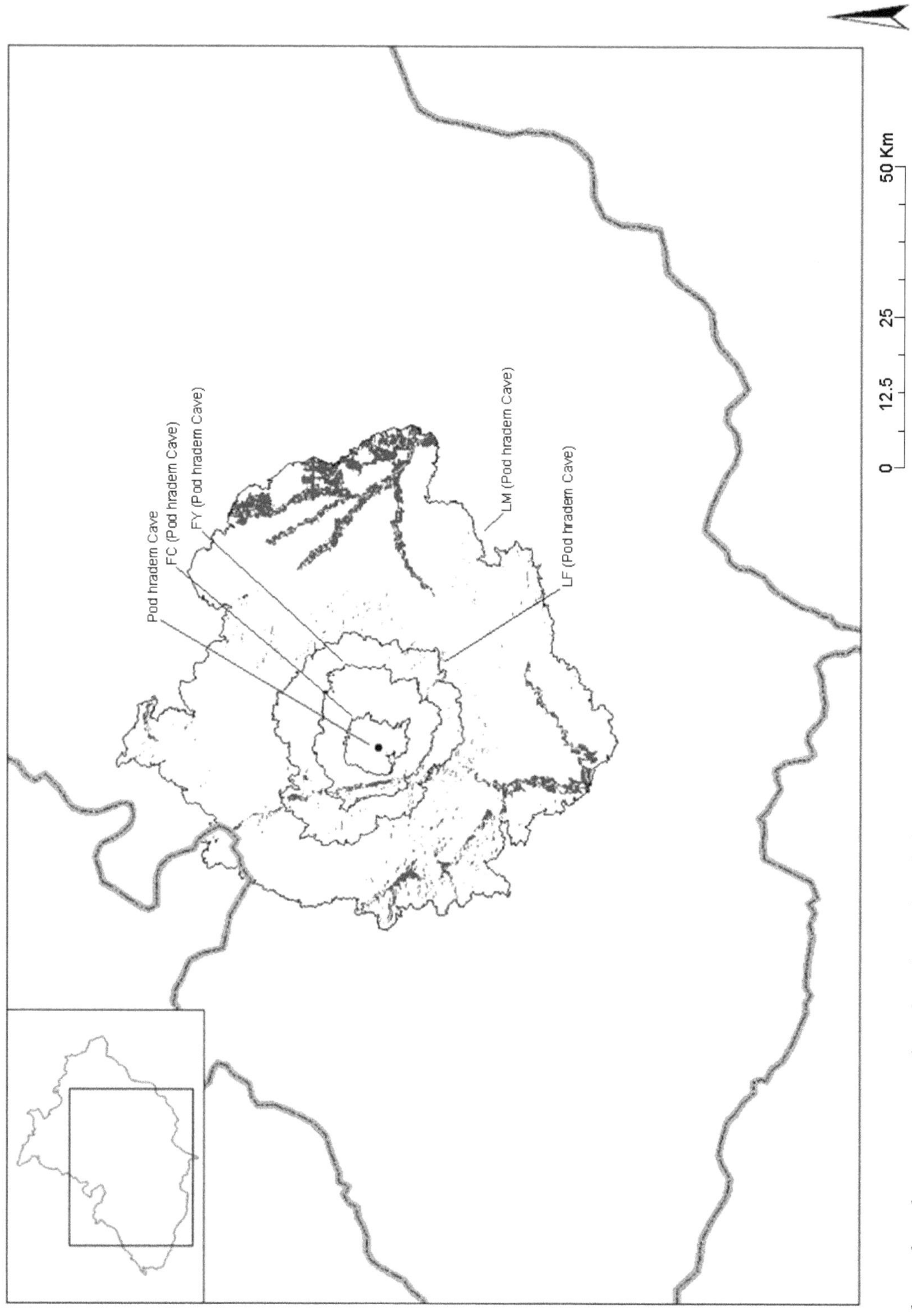

Fig. 7.20. Derived cave-bear home ranges and preferred cave-bear habitats (grey shaded areas) associated with cave-bear mating periods in cold climates. *FC = female cave bears with cubs, FY = female cave bears with yearlings, LF = lone female cave bears, and LM = lone male cave bears.*

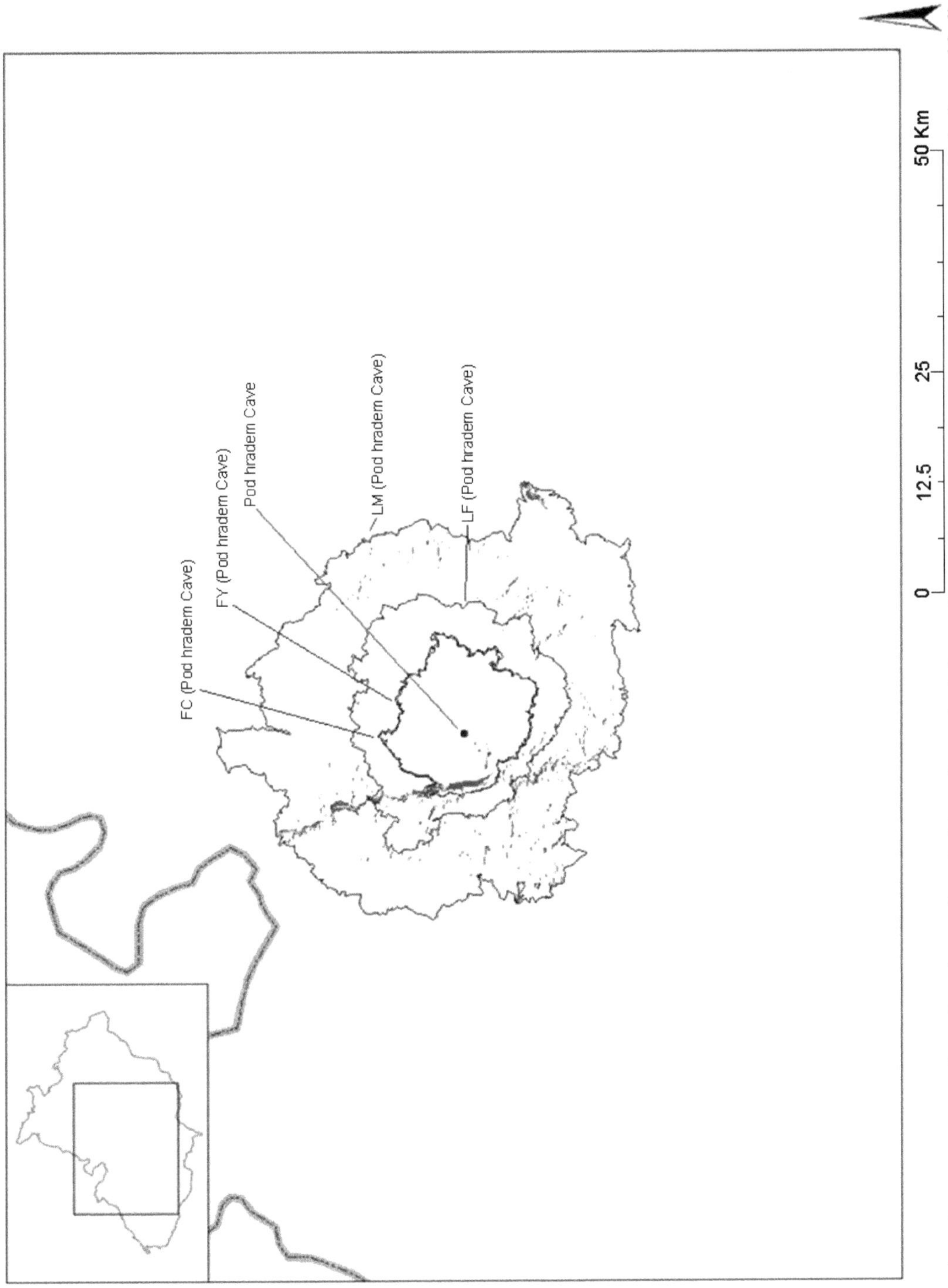

Fig. 7.21. Derived cave-bear home ranges and preferred cave-bear habitats (grey shaded areas) associated with cave-bear post-mating periods during cold climates. *FC = female cave bears with cubs, FY = female cave bears with yearlings, LF = lone female cave bears, and LM = lone male cave bears.*

	Climate	Cave-bear season	Females with cubs	Females with yearlings	Lone females	Lone males
Pod hradem Cave						
	Warm					
		Mating	1.00	0.34	0.14	0.05
		Post-mating	0.85	0.55	0.24	0.10
	Cold					
		Mating	0.19	0.07	0.03	0.02
		Post-mating	0.16	0.11	0.06	0.03
Šipka Cave						
	Warm					
		Mating	1.00	0.34	0.14	0.05
		Post-mating	0.85	0.55	0.24	0.10
Barová Cave						
	Warm					
		Mating	1.00	0.34	0.14	0.05
		Post-mating	0.85	0.55	0.24	0.10

Table 7.2. Derived relative cave-bear densities (RCD) for cave bears (females with cubs, females with yearlings, lone females, and lone males) associated with Pod hradem Cave, Šipka Cave, and Barová Cave during cave-bear mating and post-mating periods in warm and cold climates. Maximum RCD value is 1.00.

km in an east-west direction. The RCD value associated with lone female cave bears is 0.24 (Table 7.2). Home ranges for lone male cave bears are considerably larger than those associated with lone female cave bears. These measure about 30-35 km in a north-south direction, and about 25 km in an east-west direction; however, the southern region of this home range is only about 10 km larger than home ranges for lone female cave bears, whereas in the north they are about 20 km larger. The RCD value for lone male cave bears during this period is 0.10 (Table 7.2). Preferred cave-bear habitat areas associated with Šipka Cave during this period are mostly sporadic, but areas to the north, northwest and southeast are particularly densely occupied.

Results associated with Pod hradem Cave during the post-mating period in warm climates (Fig. 7.19) indicate that home ranges for female cave bears with cubs are about 7 km in a north-south and east-west direction, and home ranges for female cave bears with yearlings only slightly larger, approximately 1 km larger in all directions. The RCD value for females with cubs is 0.85, and the RCD value for females with yearlings is 0.55 (Table 7.2). Home ranges for lone females are not considerably larger than those for female cave bears with cubs and female cave bears with yearlings, and measure approximately 10-12 km in both a north-south and east-west direction. The RCD value for lone cave-bear females is 0.24. Home ranges associated with lone males are considerably larger than those for lone female cave bears, and are about 20-25 km in a north-south direction, and about 15-20 km in an east-west direction. The RCD value for lone male cave bears is 0.10 (Table 7.2). Preferred cave-bear habitat areas associated with Pod hradem Cave during this period are mostly sporadic in nature, but some areas to the west of Pod hradem Cave are more densely occupied.

Results associated with Barová Cave during the post-mating period in warm climates (Fig. 7.19) indicate that home ranges for female cave bears with cubs are virtually the same as they are for female cave bears with yearlings, and measure approximately 5-7 km in a north-south and east-west direction. The RCD value for female cave bears with cubs is 0.85, and the RCD for female cave bears with yearlings is 0.55 (Table 7.2). Home ranges for lone female cave bears from Barová Cave are generally about 3-4 km larger in most directions, except for the southwest region where the border of the home range are almost in line with those for female cave bears with cubs and yearlings. The RCD for lone female cave bears is 0.24 (Table 7.2). Home ranges associated lone males are considerably larger than those for lone female cave bears, and measure approximately 15 km in an east-west direction, and about 20 km in a north-south direction. Home ranges associated with lone male cave bears are particularly irregular in shape, and significantly elongated towards the northwest region of the home range. The RCD value for lone male cave bears is 0.10 (Table 7.2). Preferred cave-bear habitat areas associated with Barová Cave during this period are mostly sporadic in nature, with the most densely occupied regions occurring immediately to the south, west, and northwest of the site.

Results associated with Pod hradem Cave during the mating period in cold climates (Fig. 7.20) indicate that home ranges associated with female cave bears with cubs are about 10 km in a north-south and east-west direction. The RCD value associated with female cave bears with cubs is 0.19 (Table 7.2). Home ranges associated with female cave bears with yearlings are larger than those for female cave bears with cubs, and measure approximately 20 km in a north-south and east-west direction. The RCD value associated with female cave bears with yearlings is 0.07 (Table 7.2). Home ranges for lone female cave bears are larger than those for female cave bears with yearlings, and measure approximately 25-30 km in north-south and east-west direction. The RCD associated with lone female cave bears is 0.03 (Table 7.2). Home ranges for lone male cave bears are significantly larger than those associated with lone female cave bears, and are approximately 70-75 km in both a north-south and an east-west direction. The RCD associated with lone male cave bears is 0.02 (Table 7.2). Preferred cave-bear habitat areas associated with Pod hradem Cave during this period are distributed to the west, south-west, and south, but the most densely occupied regions are in the far east of the region.

Results associated with Pod hradem Cave during the post-mating period in cold climates (Fig. 7.21) show that home ranges for female cave bears with cubs are virtually the same as they are for female cave bears with yearlings, and measure approximately 20 km in both a north-south and an east-west direction. The RCD for female cave bears with cubs is 0.16, and the RCD for female cave bears with yearlings is 0.11 (Table 7.2). Home ranges associated with lone female cave bears are generally about 5-10 km larger in all directions than those associated with female cave bears with cubs and yearlings. The RCD for lone females is 0.06 (Table 7.2). Home ranges for lone male cave bears are larger than those for lone female cave bears, and mostly measure about 10-15 km larger in all directions, apart from in the south-east where the home range for lone males extends about 20 km further than that for lone female cave bears. The RCD for lone male cave bears is 0.03 (Table 7.2). Preferred cave-bear habitat areas associated with Pod hradem Cave during this period are sporadic in nature, and mostly focused in areas to the west of Pod hradem Cave.

Summary and conclusions

In this chapter I reconstructed the distribution patterns of cave bears associated with the cave-bear case study site. I used cave-bear teeth to gain insights into the demographic nature of cave bears in the study region. I modelled cave-bear home ranges, using digitally mapped cave-bear site locations, together with contemporary studies on brown bears and associated home ranges. RCD values were calculated using modern-day brown bear demography and associated home range surface areas. Preferred cave-bear habitats were derived using cave-bear dietary habits and already modelled digital palaeovegetation maps, together with ArcGIS9.2. Overall, this chapter has shown that the distribution of cave bears in the study region changed significantly according to climate, cave-bear mating and post-mating periods, and cave bear demographic group. In the next chapter, I put the results of this chapter together with the results of Chapter 6 to identify potential interactions between hominins and cave bears.

Chapter 8: Identifying interactions

Introduction

The objective of this chapter is to identify potential interactions between hominins and cave bears. I begin by briefly describing the methodology used, and then present and summarise the results in terms of Neanderthals and AMH and their potential for interactions with cave bears.

Identifying interactions

Potential interactions between hominins and cave bears are identified by overlaying the results of contemporaneous hominin and cave-bear distribution patterns derived in Chapters 6 and 7. Results associated with warm (Šipka Cave, Kůlna Cave, Vedrovice V, and Bohunice-kejbalý I and II) and cold (Stránská skála IIA and IIIA, Dolní Věstonice I, II, III, and Pavlov I) climates are presented for cave-bear mating, post-mating, and hibernation periods. In cases where hominin and cave-bear distribution patterns overlap, all relevant information such as hominin distribution patterns, prey species diversity, cave-bear home ranges, and cave-bear preferred habitats are shown, with associated RCD values included in the text. In cases where no overlap is apparent only minimal information is provided, i.e. the outer perimeter of hominin and cave-bear distribution areas. In cases where cave bears are hibernating, the location of the cave-bear site is shown together with the outer perimeter of the hominin distribution area.

Results associated with Šipka Cave during the cave-bear mating period in warm climates (Fig. 8.1) indicate that direct overlap regions occur. Overlap areas are associated with all of the hominin home range, and correspond mostly with the home range of cave-bear females with cubs (RCD = 1.00). Hominin distribution regions corresponding with these areas have prey species diversity values of 1 (reindeer) and 2 (reindeer and horse), and there are two small cave-bear preferred habitat regions located close to the site of Šipka Cave. At the far northeast and south of the hominin distribution area, two small areas overlap with the home ranges of cave-bear females with yearlings (RCD = 0.34), and these have a prey species diversity value of 1 (reindeer). Results associated with Šipka Cave during the cave-bear post-mating periods again indicate that potential overlap regions occur (Fig. 8.2). The major difference, in comparison with results associated with cave-bear mating periods, is that hominin distribution areas occupy a much smaller region of home ranges associated with female cave bears with cubs (RCD = 0.85). Results associated with Šipka Cave during cave-bear hibernation periods simply demonstrate that hominin distribution patterns overlap with the cave-bear hibernation site, as would be expected (Fig. 8.3).

Results associated with Kůlna Cave during cave-bear mating periods in warm climates demonstrate that potential areas of overlap occur between hominin distribution areas and cave-bear distribution patterns (Pod hradem Cave and Barová Cave) (Fig. 8.4). Overlap regions correspond with all hominin distribution areas, and with the home ranges of all cave-bear demographic groups associated with Pod hradem Cave except for female cave bears with cubs, and with lone male cave-bear home ranges associated with Barová Cave. The spatial extents of the home range associated with female cave bears with yearlings (RCD = 0.34) (Pod hradem Cave) coincide with a relatively small region of the hominin distribution area to the southwest of Kůlna Cave, and this area mostly has a prey species diversity value of 2 (reindeer and mammoth). Cave-bear home ranges associated with lone female cave bears (RCD = 0.14) (Pod hradem Cave) extend about half way across hominin distribution areas, and these mostly have a prey species diversity value of 2 (reindeer and mammoth). Cave-bear home ranges associated with lone male cave bears (RCD 0.05) (Pod hradem Cave and Barová Cave) extend across all of the hominin distribution area, and these areas have a prey species diversity value of 1 (reindeer) and 2 (reindeer and mammoth). This region also coincides with cave-bear preferred habitat areas, located towards the western region of the hominin distribution area. Results associated with Kůlna Cave during cave-bear post-mating periods (Fig. 8.5) show that areas of potential overlap occur with cave bears from both Pod hradem Cave and Barová Cave, but home ranges associated with female cave bears with cubs (RCD = 0.85) (Pod hradem Cave) do not coincide with hominin distribution patterns. Home ranges associated with female cave bears with yearlings (RCD = 0.55) (Pod hradem Cave) coincide with the far southeast region of the hominin distribution area, which has a prey species diversity value of 2 (reindeer and mammoth). This is similarly the case for home ranges associated with lone female cave bears (RCD = 0.24) (Pod hradem Cave). Home ranges associated with lone male cave bears (RCD = 0.10) (Pod hradem Cave) cover the whole region of the hominin distribution area, coinciding with areas that have prey species diversity values of 1 (mammoth) and 2 (reindeer and mammoth), and areas of cave-bear preferred habitat. Home ranges associated with lone male cave bears (Barová Cave) occupy both the south-east part of the hominin distribution area, and a large proportion of the far west area of the hominin distribution region, coinciding with prey species diversity values of 1 (mammoth) and 2 (reindeer and mammoth), and cave-bear preferred habitat areas. During the winter period, hominin distribution patterns do not coincide with any of the cave-bear hibernation sites. The nearest cave-bear site is Pod hradem Cave, located about 5 km south of Kůlna Cave (Fig. 8.6).

Results associated with Bohunice-kejbalý (I and II) during cave-bear mating periods in warm climates indicate that potential areas of overlap occur (Fig. 8.7).

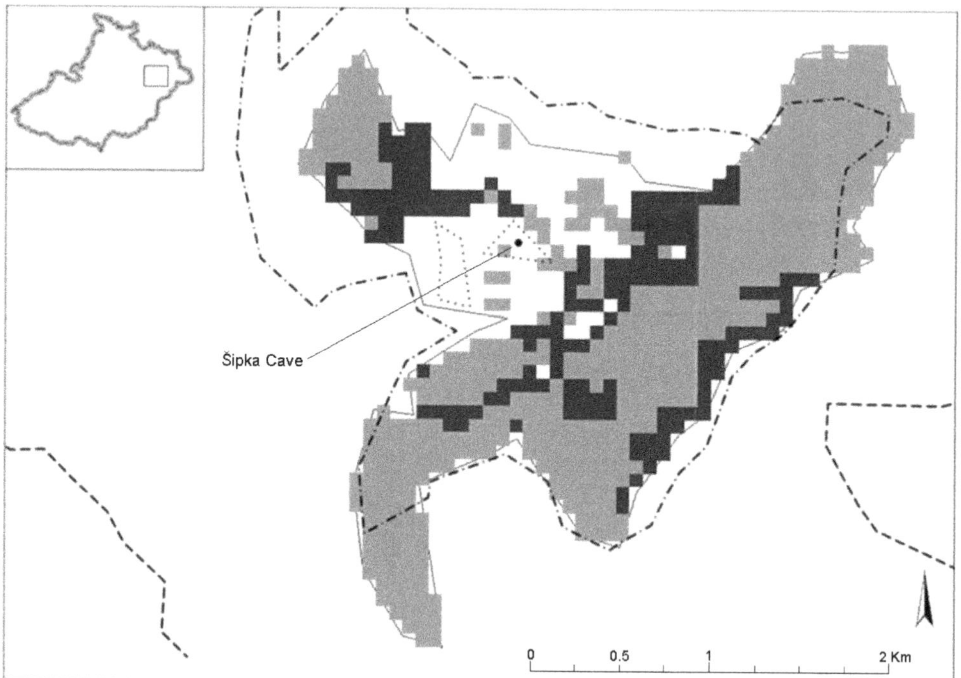

Fig. 8.1. Hominin and cave-bear distribution areas, and associated prey species diversity for Šipka Cave during cave-bear mating periods. Species diversity – light grey = 1 (reindeer), and dark grey = 2 (reindeer and horse). Dash-dot-dash = home range of female cave bears with cubs (RCD = 1), dash = home range of female cave bears with yearlings (RCD = 0.34), dotted = cave-bear preferred habitat.

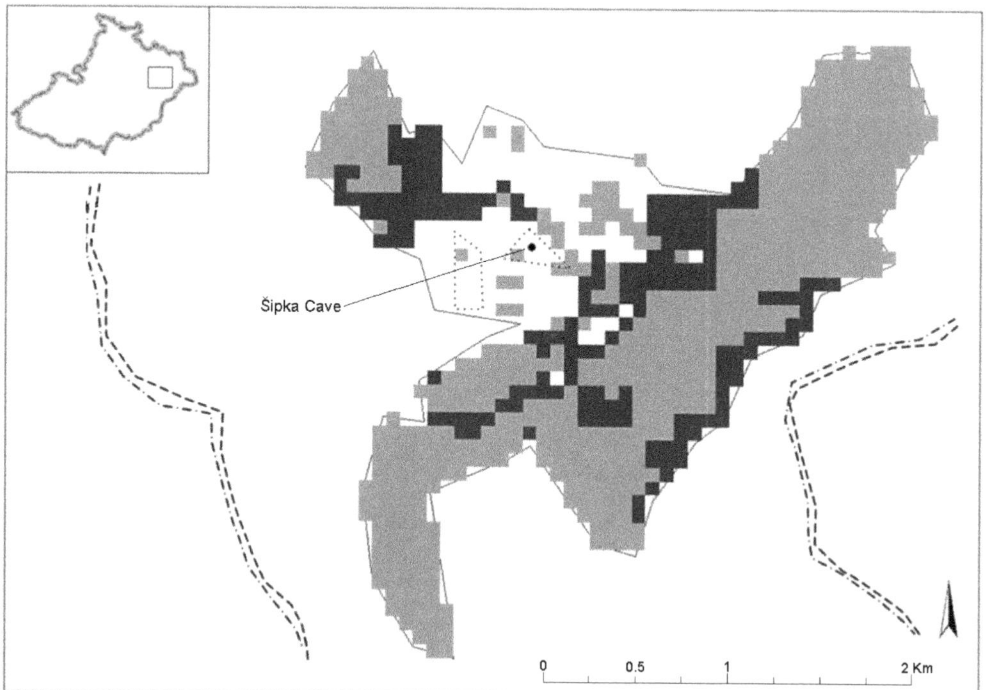

Fig. 8.2. Hominin and cave bear distribution areas, and associated prey species diversity for Šipka Cave during cave-bear post-mating periods. Species diversity – light grey = 1 (reindeer), and dark grey = 2 (reindeer and horse). Dash = home range of female cave bears with cubs (RCD = 0.85), dash-dot-dash = home range of female cave bears with yearlings (RCD = 0.55), dotted = cave-bear preferred habitats.

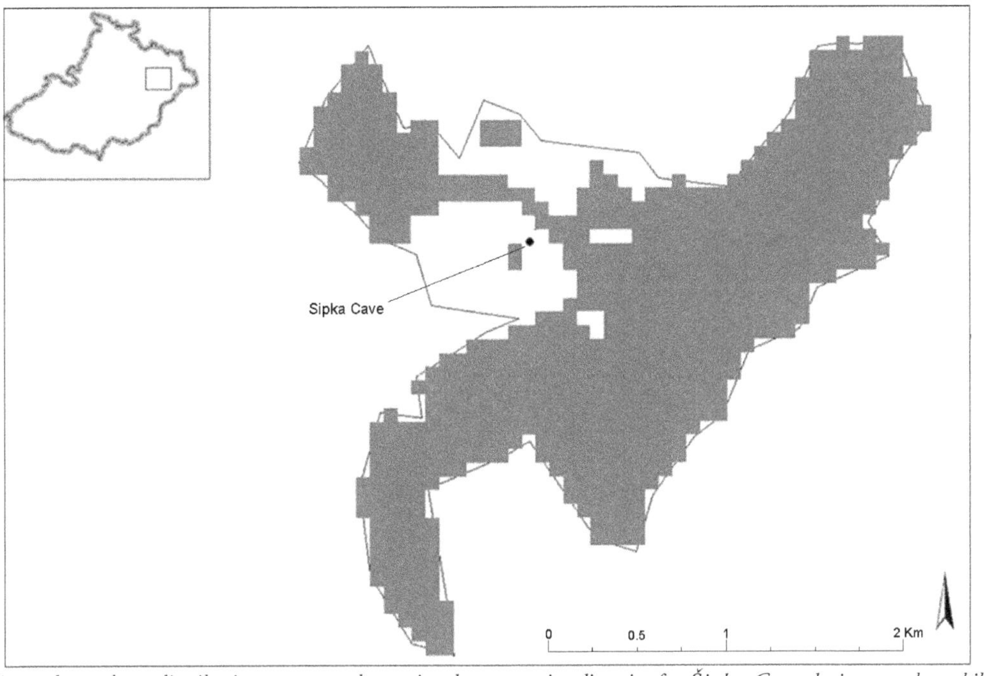

Fig. 8.3. Hominin and cave-bear distribution areas, and associated prey species diversity for Šipka Cave during cave-bear hibernation periods. Species diversity – grey = 2 (horse and reindeer).

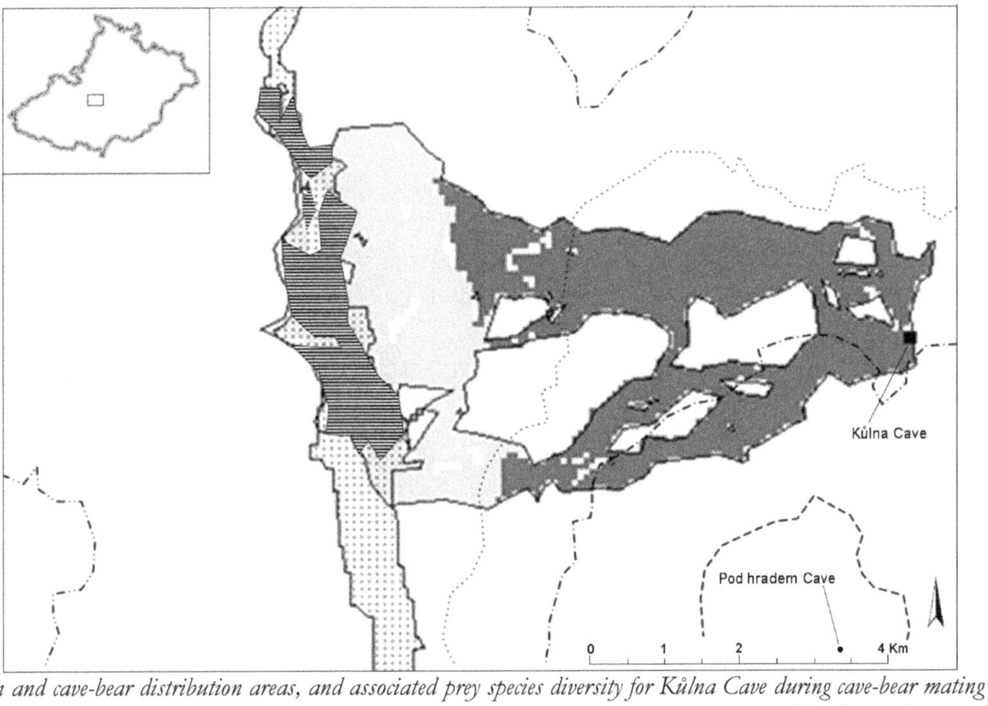

Fig. 8.4. Hominin and cave-bear distribution areas, and associated prey species diversity for Kůlna Cave during cave-bear mating periods. Species diversity – light grey = 1 (mammoth), and dark grey = 2 (reindeer and mammoth). Dash = home range of female cave bears with cubs (RCD = 1.00), dash-dot-dash = home range of female cave bears with yearlings (RCD = 0.34), dotted = home range of lone female cave bears (RCD = 0.14), dash-dot-dot-dash = home range of lone male cave bears (RCD = 0.05), horizontal stripes = cave-bear preferred habitat areas, and dotted area = Cretaceous chert.

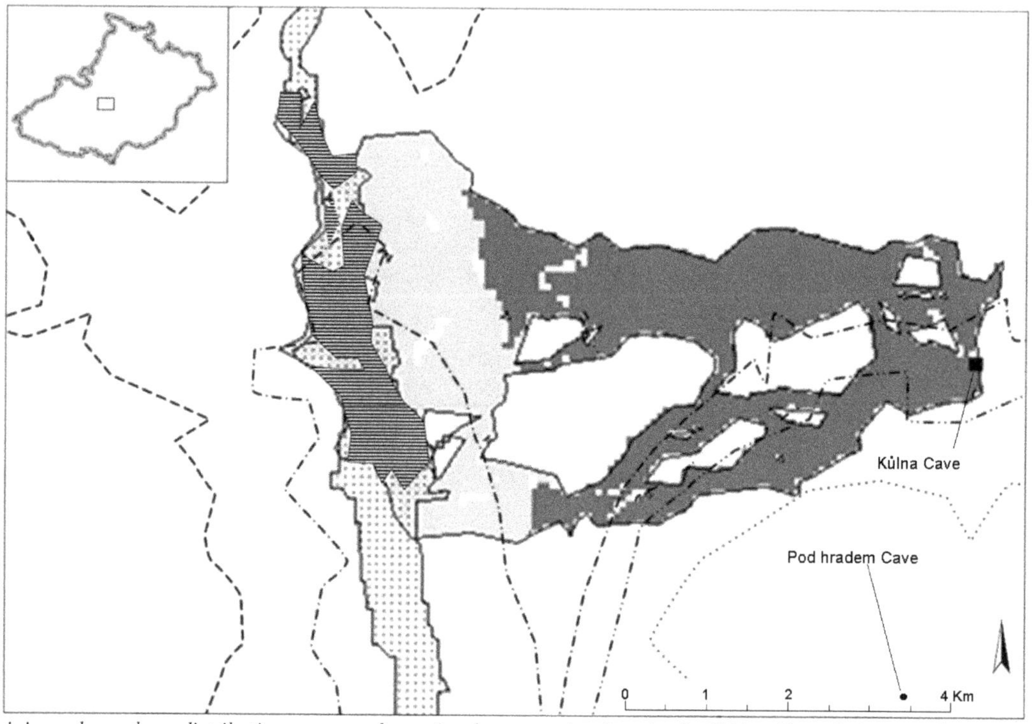

Fig. 8.5. Hominin and cave-bear distribution areas, and associated prey species diversity for Kůlna Cave during cave-bear post-mating periods. Species diversity – light grey = 1 (mammoth), and dark grey = 2 (reindeer and mammoth). Dotted = home range of female cave bears with cubs (Pod hradem Cave) (RCD = 0.85), Dash-dot-dot-dash = home range of female cave bears with yearlings (Pod hradem Cave) (RCD = 0.55), dash-dot-dash = home range of lone male cave bears (Barová Cave) (RCD = 0.10), dash = home range of lone female cave bears (Pod hradem Cave) (RCD = 0.24), horizontal stripes = cave-bear preferred habitat areas, and dotted area = Cretaceous chert.

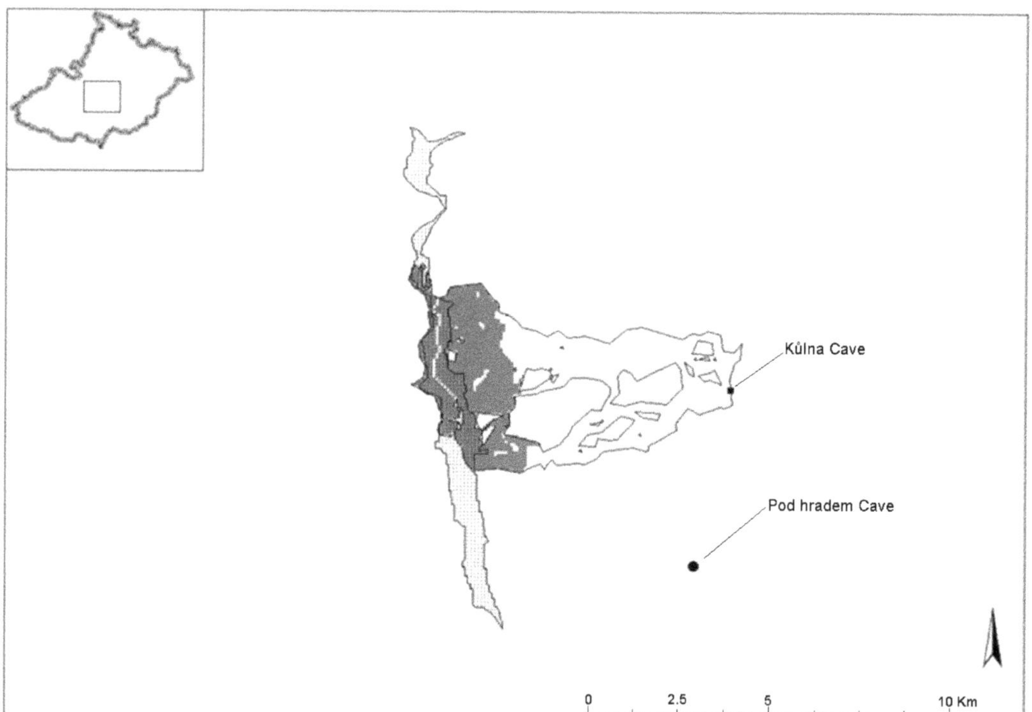

Fig. 8.6. Hominin and cave-bear distribution areas, and associated prey species diversity for Kůlna Cave during cave-bear hibernation periods. Species diversity – grey = 2 (reindeer and mammoth). Dotted area = Cretaceous chert.

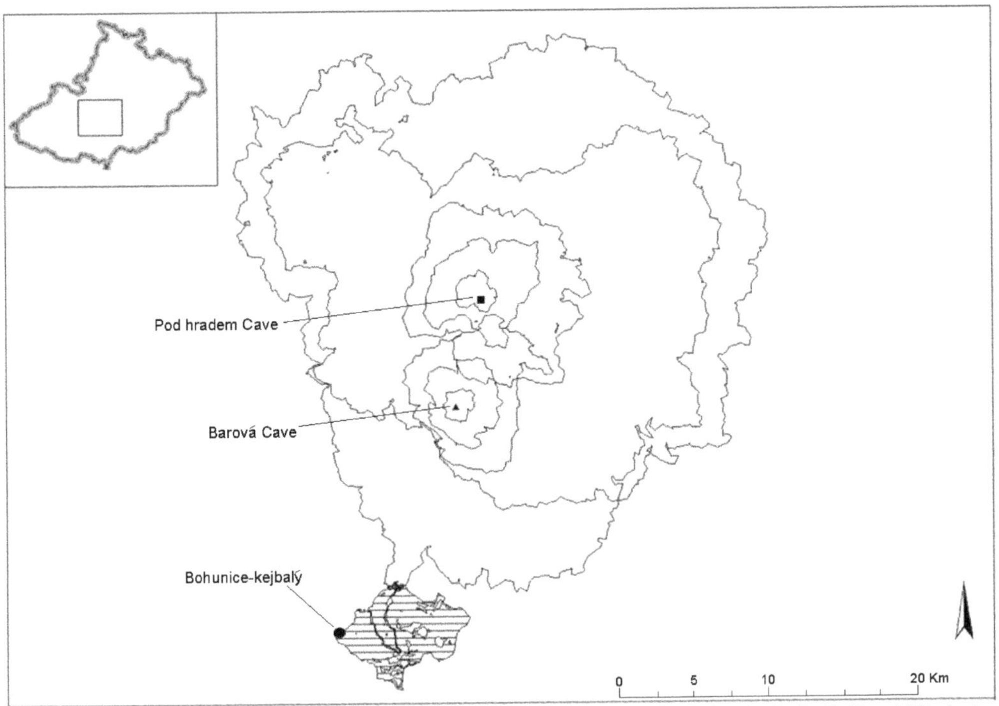

Fig. 8.7. Hominin and cave-bear distribution areas, and associated prey species diversity for Bohunice-kejbalý (I and II) during cave-bear mating periods. Cave-bear home range perimeter overlapping with hominin distribution area = home range of lone male cave bears (Barová Cave) (RCD = 0.05), with prey species diversity of 2 (horse and mammoth), and populated with cave-bear preferred habitat area.

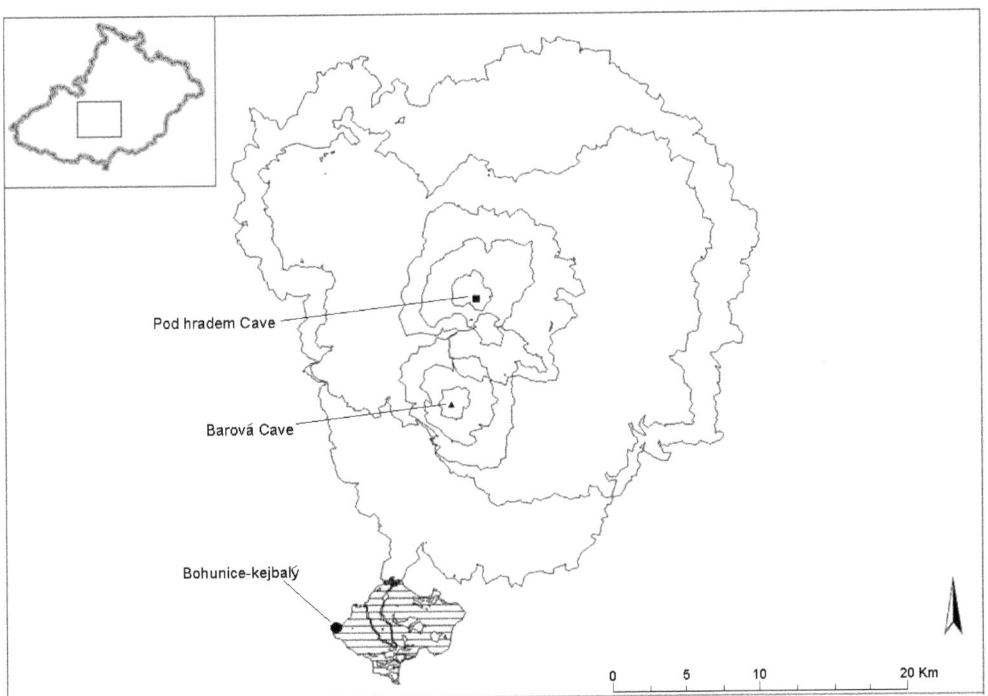

Fig. 8.8. Hominin and cave-bear distribution areas, and associated prey species diversity for Bohunice-kejbalý (I and II) during cave-bear post-mating periods.

Fig. 8.9. Hominin and cave-bear distribution areas for Bohunice-kejbalý (I and II) during cave-bear hibernation periods.

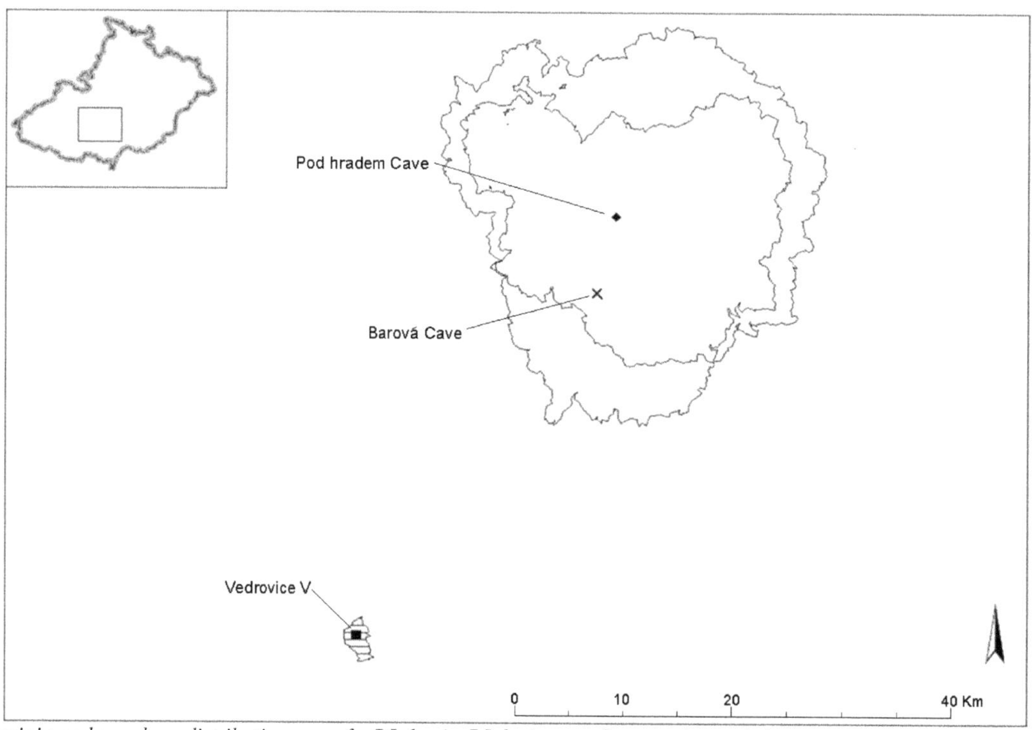

Fig. 8.10. Hominin and cave-bear distribution areas for Vedrovice V during cave-bear mating periods.

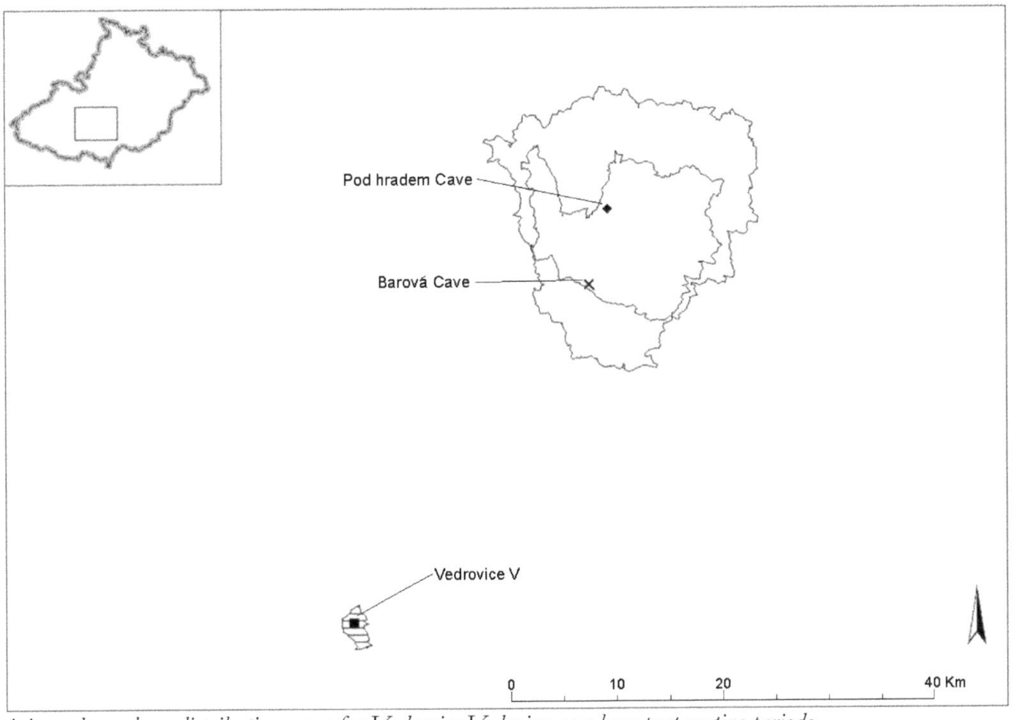

Fig. 8.11. Hominin and cave-bear distribution areas for Vedrovice V during cave-bear post-mating periods.

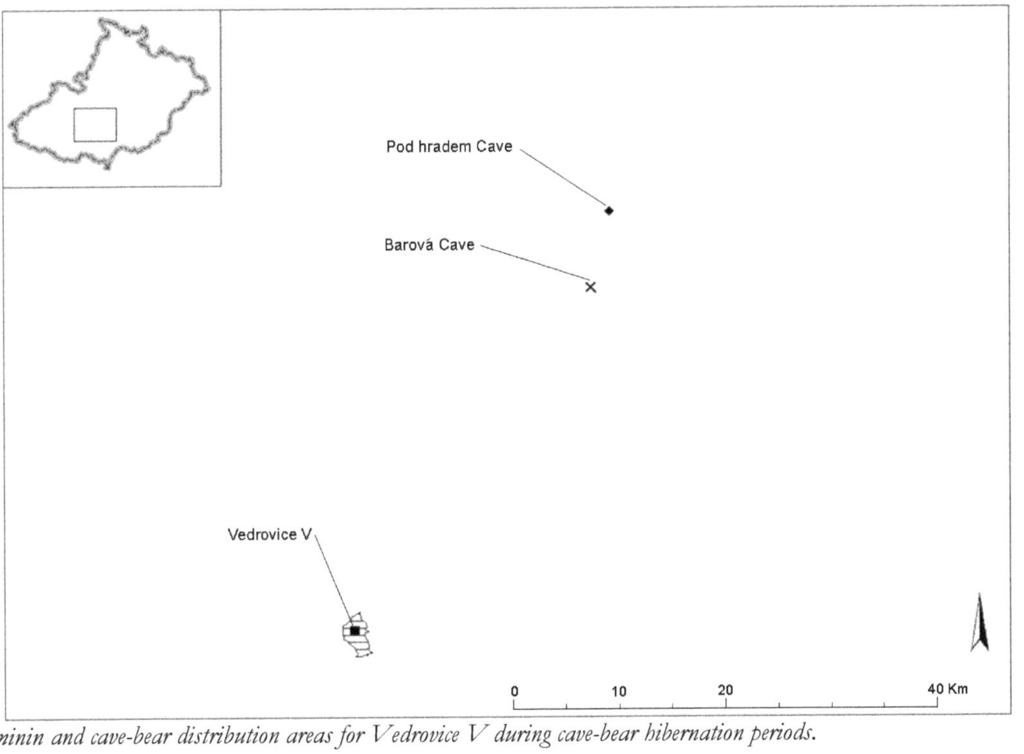

Fig. 8.12. Hominin and cave-bear distribution areas for Vedrovice V during cave-bear hibernation periods.

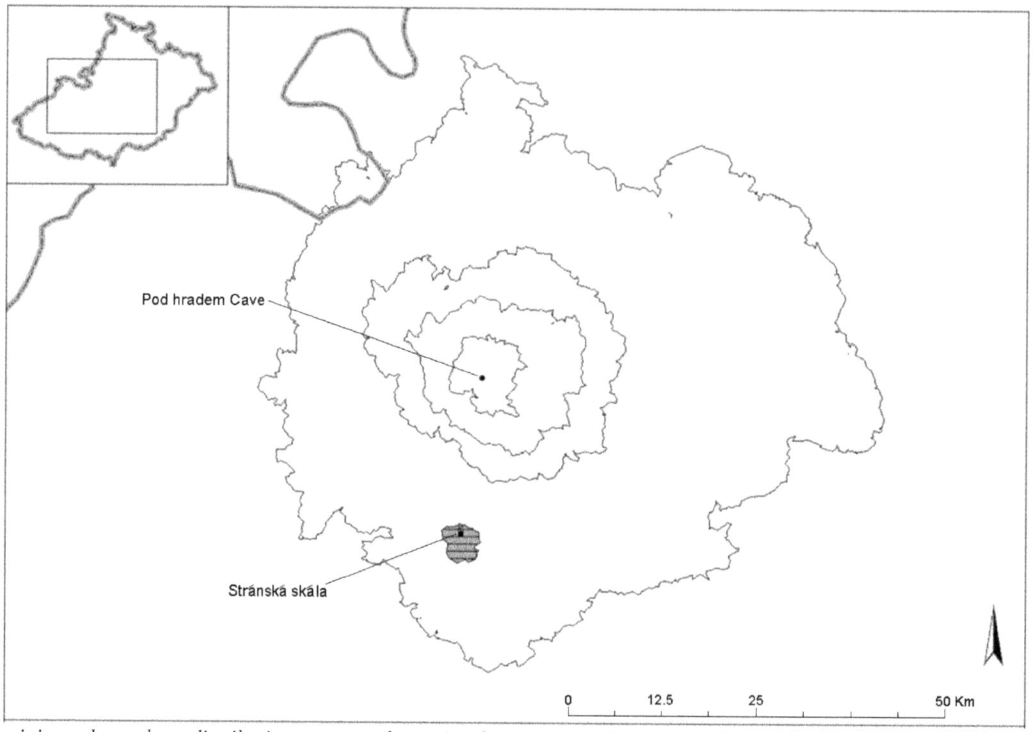

Fig. 8.13. Hominin and cave-bear distribution areas, and associated prey species diversity for Stránská skála (IIA and IIIA) during cave-bear mating periods. Species diversity – grey = 2 (horse and mammoth). Cave-bear home range area overlapping with hominin distribution area = home range of lone male cave bears (Pod hradem Cave) (RCD = 0.05).

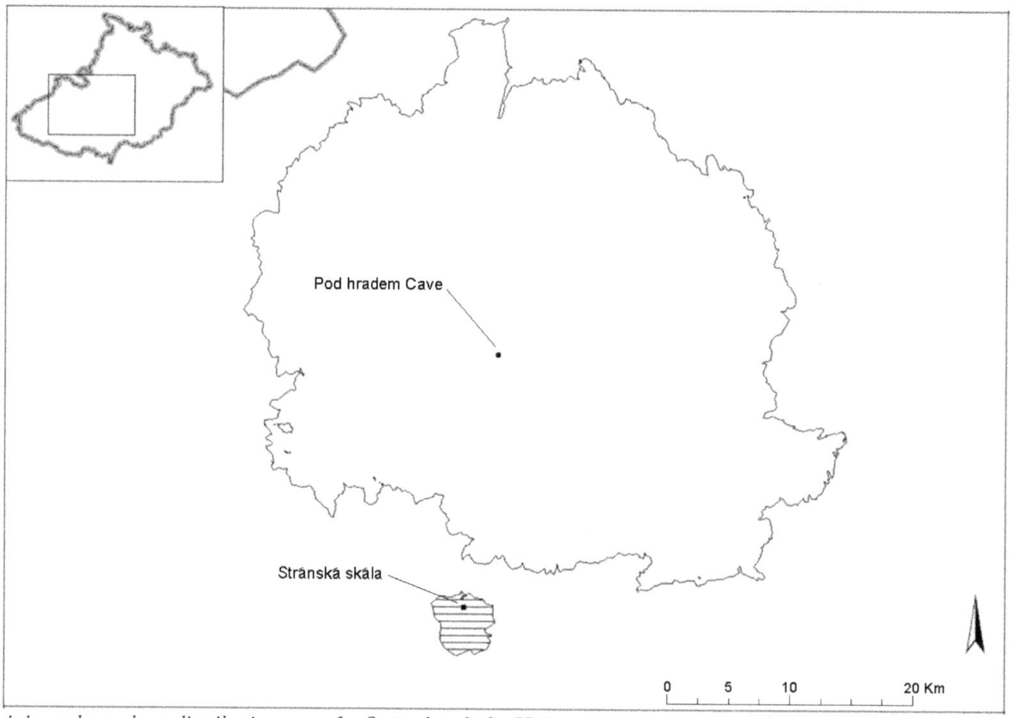

Fig. 8.14. Hominin and cave-bear distribution areas for Stránská skála (IIA and IIIA) during cave-bear post-mating periods.

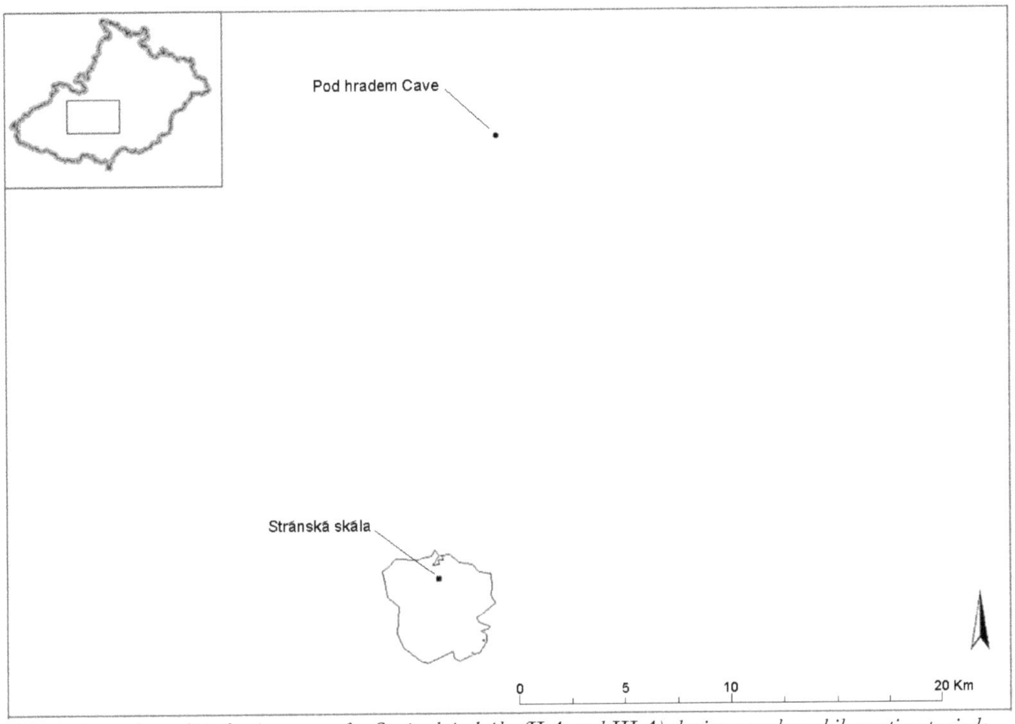

Fig. 8.15. Hominin and cave-bear distribution areas for Stránská skála (IIA and IIIA) during cave-bear hibernation periods.

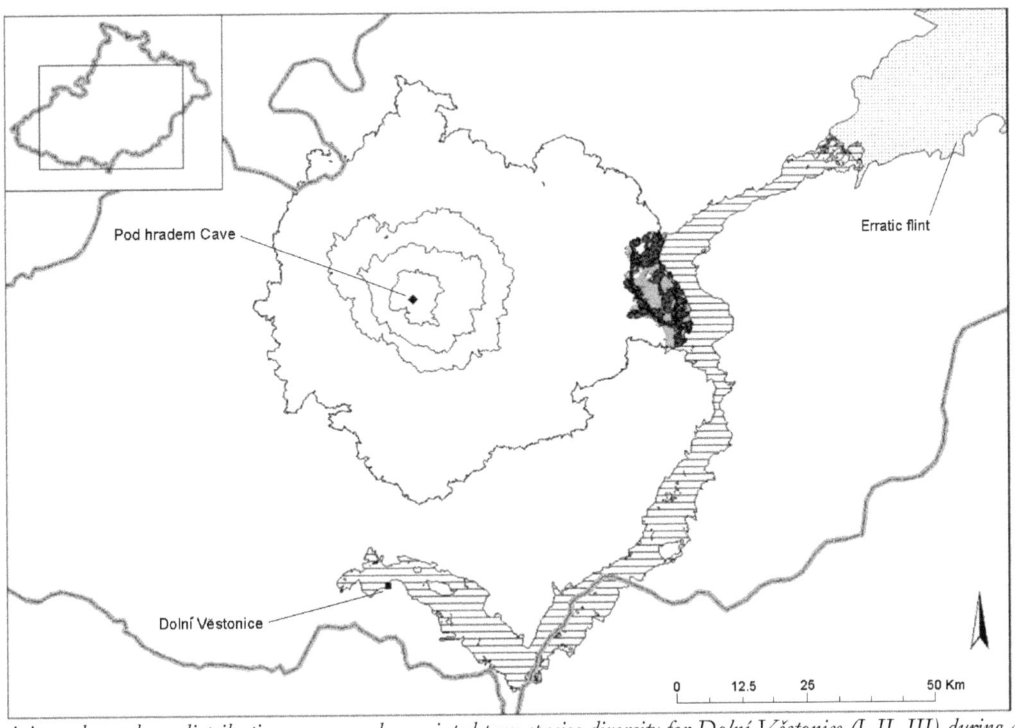

Fig. 8.16. Hominin and cave-bear distribution areas, and associated prey species diversity for Dolní Věstonice (I, II, III) during cave-bear mating periods. Species diversity – light grey = 2 (Arctic fox and horse), and dark grey = 4 (mammoth, Arctic fox, hare, wolf). Cave-bear home range overlapping with hominin distribution areas = home range of lone male cave bears (RCD = 0.05).

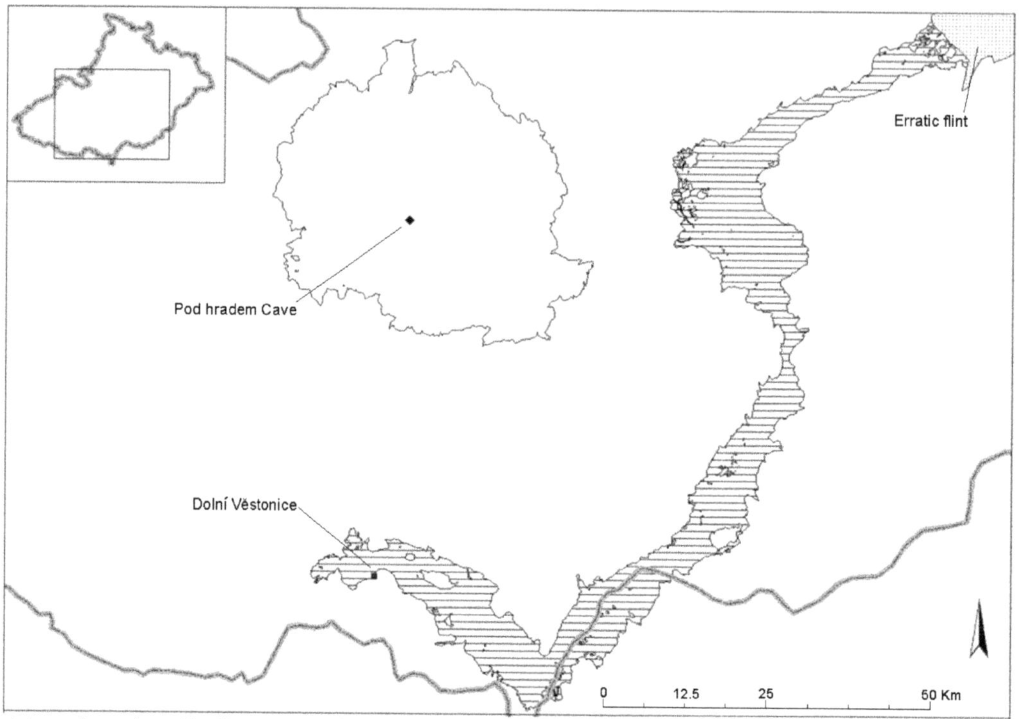

Fig. 8.17. Hominin and cave-bear distribution areas for Dolní Věstonice (I, II, III) during cave-bear mating periods.

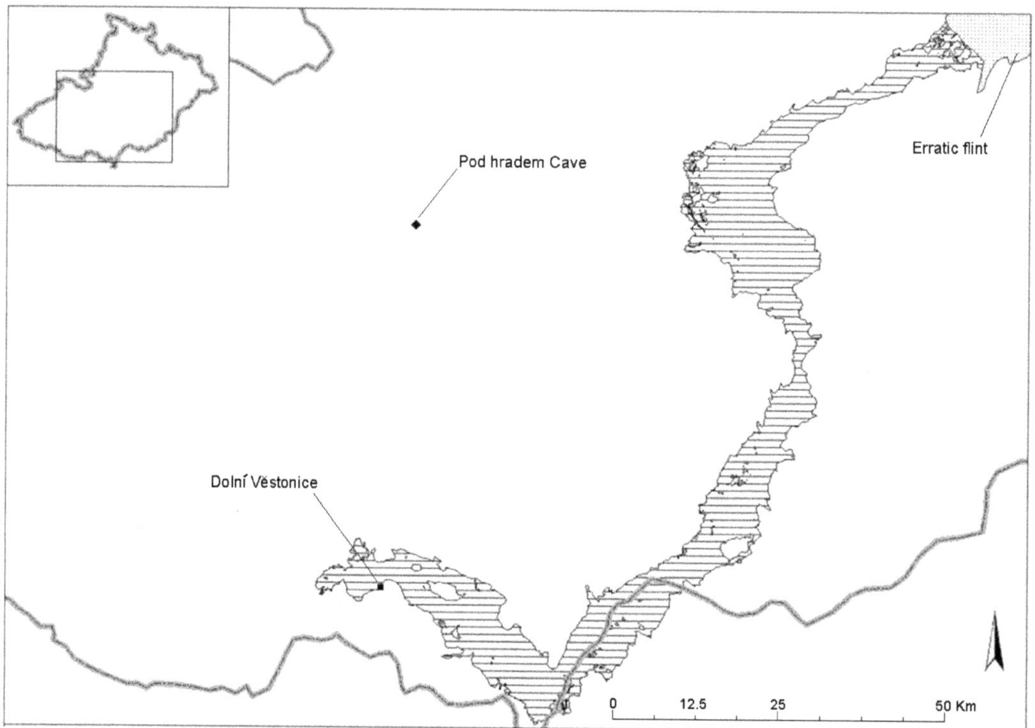

Fig. 8.18. Hominin and cave-bear distribution areas for Dolní Věstonice (I, II, III) during cave-bear hibernation periods.

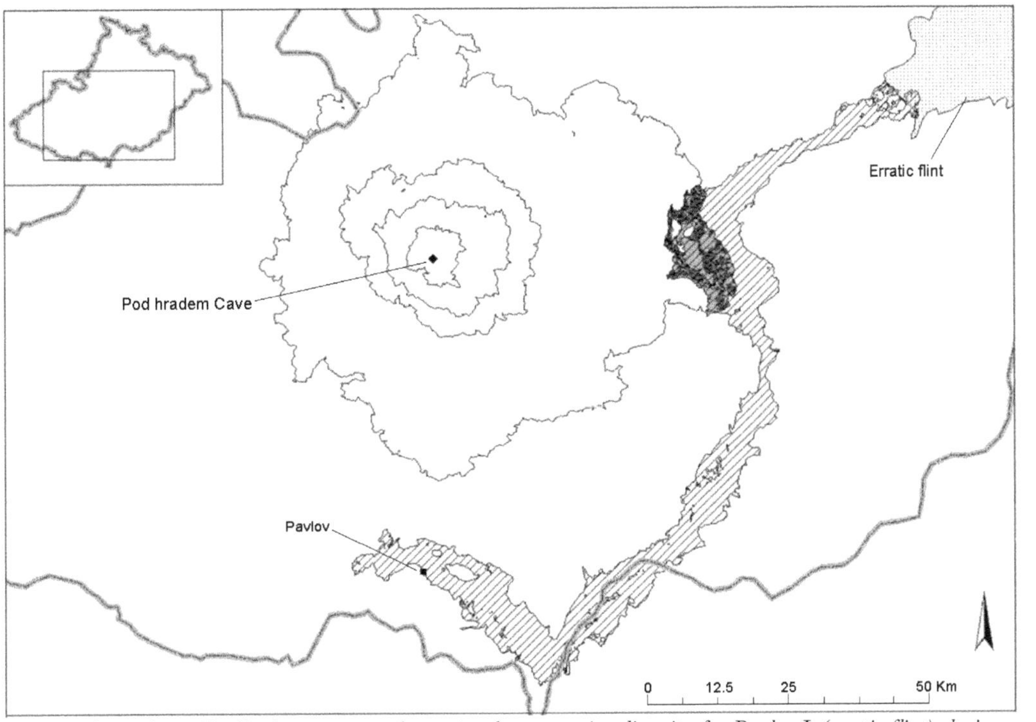

Fig. 8.19. Hominin and cave-bear distribution areas, and associated prey species diversity for Pavlov I (erratic flint) during cave-bear mating periods. Prey species diversity – dark grey = 2 (Arctic fox and horse) and cave-bear preferred habitat, and light grey = 4 (wolf, mammoth, Arctic fox, and hare). Cave-bear home range overlapping with hominin distribution areas = home range of lone male cave bears (RCD = 0.05).

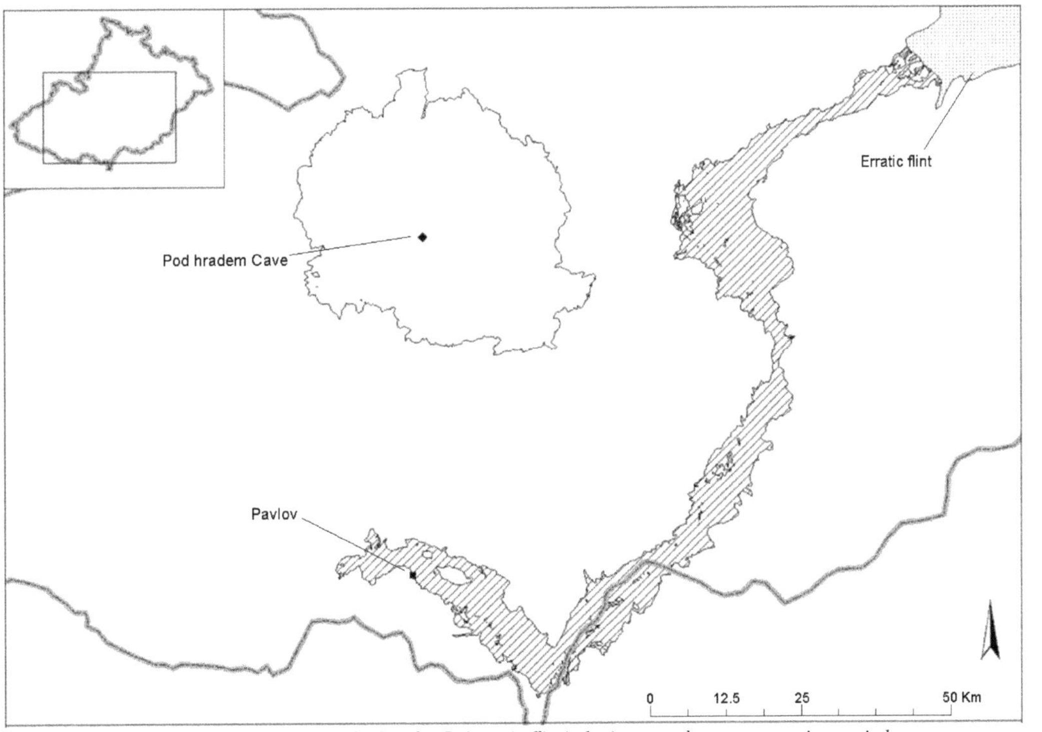

Fig. 8.20. Hominin and cave-bear distribution areas for Pavlov I (erratic flint) during cave-bear post-mating periods.

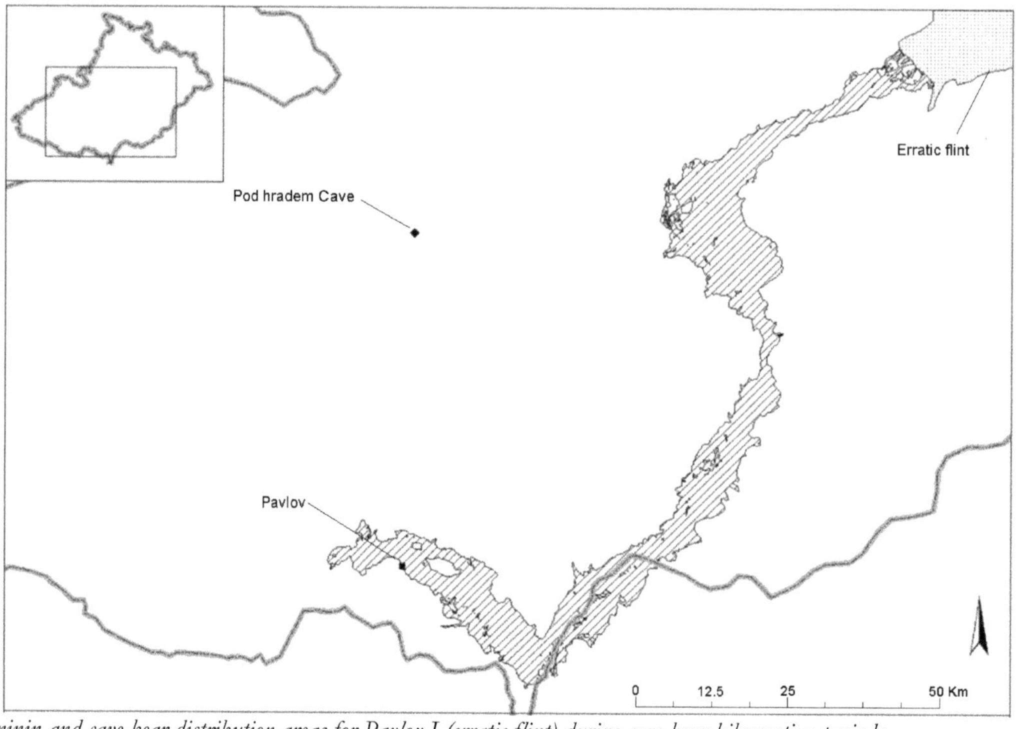

Fig. 8.21. Hominin and cave-bear distribution areas for Pavlov I (erratic flint) during cave-bear hibernation periods.

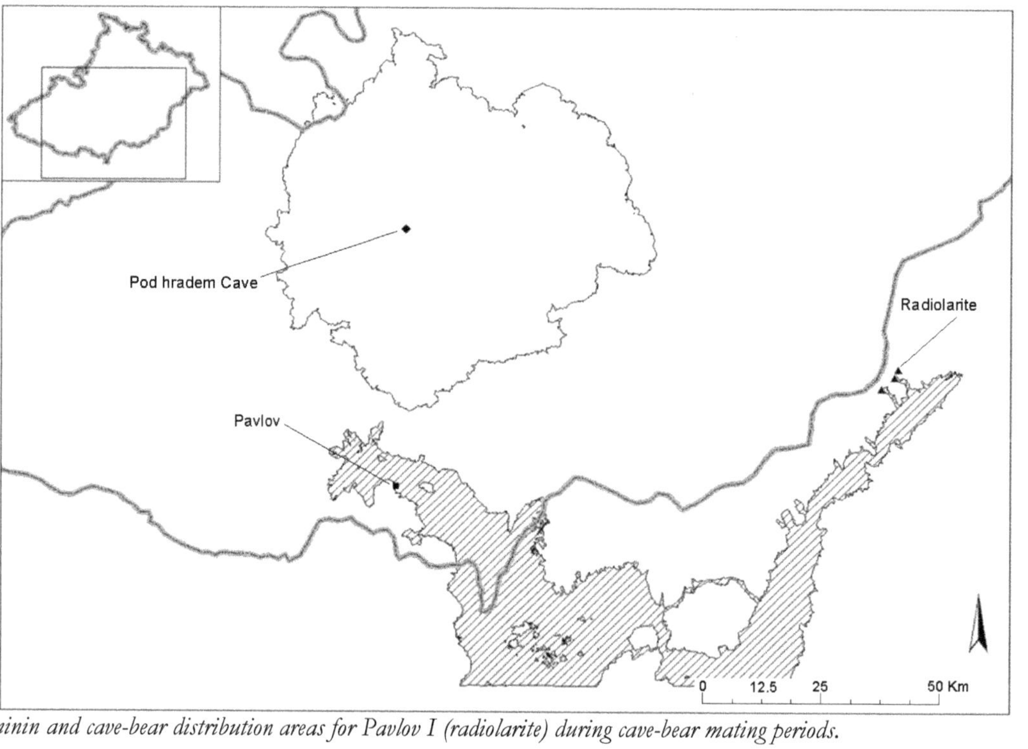

Fig. 8.22. Hominin and cave-bear distribution areas for Pavlov I (radiolarite) during cave-bear mating periods.

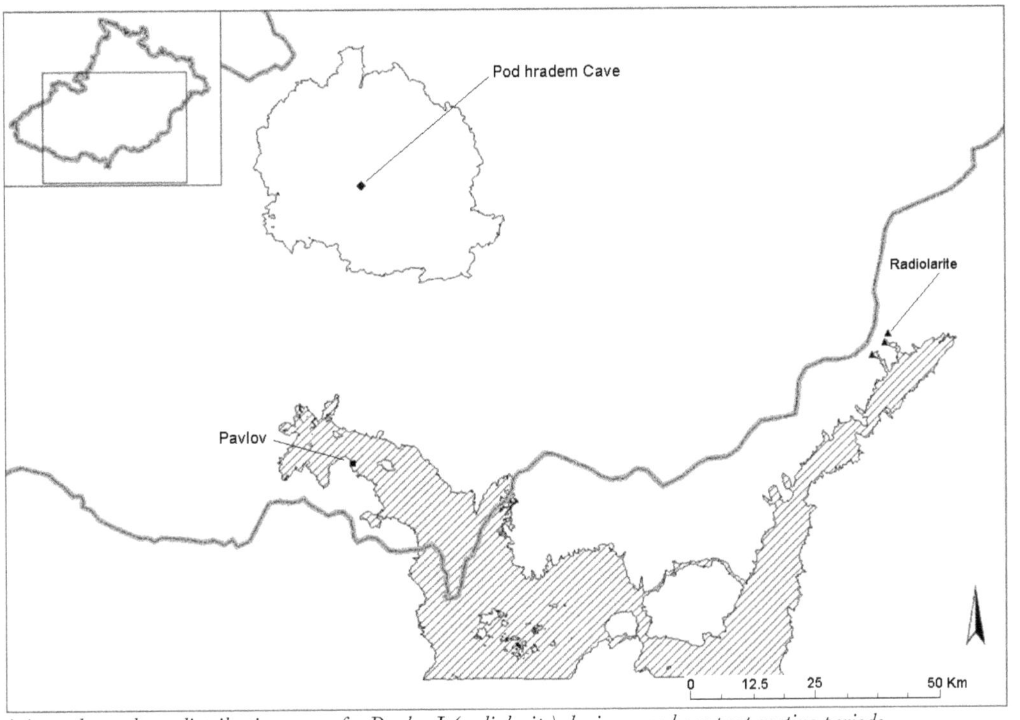

Fig. 8.23. Hominin and cave-bear distribution areas for Pavlov I (radiolarite) during cave-bear post-mating periods.

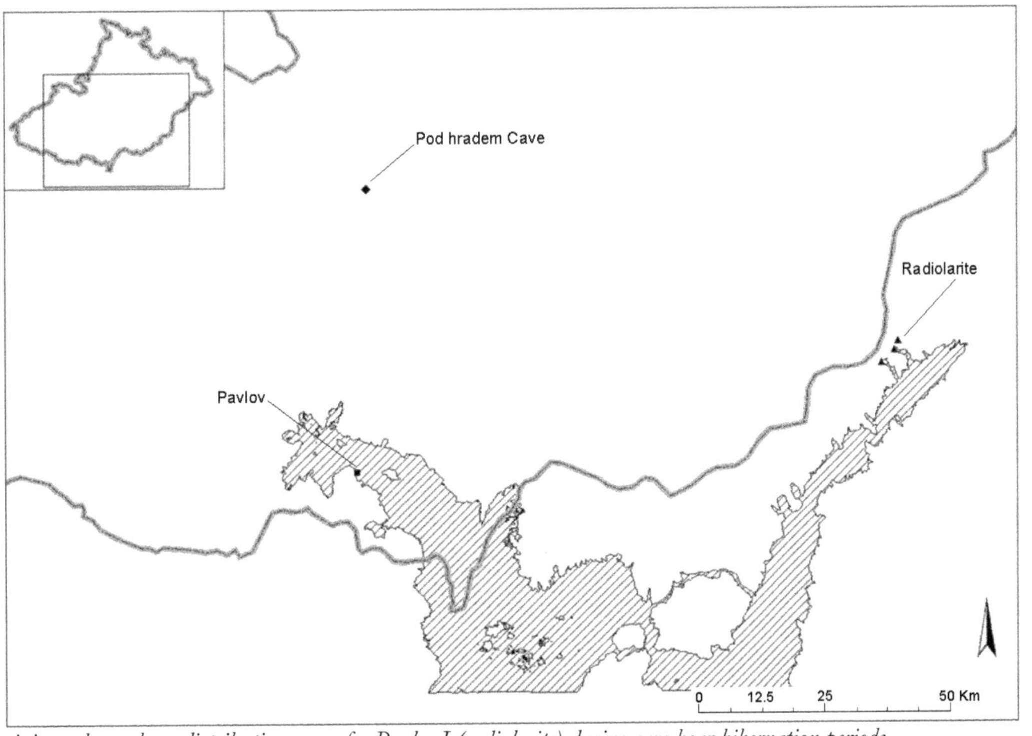

Fig. 8.24. Hominin and cave-bear distribution areas for Pavlov I (radiolarite) during cave-bear hibernation periods.

The overlap region is located at the very southern tip of the cave-bear distribution area and associated with lone male cave bears (Barová Cave) (RCD = 0.05), and the most northerly region of the hominin distribution region. The area has a prey species diversity value of 2, and measures about 2 km in an east-west direction and 1 km in a north-south direction. During cave-bear post-mating periods there are no areas of potential overlap between hominin and cave-bear distribution patterns (Fig. 8.8). The closest regions occur towards the northern edge of the hominin distribution area and towards the southern edge of cave-bear distribution areas, and are located about 7 km from each other. Results associated with winter periods (Fig. 8.9) indicate that there are again no areas of potential overlap. The closest cave-bear site is Barová Cave, located about 15 km north of Bohunice-kejbalý (I and II), with Pod hradem Cave located a further 7 km north of Barová Cave.

Results associated with Vedrovice V during cave-bear mating periods in warm climates (Fig. 8.10) indicate that there are no potential areas of overlap between hominin distribution patterns and cave-bear distribution areas. The closest cave-bear regions occur about 25 km northeast of hominin distribution areas. During cave-bear post-mating periods no areas of overlap occur (Fig. 8.11), and the closest cave-bear areas are located about 35 km northeast of hominin distribution areas. During cave-bear hibernation periods (Fig. 8.12) there are no areas of overlap, and the closest cave-bear site is Barová Cave, located about 40 km northeast of hominin distribution areas, with Pod hradem Cave located a further 7 km north of Barová Cave.

Results associated with Stránská skála (IIA and IIIA) during cave-bear mating periods in cold climates indicate that areas of potential overlap occur. Overlap regions are associated with lone male cave bears (Pod hradem Cave) (RCD = 0.02) and all hominin distribution areas, and these have a uniform prey species diversity value of 2 (horse and mammoth) (Fig. 8.13). During cave-bear post-mating periods there are no areas of potential overlap, and the closest cave-bear region occurs about 5 km north of the hominin distribution areas (Fig. 8.14). Results associated with cave-bear hibernation periods indicate that there are no areas of potential overlap, and the nearest cave-bear hibernation site (Pod hradem Cave) is located about 25 km north of the hominin distribution area (Fig. 8.15).

Results associated with Dolní Věstonice (I, II, III) during cave-bear mating periods in cold climates indicate that potential overlap occurs (Fig. 8.16). Overlap regions occupy only a relatively small area (25 km x 12 km) of the hominin distribution area, located about 75 km northeast of Dolní Věstonice, and associated with lone male cave-bear home ranges (Pod hradem Cave) (RCD = 0.02). In this region, prey species diversity values vary between 1 (Arctic fox) and 4 (mammoth, Arctic fox, hare, and wolf), with cave-bear preferred habitat regions coinciding with a prey species diversity value of 4. Results associated with cave-bear post-mating periods demonstrate no areas of potential overlap (Fig. 8.17). The closest cave-bear areas are located about 15 km west of

			Cave-bear demographic group				Relative frequency of interaction
		Cave-bear season	**FC**	**FY**	**LF**	**LM**	
Neanderthal	**Šipka Cave**	**Hibernation**	x	x	x	x	**High**
		Mating	x	x	x	x	**High**
		Post-mating	x	x	x	x	**High**
	Kůlna Cave	**Hibernation**	-	-	-	-	-
		Mating	-	x	x	x	**High**
		Post-mating	-	x	x	x	**High**
	Bohunice-kejbalý (I and II)	**Hibernation**	-	-	-	-	-
		Mating	-	-	-	x	Low
		Post-mating	-	-	-	-	-
	Vedrovice V	**Hibernation**	-	-	-	-	-
		Mating	-	-	-	-	-
		Post-mating	-	-	-	-	-
AMH	**Stránská skála (IIA and IIIA)**	**Hibernation**	-	-	-	-	-
		Mating	-	-	-	x	Low
		Post-mating	-	-	-	-	-
	Dolní Věstonice (I, II, III)/ Pavlov I (erratic flint)	**Hibernation**	-	-	-	-	-
		Mating	-	-	-	x	Low
		Post-mating	-	-	-	-	-
	Pavlov I (radiolarite)	**Hibernation**	-	-	-	-	-
		Mating	-	-	-	x?	Low?
		Post-mating	-	-	-	-	-

Table 8.1. Summary of hominin-cave bear interaction results, showing Neanderthals and AMH and their associated archaeological sites, cave-bear seasonal periods (i.e. cave-bear hibernation, mating, and post-mating periods), cave-bear demographic groups, and relative frequency of potential interaction. FC = female cave bears with cubs, FY = female cave bears with yearlings, LF = lone female cave bears, LM = lone male cave bears, ? = no overlap between hominin and cave-bear distribution patterns, but interactions possible.

the hominin distribution regions, and associated with the home ranges of lone male cave bears (Pod hradem Cave) (RCD = 0.03). During cave-bear hibernation periods, no areas of potential overlap occur. Pod hradem Cave is located about 40 km west of the nearest hominin distribution area (Fig. 8.18).

Results associated with Pavlov I (erratic flint raw material source) during cave-bear mating periods in cold climates (Fig. 8.19) are largely similar but not the same as those associated with Dolní Věstonice (I, II, and III) during this period. Overlap regions occur in a relatively small area of the derived hominin least-cost corridor, located about 75 km northeast of Pavlov, and measure about 25 km in a north-south direction and between about 2 km and 12 km in an east-west direction. In this region, prey species diversity values vary between 2 (Arctic fox and horse) and 4 (wolf, mammoth, Arctic fox, and hare). Areas with a prey species diversity value of 4 also correspond with cave-bear preferred habitat areas. Results associated with cave-bear post-mating periods indicate that no potential areas of overlap (Fig. 8.20). The closest cave-bear regions occur about 15 km west of the nearest hominin distribution region, and these areas are associated with lone male cave bears (RCD = 0.03). During cave-bear hibernation periods there are no areas of potential overlap. Pod hradem Cave is located about 40 km west of the hominin distribution area (Fig. 8.21).

Results associated with Pavlov I (radiolarite raw material source) during cave-bear mating periods in cold climates suggest no potential areas of overlap (Fig. 8.22). The closest cave-bear regions are associated with lone male cave bears (RCD = 0.02), located less than 10 km north of the closest hominin distribution areas. Results associated with cave-bear post-mating periods again indicate no areas of potential overlap (Fig. 8.23). The closest cave-bear regions are associated with lone male cave bears (RCD = 0.03). These are located about 30 km north of the closest hominin distribution area. Results associated with cave-bear hibernation periods indicate no areas of overlap. The closest hominin distribution areas are located between 40 km and 50 km from Pod hradem Cave (Fig. 8.24).

Overall the results of this chapter demonstrate interesting and contrasting results, particularly in terms of potential interactions associated with Neanderthals and AMH (Table 8.1). Neanderthals associated with Šipka Cave would have had the highest potential for interactions with cave bears, and these could have occurred with cave bears of all demographic groups, and during all periods. For those associated with Kůlna Cave, interactions are likely to have been relatively high frequency and mostly with cave-bear females with young, and where interactions did occur, they were associated with lone male cave bears during cave-bear mating and post-mating periods, but not cave-bear hibernation periods. Neanderthals associated with Bohunice-kejbalý (I and II) could only have interacted with male cave bears, and these interactions are likely to have occurred only during cave-bear mating periods, and relatively infrequently. In the case of AMH, potential interactions could have occurred for hominins occupying Stránská skála (IIA and IIIA), Dolní Věstonice (I, II, and II), and Pavlov I (erratic flint), but this is less likely for those associated with Pavlov I (radiolarite). In all cases, AMH could only have interacted with lone male cave bears, during cave-bear mating periods, and any interactions that took place would have been infrequent.

Summary and conclusions

This chapter has attempted to identify potential interactions between hominins and cave bears. I briefly discussed the methodology used for this chapter, and presented the results. The main points to take from this chapter are: Neanderthals potentially interacted with cave bears much more frequently than AMH; Neanderthals were much more likely to have interacted with female cave bears with young than cave bears associated with other demographic groups; AMH were much more likely to encounter lone male cave bears than other cave bears; and AMH interacted with cave bears much less frequently than Neanderthals.

Chapter 9: Understanding the significance of interactions

Introduction

The objective of this chapter is to discuss the significance of the results of the book. I first provide an overview of the book, restate the working hypothesis, and define relational cohesion values for different hominin-cave bear interaction contexts. I discuss the results in terms of opportunity for relational cohesion through interactions with cave bears, expected and archaeological evidence of the use of cave-bear remains and depictions of cave bears, and attempt to answer the overriding question of the book in terms of Neanderthals and AMH: what was the role of interactions for these peoples' sense of identity? Finally, I provide pointers for future work.

Interactions and relational cohesion

In Chapter 1, I described what this book was about, what interactions are, their place in archaeological inquiry, and how I became interested in the topic. I provided an overview of human-animal relations in the context of modern-day Western, non-Western, historic, and prehistoric societies, and discussed ways in which animals have been approached in archaeology, focusing on 'Western' dualism. I proposed an alternative singular-relational approach and discussed the consequences of this for understanding and investigating the significance of interactions with animals for past people, and provided the rationale for choosing cave bears in Moravia and Silesia during OIS3 as a case study context. In Chapter 2, I explored the topic of interactions, and I discussed the main issues and approaches, addressing three key and associated subsidiary questions. In Chapter 3, I provided an overview of cave bears, looking at their history of research, phylogeny, geographical and chronological distribution, evidence of their presence in the past, demography, important elements of their ecology, physiology, biology, and ethology, and evidence of human use of their remains and depictions of cave bears. In Chapter 4, I provided an overview of the case study, discussing the geographical and chronological context of the study region, geology, topography, palaeohydrology, climate, palaeovegetation, palaeofauna, hominins and cave bears in the study region, and the chosen case study sites. In Chapter 5, I built a GIS framework, mapping the locations of the case study sites, lithic raw material outcrops, topography, palaeohydrology, palaeovegetation, friction maps, and prey species preferred habitats and diversity. In Chapter 6 and Chapter 7, I reconstructed hominin and cave-bear distribution patterns, and in Chapter 8 I identified potential place/times of interactions between the two.

At the heart of this book is the idea or hypothesis that the frequency at which human beings interact with things in the world plays an important role in how they subsequently behave towards those things. In the context of human interactions with animals, I proposed that interactions with important animals, i.e. those animals that are particularly anthropomorphic and have apparent 'super-human' qualities (e.g. bears), would have been pivotal for past people's ability to maintain a sense of relational cohesion with the world: regular interactions with such animals allowed for the perpetuation and enchainment of important associated identities, therefore providing the means by which relational cohesion could be established and maintained. In cases where regular interactions with such animals were rare or sporadic, relational cohesion had to be maintained through interactions with other third-party objects, such as associated animal remains and/or depictions of animals. As a result, use of the remains and depictions of these animals should be common. On the other hand, if interactions with such animals were frequent then associated relational cohesion could be established and maintained through direct interactions, and it would not be necessary to use third-party objects to achieve a sense of relational cohesion. In these circumstances, use of associated animal remains and/or depictions of animals should be relatively rare.

In order to test the working hypothesis, it is necessary to establish not only the frequency at which humans interacted with animals, but also the value of such interactions in maintaining a sense of relational cohesion: although increases in frequency of interactions are likely to have increased the potential for relational cohesion, different interaction contexts are also likely to have provided different relational cohesion values. For instance, interactions with hibernating cave bears are likely to have been particularly important, as it is during periods of hibernation that cave bears are demonstrating their apparent super-human qualities, i.e. their ability to go without food or water for months on end. Also, interactions with cave-bear females with young are likely to have offered increased relational cohesion value: witnessing a female with young would have confirmed the human-like tendency of nurturing in cave bears, and the particular aggressive nature of cave-bear mothers keen to defend their offspring, highlighting aggressive elements in cave bears similar to those of humans. On the other hand, interactions with lone female and lone male cave bears are likely to have offered less relational cohesion value: unless these animals were surprised or encountered when feeding, manifestations of human-like or super-human qualities are likely to have been less commonly observed than in the case of hibernating cave bears or cave-bear females with young. Thus, in an attempt to provide some kind of understanding of the relational cohesion value of hominin interactions with cave bears, I assign quantitative values to different interaction frequencies and contexts. High frequency interactions are given a value of 3 out of a maximum of 3, medium frequency interactions 2, and low frequency interactions 1. Interactions with hibernating cave bears are given a value of 9 out of a maximum of 10, female cave bears with cubs are given a value of 7, female cave bears with yearlings a value of 6, and lone female cave bears and lone male cave bears are given a value of 3. These are combined with the results of Chapter 8 to provide overall relational cohesion values for hominins and associated expected interactions with cave bears, and are used to describe the opportunity that hominins had to maintain their sense of relational cohesion through interactions with cave bears, and to test the working hypothesis (Table 9.1).

For Mousterians associated with Šipka Cave, the potential to maintain relational cohesion through direct interactions with cave bears was possible during the whole year (Table 9.1). During cave-bear hibernation periods, the potential for relatively frequent interactions with hibernating cave bears, and the associated high relational cohesion value of these animals, provided these people with the opportunity to establish and maintain a sense of relational cohesion during this period. During cave-bear mating periods, potentially high frequency interactions, in particular with female cave bears with cubs and yearlings, would have again allowed for a sense of relational cohesion to be established through direct interactions, but in these cases relational cohesion was mediated via the observation of human-like rather than apparent super-human behavioural qualities of these animals. Although interactions may have occurred with lone male and female cave bears during these periods, the relatively low relational cohesion value of these animals and the low frequency of such interactions would have meant that these animals played only a small part in human sense of relational cohesion. The situation changed little during cave-bear post-mating periods: although the overall distribution of cave-bear groups may have changed slightly as a result of increases and decreases in home range extents and population density, the potential for high frequency interactions with cave-bear females with young still remained, and together with their high relational cohesion value would have meant that these animals were most important in terms of maintaining a sense of relational cohesion. Other cave-bear groups (lone males and lone females) would have played only a minor role. Overall, Mousterians occupying Šipka Cave during OIS3 potentially had the opportunity to establish a sense of relational cohesion through direct interactions with cave bears and by observing both their apparent super-human and human-like qualities. Although relational cohesion could be maintained during the winter periods by interacting with hibernating cave bears, it was during the summer that cave bears offered the greatest potential and value for humans (Table 9.1).

For Micoquians associated with Kůlna Cave, the potential to maintain relational cohesion through direct interactions with cave bears largely occurred only during cave-bear mating and post-mating periods. During cave-bear mating periods, potentially frequent interactions with female cave bears with yearlings, but not with female cave bears with cubs, provided the means by which people established and maintained a sense of relational cohesion. As in the case of Šipka Cave, the particular human-like nurturing qualities and defensive behaviour of cave-bear mothers provided relatively high relational cohesion values. Again, interactions may have occurred with lone male and lone female cave bears, but the low relational cohesion value of these animals and the low interaction rate would have meant that they provided little significance for humans. During cave-bear post-mating periods, Micoquians again maintained a sense of relational cohesion mostly through interacting with female cave bears with yearlings, but the reduction in home range extents associated with lone male and lone female cave bears may have meant that these animals played a slightly larger role in human sense of relational cohesion. Overall, Micoquians occupying Kůlna Cave had the opportunity to establish and maintain a sense of relational cohesion during most of the year, and this was largely achieved by interacting with cave-bear females with yearlings and through observation of their apparent human-like behaviours (nurturing and aggression).

In the case of hominins associated with Šipka Cave and Kůlna Cave, and according to the working hypothesis, there should be little or no evidence that cave-bear remains were commonly used: the potential for high frequency interactions with cave bears was sufficient for maintaining a sense of relational cohesion. It is difficult to assess whether or not this was the case for Mousterian occupants of Šipka Cave, as the faunal remains associated with this site are yet to be fully examined. Archaeological evidence associated with Micoquian occupation of Kůlna Cave during OIS3 indicates that animal remains such as mammoth ivory and bone and reindeer antler were used (Oliva 2005, 68; Rink *et al.* 1996, 981); however, there is little or no evidence that cave-bear remains were used by these people. This is surprising: it is difficult to imagine that Mousterian and Micoquian people would not have at least been aware of the abundance of cave-bear remains in nearby caves (e.g. Pod hradem Cave), or in the case of Mousterians, the same cave, given the degree to which cave-bear remains are apparent in these caves even today. Also, it is interesting to note that evidence regarding Mousterian and Micoquian use of animals in other regions of Europe (Holgueras 2009) indicates, as in the case of Kůlna Cave, that animal remains were at least used to some extent, but again there is little or no evidence that cave-bear remains were commonly used. This might suggest that these people are actively avoiding the use of cave-bear remains in favour of other animal remains, possibly supporting the idea that these people had different relationships with particular animals, i.e. those animals that demonstrate evidence of super-human or human-like qualities such as bears were not treated or thought about in the same way as other animals. On the other hand, the use of animal remains may have been driven, not by the need or desire to maintain a sense of relational cohesion with the associated animal, but primarily by other factors such as the material qualities of animal remains that allowed them to be used in a certain

			Cave-bear interaction context and relational cohesion value									
	Hominin sites	**Cave-bear season**	**HB (9)**	**FC (7)**	**FY (6)**	**LF (3)**	**LM (3)**	**Summed relational cohesion values**	**Opportunity for relational cohesion through cave-bear interactions**	**Expected use of cave-bear remains/ depictions of cave bears**	**Actual use of cave bear remains**	**Actual depictions of cave bears**
Neanderthal	**Šipka Cave**	**Hibernation**	High (27)	-	-	-	-	27	**Medium**	**Low**	**None?**	**None?**
		Mating	-	High (21)	High (18)	Medium (6)	Low (3)	48	**High**			
		Post-mating	-	High (21)	High (18)	Medium (6)	Low (3)	48	**High**			
	Kůlna Cave	**Hibernation**	-	-	-	-	-	0	-	**Low**	**None**	**None**
		Mating	-	-	High (18)	Medium (6)	Low (3)	27	**Medium**			
		Post-mating	-	-	High (18)	Medium (6)	Low (3)	27	**Medium**			
	Bohunice-kejbalý (I and II)	**Hibernation**	-	-	-	-	-	0	-	**High**	**None?**	**None?**
		Mating	-	-	-	-	Low (3)	3	**Low**			
		Post-mating	-	-	-	-	-	0	-			
	Vedrovice V	**Hibernation**	-	-	-	-	-	0	-	**High**	**None?**	**None?**
		Mating	-	-	-	-	-	0	-			
		Post-mating	-	-	-	-	-	0	-			
AMH	**Stránská skála (IIA and IIIA)**	**Hibernation**	-	-	-	-	-	0	-	**High**	**None?**	**None?**
		Mating	-	-	-	-	Low (3)	3	**Low**			
		Post-mating	-	-	-	-	-	0	-			
	Dolní Věstonice (I, II, III)/ Pavlov I (erratic flint)	**Hibernation**	-	-	-	-	-	0	-	**High**	**Low-medium**	**High**
		Mating	-	-	-	-	Low (3)	3	**Low**			
		Post-mating	-	-	-	-	-	0	-			
	Pavlov I (radiolarite)	**Hibernation**	-	-	-	-	-	0	-			
		Mating	-	-	-	-	Low ? (3)	3?	**Low?**			
		Post-mating	-	-	-	-	-	0	-			

Table 9.1. Summary of hominin opportunity for relational cohesion through interactions with cave bears, and expected and actual use of cave-bears remains and depictions of cave bears. HB = hibernating cave bears, FC = female cave bears with cubs, FY = female cave bears with yearlings, LF = lone female cave bears, LM = lone male cave bears.

way: the enchainment of identities associated with animal remains may have been a consequence rather than a driving force of behaviour.

For Bohunicians associated with Bohunice-kejbalý (I and II), the potential to maintain a sense of relational cohesion through direct interactions with cave bears was only possible during cave-bear mating periods. During these times, interactions with lone male cave bears were possible; however, given the low relational cohesion value of lone male cave bears and the infrequent basis on which they were encountered, it is unlikely that such interactions were able to maintain a sense of relational cohesion for these people. In the case of Szeletians associated with Vedrovice V, interactions did not occur, at least in terms of the reconstructed hominin and cave-bear distribution patterns in this book. These people, then, would not have been able to maintain their sense of relational cohesion through interactions. In both cases, however, it is possible that excursions were made into areas that were more commonly populated with cave bears (i.e. the Moravian Karst), and possible cut marks found on a cave-bear radius in association with Szeletian stone tools at the site of Pod hradem Cave may indicate that this happened. However, even if these people did make occasional trips to such regions, potential interactions with cave bears would still have been rare. As a result, and according to the working hypothesis, it would be expected that these people would have made regular use of cave-bear remains. However, apart from a 'harpoon' made from a mammoth tooth root, found in association with Szeletian occupation of the study region (Oliva 2005, 31), there are virtually no other faunal remains associated with Bohunicians and Szeletians either in the context of the case study or within a wider spatial context. However, it is interesting to note that at the site of Üçağizli Cave, Turkey, where the presence of Bohunician-like technologies are found (Layers F-H), more than 100 perforated shell beads or small pendants have been recovered (Kuhn *et al.* 2004, 124), suggesting that these people possibly had the capability and inclination to use objects as a means for establishing a sense of relational cohesion.

For AMH associated with Stránská skála (IIA and IIIA), the opportunity to establish and maintain a sense of relational cohesion through direct interactions with cave bears occurred only during cave-bear mating periods and with lone male cave bears. Given the relatively infrequent nature of such interactions and the low relational cohesion value of lone male cave bears, it is unlikely that such interactions were sufficient to maintain an associated sense of relational cohesion. As in the case of Bohunicians and to a lesser extent Szeletians, however, the relatively close proximity of the Moravian Karst would have meant that it was possible for these people to make occasional trips to this area, and therefore increase the potential frequency at which cave bears were encountered. However, this is unlikely to have made a significant difference in the overall frequency of interaction, and it is still probable that these people were not able to maintain a sense of relational cohesion through direct interactions alone. In such a scenario, and according to the working hypothesis, it would be expected that some attempt was made by these people to compensate for the apparent lack of direct interactions with these animals, and that either their remains were used or cave bears were depicted in some way. Unfortunately, as only stone tools were recovered from this site it is not possible to test the hypothesis further, at least within the context of this site. However, the presence of animal remains, in particular the presence and use of bear canines for pendants, found at other Aurignacian sites in the study region such as Mladeč indicates that Aurignacian people in the study region were attempting to enchain important identities through the use of such items, but it is possible that this was associated with brown bears and not cave bears. Moreover, the particular prevalence of canines found associated with Mladeč suggests that it was not just bears that were important for maintaining a sense of relational cohesion, but also other carnivorous animals. However, the fact that Aurignacian people in the study region did not depict cave bears or any other animals suggests that if important identities associated with animals were enchained, this was achieved mostly through the use of their remains. This is not the case elsewhere in Europe. Much of the parietal art found at Chauvet Cave in France is associated with Aurignacian people. Indeed, at this site a large number of animal depictions, including cave bears, not only suggest that these people utilised animal depictions in the form of parietal art as a means of establishing and maintaining a sense of relational cohesion with these animals; they also demonstrate an attempt to actively accumulate identities in single locations, producing places that were crucial for the enchainment of important identities and associated sense of relational cohesion. Also, the presence of mobile animal figurines found in association with Aurignacian sites in places such as Germany demonstrates that these people were not only able to enchain important identities in such objects, but the fact that they could be carried meant that relational cohesion was not bound by the behavioural tendencies of animals and associated interactions, or by the spatially static character of parietal art: enchainment of identities and associated relational cohesion was a matter of human governance.

For Gravettians associated with both Dolní Věstonice (I, II, III) and Pavlov I (erratic flint and radiolarite), the potential to establish and maintain a sense of relational cohesion through direct interactions with cave bears could only have occurred during cave-bear mating periods. Moreover, as such interactions are likely to have been relatively infrequent and lone male cave bears had a relatively low relational cohesion value, it is unlikely that such interactions were sufficient for maintaining a sense of relational cohesion. Thus, it would be expected that these people used cave-bear remains or depicted their form to compensate for an apparent lack of relational cohesion established through interactions. The presence of cave-bear remains such as penis bones, and cave-bear teeth used to make pendants found at these sites suggests that some attempt was made to do this. Moreover, the presence of bear figurines seems to indicate that, as with some Aurignacian sites in Germany, these people not only had the capability and inclination to enchain important identities through such objects, but also had governance over where and when enchainment of identities and associated confirmation of relational cohesion took place. The presence and use of animal remains such as mammoth bones and ivory, and depictions of these animals and other animals such as lions in the form of figurative art, suggest that these people were engaged in similar relations with other animals. But the fact that some of these animals were also commonly hunted (mammoth) suggests that the relationship between human interactions with animals, use of their remains, and the degree to which they are depicted is multifaceted.

This discussion has demonstrated that the role that interactions with animals played in the lives of past humans was complex. Relational cohesion was not simply a matter of frequency of interaction with 'special' animals, but also to do with the active avoidance and material qualities of their remains, and the manner in which important identities were enchained (e.g. parietal or figurative art). However, one clear point arises from this discussion and book as a whole. If relational cohesion was a central part of the lives of Neanderthals and AMH, Neanderthals and AMH had very different ways of achieving this. On the whole, Neanderthals were generally not using any third-party mechanisms as a means of achieving relational cohesion, whereas AMH were. This may be because Neanderthals were 'local people', commonly interacting with important elements of their world on a daily basis, and so had no need to enchain important identities via objects. However, although Neanderthals do seem to have been more local in their behaviour, it is inevitable that some things would have been encountered more often than others, and so the localised behaviour in itself does not explain an apparent lack of an attempt to enchain important identities. An alternative explanation is that Neanderthals and AMH had distinct ways of understanding the world. Although both Neanderthals and AMH may both have understood the world as consisting of a multitude of identities, fluid and unbound by physicality, for Neanderthals deposition and subsequent enchainment of identity and relational cohesion was a matter of perceptual immediacy: only those things that could be interacted with through direct perception were involved in the making of the Neanderthal world. Relational cohesion for Neanderthals was governed primarily not by memory of dwelling, but by dwelling itself. In the case of AMH, however, memory of the perceived world was part and parcel of the dwelt-in world, and only by reconciling the two, i.e. using objects to enchain spatially and temporally distant identities, was it possible for these people to maintain a sense of relational cohesion. To sum up, Neanderthals existed in places where they dwelt, but for AMH the past dwelt in places where they existed. Thus, in answer to the main question of the book, what was the role of interactions for past human sense of identity? For Neanderthals the answer is 'dominant': the perceived world took precedence over memory of the perceived world. For AMH the answer is 'secondary': memory of the world took precedence over the immediacy of the perceived world.

As a final summary, human relations with animals are not just a matter for prehistoric hunter-gatherers: as was highlighted at the start of this book, whatever the character of society, (prehistoric farming, historic, modern-day hunter-gatherer, and modern-day Western), animals are part and parcel of people's everyday lives, and central to the way we as humans understand ourselves and the world around us. Moreover, it seems that, like Palaeolithic AMH of Europe, modern-day stories and myths of animals and their significance for people today suggest that it is not what is directly perceived that is most important for people's understanding of the world, but rather memory and imagination of the perceived world.

Further work

There are a number of areas of research that could build on the work undertaken here. One obvious line of inquiry is to continue with the bear theme and investigate contemporary societies and their relations with bears. In Hokkaido in Japan, bears still play a major role in the lives of the Ainu people: bears are commonly hunted, and they still regularly undertake bear ceremonies. Moreover, there have been few if any investigations concerning such behaviours and their significance for understanding prehistoric hunter-gatherer relations with bears and other animals. An alternative approach could be to widen the zoological, geographical, and spatial context and investigate late Palaeolithic human interactions with prey and predatory animals (i.e. herbivores and carnivores), and associated use of their remains and/or depictions of these animals. In south-western France, evidence of remains of herbivorous and carnivorous animals and depictions of these animals are commonly found in association with human occupation during this period. Alternatively, there is no reason why the idea of enchainment of identities through objects and associated establishment of relational cohesion could not be applied to other areas of research, such as lithic raw material procurement. The apparent tendency for Neanderthals and AMH, at least in Europe during OIS3, to procure lithic raw materials from very different proximities suggests that the procurement of these materials may have played a very different role in the lives of people and their overall understanding of themselves and the world around them.

Summary and conclusion

In this book, I hoped to have provided food for thought, and offered some new insights into not only how it might be possible to investigate ideas of relational cohesion, but also how new understandings into the role of interactions with animals for past humans can be achieved. To some extent this has been a partially experimental and pragmatic process, but has demonstrated that it is only by undertaking such journeys, built on the theoretical and methodological work of others, that advancement in archaeology and a better understanding of past people and ultimately ourselves can be achieved.

Appendix 1

ID	Tooth ID	Storage ID	Site ID	Tooth Type ID	Left/Right	Measurement Type Length	Length1 (mm)	Length2 (mm)	Length3 (mm)	Measurement Type Breadth	Breadth1 (mm)	Breadth2 (mm)	Breadth3 (mm)	Wear Stage	Articulation ID	Layer	Box Number	Medium Box Number Suffix	Small Box Number Suffix	Log Book Number	Log Book Page Number
834	837	Budišov	Pod hradem Cave	C		No measurement				Close estimate	19.09	19.09	19.09		D	5	30	A		1	185
835	838	Budišov	Pod hradem Cave	C		No measurement				No measurement					D	5	30	A		1	185
839	842	Budišov	Pod hradem Cave	C		Exact	30.14	30.14	30.09	Exact	21.98	21.98	21.94		D	5	30		b	1	185
840	843	Budišov	Pod hradem Cave	C		No measurement				No measurement					D	5	29	A		1	185
845	848	Budišov	Pod hradem Cave	C		No measurement				No measurement					M	5	28		a	1	187
863	866	Budišov	Pod hradem Cave	C		Close estimate	26.13	26.13	26.13	Exact	17.46	17.33	17.48		D	5	26	B		1	190
865	868	Budišov	Pod hradem Cave	C		Exact	32.84	33	33.12	Exact	23.63	23.48	23.15		D	5	25			1	191
875	888	Budišov	Pod hradem Cave	C		Exact	24.25	24.88	24.98	Exact	16.15	16.14	16.11		D	5	24	C		1	192
846	849	Budišov	Pod hradem Cave	C (J)		Exact	17.91	17.91	17.91	Exact	13.44	13.44	13.44		D	5	28		a	1	187
851	854	Budišov	Pod hradem Cave	C (J)		No measurement				No measurement					D	5	27	B		1	188
864	867	Budišov	Pod hradem Cave	C (J)		No measurement				No measurement					M	5	26	C		1	190
866	879	Budišov	Pod hradem Cave	C (J)		Exact	22.84	22.84	22.84	Exact	16.17	16.17	16.17		D	5	25			1	191
836	839	Budišov	Pod hradem Cave	LM1	R	Exact	29.41	29.4	29.4	Exact	13.41	13.41	13.32	I	D	5	30	A		1	185
837	840	Budišov	Pod hradem Cave	LM1	R	Close estimate	29.58	29.58	29.58	Close estimate	13.94	13.94	13.94	II	D	5	30	A		1	185
843	846	Budišov	Pod hradem Cave	LM1	R	Exact	29.87	29.86	29.87	Exact	13.85	13.8	13.82	I	D	5	28	B		1	186
847	850	Budišov	Pod hradem Cave	LM1	R	Exact	29.83	29.81	29.8	Exact	13.55	13.33	13.55	II	D	5	28		a	1	187
852	855	Budišov	Pod hradem Cave	LM1	R	Exact	29.36	29.43	29.34	Exact	15.08	15.02	15.06	IX	M	5	27	D		1	188
855	858	Budišov	Pod hradem Cave	LM1	R	Exact	30.75	30.76	30.75	Exact	15.09	15.14	15.17	II	D	5	27	D		1	188
858	861	Budišov	Pod hradem Cave	LM1	R	Exact	31.88	31.81	31.89	Exact	14.98	14.94	15.02	II	D	5	26	B		1	190
859	862	Budišov	Pod hradem Cave	LM1	R	Close estimate	28.64	28.64	28.64	Exact	13.15	13.19	13.04	II	D	5	26	B		1	190
867	880	Budišov	Pod hradem Cave	LM1	R	Exact	29.25	29.25	29.26	Exact	13.79	13.82	13.72	IV	D	5	25	C		1	191
870	883	Budišov	Pod hradem Cave	LM1	R	Exact	29.43	29.44	29.45	Exact	13.7	13.7	13.74	I	D	5	25	D		1	191
871	884	Budišov	Pod hradem Cave	LM1	L	Exact	31.32	31.33	31.54	Exact	16.07	16.04	16.05	IX	D	5	24	A		1	191

Table. A1.1. Odontometric and tooth-wear analyses results, organised according to cave-bear site, layer and tooth type. Tooth Type ID – C = canine, C(j) = juvenile canine, LM1 = lower molar 1, LM2 = lower molar 2, LM3 = lower molar 3, UM1 = upper molar 1, UM2 = upper molar 2, Articualtion ID – D = disarticulated, M = tooth attached to mandible, C = tooth attached to maxilla.

ID	Tooth ID	Storage ID	Site ID	Tooth Type ID	Left/Right	Measurement Type Length	Length1 (mm)	Length2 (mm)	Length3 (mm)	Measurement Type Breadth	Breadth1 (mm)	Breadth2 (mm)	Breadth3 (mm)	Wear Stage	Articulation ID	Layer	Box Number	Medium Box Number Suffix	Small Box Number Suffix	Log Book Number	Log Book Page Number
872	885	Budišov	Pod hradem Cave	LM1	L	Exact	30.17	30.19	30.24	Exact	13.69	13.6	13.82	I	D	5	24	A		1	191
873	886	Budišov	Pod hradem Cave	LM1	L	Exact	30.17	30.15	30.17	Exact	14.38	14.32	14.32	II	D	5	24	C		1	191
874	887	Budišov	Pod hradem Cave	LM1		No measurement				No measurement				IX	D	5	24	C		1	192
848	851	Budišov	Pod hradem Cave	LM2	R	No measurement				Exact	21.98	22.14	22.09	IX	D	5	27	C		1	187
853	856	Budišov	Pod hradem Cave	LM2	R	Exact	30.89	30.86	30.8	Exact	19.71	20.01	19.8	IX	M	5	27	D		1	188
860	863	Budišov	Pod hradem Cave	LM2	R	Exact	32.65	32.83	32.75	Exact	18.89	18.93	19.02	II	D	5	26	B		1	190
861	864	Budišov	Pod hradem Cave	LM2	R	Exact	31.32	31.31	31.29	Exact	18.55	18.61	18.51	II	D	5	26	B		1	190
842	845	Budišov	Pod hradem Cave	LM3	R	Exact	27.75	27.73	27.74	Exact	18.98	18.97	18.95	III	D	5	29	E		1	186
854	857	Budišov	Pod hradem Cave	LM3	R	Exact	27.53	27.55	28.09	Exact	19.99	20.02	20.22	IX	M	5	27	D		1	188
856	859	Budišov	Pod hradem Cave	LM3	R	Exact	28.07	28.07	28.07	Exact	18.89	18.89	18.87	III	D	5	27		a	1	189
841	844	Budišov	Pod hradem Cave	UM1	L	No measurement				No measurement				II	D	5	29	A		1	185
844	847	Budišov	Pod hradem Cave	UM1	L	Exact	27.68	27.64	27.67	Exact	16.61	16.53	16.78	II	D	5	28	B		1	186
862	865	Budišov	Pod hradem Cave	UM1	L	Exact	29.36	29.27	29.32	Exact	18.64	18.44	18.7	II	D	5	26	B		1	190
868	881	Budišov	Pod hradem Cave	UM1	L	Exact	30.06	29.85	30.08	Estimate	19.63	19.63	19.63	II	D	5	25	C		1	191
869	882	Budišov	Pod hradem Cave	UM1		No measurement				No measurement					D	5	25	C		1	191
838	841	Budišov	Pod hradem Cave	UM2	L	Exact	42.32	42.34	42.33	Exact	20.18	20.14	20.18	III	D	5	30	A		1	185
849	852	Budišov	Pod hradem Cave	UM2	R	Close estimate	40.94	40.94	40.94	Close estimate	24.73	24.73	24.73	VII	D	5	27	B		1	188
850	853	Budišov	Pod hradem Cave	UM2	L	Close estimate	44.57	44.57	44.57	Close estimate	25.7	25.7	25.7	VIII	D	5	27	B		1	188
857	860	Budišov	Pod hradem Cave	UM2	R	No measurement				No measurement				IX	D	5	27		a	1	189
831	834	Budišov	Pod hradem Cave	LM1	R	Exact	29.9	29.89	29.9	Exact	14.18	14.15	14.13	I	D	5A	31	D		1	183
829	832	Budišov	Pod hradem Cave	LM2	R	Exact	30.44	30.46	30.48	Exact	17.36	17.34	17.34	I		5A	31	B		1	183
830	833	Budišov	Pod hradem Cave	UM1	L	Exact	29.21	29.24	29.25	Exact	17.59	17.79	17.87	II	D	5A	31	B		1	183
832	835	Budišov	Pod hradem Cave	UM1	R	Exact	25.73	25.71	25.72	Exact	19.5	19.53	19.51	IX	D	5A	31		e	1	184
826	829	Budišov	Pod hradem Cave	LM1	L	Exact	29.93	30.02	30.02	Exact	13.54	13.52	13.53	I	D	6	32	B		1	182
828	831	Budišov	Pod hradem Cave	LM1	L	Exact	32.15	32.12	32.16	Exact	15.15	15.36	15.55	II	D	6	32	D		1	182
827	830	Budišov	Pod hradem Cave	UM1	R	Close estimate	41.96	41.96	41.96	Exact	21.19	21.37	21.08	II	D	6	32	B		1	182
817	820	Budišov	Pod hradem Cave	C		Close estimate	32.41	32.41	32.41	Exact	23.31	23.51	23.49		D	6B	35	C		1	180
820	823	Budišov	Pod hradem Cave	LM1	L	Exact	31.54	31.58	31.58	Exact	15.39	15.38	15.33	II	D	6A	35	D		1	181
821	824	Budišov	Pod hradem Cave	LM1	R	Exact	31.47	31.47	31.47	Exact	14.98	14.99	14.88	II	D	6A	35	D		1	181
822	825	Budišov	Pod hradem Cave	LM1	R	Exact	28.77	28.77	28.77	Exact	13.83	13.79	13.79	IX	M	6A	34	D		1	181
819	822	Budišov	Pod hradem Cave	LM2	R	Exact	33.53	33.56	33.53	Exact	18.43	18.45	18.55	III	D	6A	35	D		1	181

Table. A1.1. (Continued.)

ID	Tooth ID	Storage ID	Site ID	Tooth Type ID	Left/Right	Measurement Type Length	Length1 (mm)	Length2 (mm)	Length3 (mm)	Measurement Type Breadth	Breadth1 (mm)	Breadth2 (mm)	Breadth3 (mm)	Wear Stage	Articulation ID	Layer	Box Number	Medium Box Number Suffix	Small Box Number Suffix	Log Book Number	Log Book Page Number
823	826	Budišov	Pod hradem Cave	LM2	R	Exact	26.98	27.01	26.99	Exact	17.67	17.66	17.67	IX	M	6A	34	D		1	181
818	821	Budišov	Pod hradem Cave	LM2	L	Exact	33.06	33.06	33.05	Exact	20.01	20.04	20.04	V	D	6B	35	C		1	180
824	827	Budišov	Pod hradem Cave	LM3	R	Exact	24.88	25.01	24.89	Exact	18.16	18.21	18.23	IX	M	6A	34	D		1	181
825	828	Budišov	Pod hradem Cave	LM3	L	Exact	28.45	28.49	28.49	Exact	20.65	20.74	20.81	II	D	6A	34	D		1	181
805	808	Budišov	Pod hradem Cave	C		Exact	21.09	21.08	21.24	Exact	15.28	15.34	15.3		M	7	37	B		1	178
809	812	Budišov	Pod hradem Cave	C		Exact	26.22	26.15	26.17	Exact	18.32	18.23	18.27		D	7	37	D		1	179
810	813	Budišov	Pod hradem Cave	C		No measurement				Close estimate	16.56	16.56	16.56		D	7	37	D		1	179
804	807	Budišov	Pod hradem Cave	LM1	L	Exact	29.11	29.09	29.11	Exact	13.99	13.49	13.52	II	D	7	37	B		1	178
806	809	Budišov	Pod hradem Cave	LM1	R	Exact	29.62	29.62	29.62	Exact	15.77	15.3	15.79	IX	M	7	37	C		1	178
807	810	Budišov	Pod hradem Cave	LM1	R	Exact	28.57	28.56	28.57	Exact	19.34	19.28	19.11	IX	M	7	37	C		1	178
811	814	Budišov	Pod hradem Cave	LM1	R	Exact	30.9	30.91	30.92	Exact	13.52	13.49	13.59	III	D	7	37	E		1	179
803	806	Budišov	Pod hradem Cave	LM2	L	Exact	28.84	28.85	28.84	Exact	16.53	16.53	16.53	II	D	7	37	B		1	178
812	815	Budišov	Pod hradem Cave	LM2	R	No measurement				Exact	17.94	17.86	17.87	II	D	7	37	E		1	179
813	816	Budišov	Pod hradem Cave	LM2	R	Exact	31.11	31.02	31.13	Exact	17.98	18	18	IV	M	7	37	E		1	179
808	811	Budišov	Pod hradem Cave	LM3	R	Exact	27.1	27.24	27.15	Exact	20.85	20.65	20.81	IX	M	7	37	C		1	178
814	817	Budišov	Pod hradem Cave	LM3	R	Exact	26.09	26.7	26.43	Exact	16.97	17.01	16.95	IV	M	7	37	E		1	179
816	819	Budišov	Pod hradem Cave	LM3	R	Exact	27.08	27.05	27.06	Exact	18.54	18.58	18.57	III	D	7	36		c	1	180
802	805	Budišov	Pod hradem Cave	UM1	L	Close estimate	24.62	24.62	24.62	Close estimate	18.32	18.32	18.32	IX	D	7	38			1	177
815	818	Budišov	Pod hradem Cave	UM2	R	Exact	47.07	47.1	47.08	Exact	24.87	24.24	24.71	VII	D	7	36		c	1	180
749	752	Budišov	Pod hradem Cave	C		Close estimate	23.21	23.21	23.21	Exact	15.76	15.64	15.63		D	8	43A	A		1	169
750	753	Budišov	Pod hradem Cave	C		No measurement				No measurement					D	8	43A	A		1	169
751	754	Budišov	Pod hradem Cave	C		Estimate	24.51	24.51	24.51	Close estimate	18.71	18.71	18.71		D	8	43A	C		1	169
752	755	Budišov	Pod hradem Cave	C		No measurement				No measurement					D	8	43A	C		1	169
766	769	Budišov	Pod hradem Cave	C		Exact	31.2	31.15	31.17	Exact	22.23	22.24	22.24		D	8	43	E		1	171
768	771	Budišov	Pod hradem Cave	C		Close estimate	27.36	27.36	27.36	Exact	19.79	19.78	19.55		D	8	42	C		1	172
769	772	Budišov	Pod hradem Cave	C		No measurement				No measurement					D	8	42	C		1	172
778	781	Budišov	Pod hradem Cave	C		No measurement				No measurement					D	8	41	A		1	174
798	801	Budišov	Pod hradem Cave	C		Exact	23.92	23.98	23.65	Exact	14.51	14.8	14.61		D	8	39	B		1	177
794	797	Budišov	Pod hradem Cave	C (J)		Exact	20.25	20.25	20.25	Exact	15.47	15.47	15.47		D	8	40	A		1	176
742	745	Budišov	Pod hradem Cave	LM1		Exact	31.3	31.31	31.28	Exact	16.47	16.47	16.54	IX	M	8	45	C		1	168
744	747	Budišov	Pod hradem Cave	LM1	L	Exact	28.19	28.19	28.19	Exact	13.05	13.05	13.05	II	D	8	45		b	1	168

Table. A1.1. (Continued.)

ID	Tooth ID	Storage ID	Site ID	Tooth Type ID	Left/Right	Measurement Type Length	Length1 (mm)	Length2 (mm)	Length3 (mm)	Measurement Type Breadth	Breadth1 (mm)	Breadth2 (mm)	Breadth3 (mm)	Wear Stage	Articulation ID	Layer	Box Number	Medium Box Number Suffix	Small Box Number Suffix	Log Book Number	Log Book Page Number
756	759	Budišov	Pod hradem Cave	LM1	R	Exact	27.07	27.05	27.14	Exact	13.64	13.84	13.73	III	D	8	43A	D		1	170
776	779	Budišov	Pod hradem Cave	LM1	L	Exact	30.29	30.29	30.27	Exact	14.12	14.33	14.27	II	D	8	41	C		1	173
779	782	Budišov	Pod hradem Cave	LM1	R	Exact	30.43	30.44	30.47	Exact	14.22	14.15	14.14	III	D	8	41	A		1	174
783	786	Budišov	Pod hradem Cave	LM1	L	Exact	28.59	28.62	28.59	Exact	13.28	13.28	13.26	I	D	8	41	B		1	174
785	788	Budišov	Pod hradem Cave	LM1	R	Close estimate	28.48	28.48	28.48	Close estimate	12.5	12.5	12.5		M	8	41	D		1	175
786	789	Budišov	Pod hradem Cave	LM1	R	Exact	30.35	30.35	30.34	Exact	13.56	13.58	13.59	III	M	8	41	E		1	175
801	804	Budišov	Pod hradem Cave	LM1	R	Exact	29.58	29.49	29.51	Exact	13.96	13.5	13.91	III	D	8	39		g	1	177
743	746	Budišov	Pod hradem Cave	LM2		Exact	32.94	33.01	32.99	Exact	19.58	19.65	19.59	IX	M	8	45	C		1	168
761	764	Budišov	Pod hradem Cave	LM2	L	Exact	32	32.01	32.12	Exact	18.15	18.33	18.33	VIII	D	8	43	C		1	171
781	784	Budišov	Pod hradem Cave	LM2	L	No measurement				Exact	17.39	17.45	17.51	II	D	8	41	A		1	174
787	790	Budišov	Pod hradem Cave	LM2	R	Exact	29.85	29.85	29.85	Exact	16.87	16.89	16.91	III	M	8	41	E		1	175
789	792	Budišov	Pod hradem Cave	LM2	R	Exact	30.22	30.76	30.52	Exact	17.39	17.41	17.38	II	M	8	41		c	1	175
791	794	Budišov	Pod hradem Cave	LM2	R	Close estimate	31.15	31.15	31.15	Close estimate	17.82	17.82	17.82	II	M	8	41		b	1	175
795	798	Budišov	Pod hradem Cave	LM2	R	Exact	32.61	32.71	32.66	Exact	18.8	18.75	18.8	III	D	8	40	C		1	176
745	748	Budišov	Pod hradem Cave	LM3	R	Exact	25.98	25.96	25.93	Exact	18.23	18.11	18.31	II	M	8	44	D		1	168
746	749	Budišov	Pod hradem Cave	LM3	R	Exact	29.45	29.37	29.34	Exact	20.09	20.07	20.06	IV	D	8	43A	B		1	169
757	760	Budišov	Pod hradem Cave	LM3	R	Exact	28.81	28.63	28.72	Exact	21.36	21.37	21.37	VII	D	8	43A	E		1	170
758	761	Budišov	Pod hradem Cave	LM3	R	Exact	27.91	27.93	27.93	Exact	17.81	17.86	17.89	III	D	8	43	B		1	170
762	765	Budišov	Pod hradem Cave	LM3	L	Exact	26.76	26.72	26.76	Exact	17.33	17.14	17.16	VIII	D	8	43	C		1	171
771	774	Budišov	Pod hradem Cave	LM3	L	Exact	26.29	26.32	26.24	Exact	18.14	18.2	18.29	III	D	8	42	C		1	172
773	776	Budišov	Pod hradem Cave	LM3	L	Exact	28.81	28.8	28.79	Exact	19.82	19.82	19.83	III	D	8	42	B		1	172
774	777	Budišov	Pod hradem Cave	LM3	L	Exact	28.41	28.46	28.46	Exact	12.41	12.41	12.46	III	D	8	42	D		1	172
777	780	Budišov	Pod hradem Cave	LM3	L	Close estimate	30.14	30.14	30.14	Exact	21.92	21.95	21.97	IX	D	8	41	E		1	173
782	785	Budišov	Pod hradem Cave	LM3	R	Exact	25.95	25.99	25.96	Exact	18.76	18.75	18.75	IX	D	8	41	B		1	174
788	791	Budišov	Pod hradem Cave	LM3	R	Exact	26.88	26.59	26.59	Exact	18.67	18.71	18.81	III	M	8	41	E		1	175
790	793	Budišov	Pod hradem Cave	LM3	R	Close estimate	26.13	26.13	26.13	Close estimate	19.34	1934	19.34	II	M	8	41		c	1	175
792	795	Budišov	Pod hradem Cave	LM3	R	Close estimate	25.86	25.86	25.86	Close estimate	19.37	19.37	19.37	II	M	8	41		b	1	175
797	800	Budišov	Pod hradem Cave	LM3	L	Exact	29.68	29.57	29.67	Exact	20.27	20.33	20.26	IX	D	8	39	A		1	177
748	751	Budišov	Pod hradem Cave	UM1	L	Exact	29.01	29.02	29.01	Exact	18.62	18.54	18.6	IV	D	8	43A	B		1	169
754	757	Budišov	Pod hradem Cave	UM1	R	Exact	31.1	31.13	31.1	Exact	21.36	21.38	21.38	VIII	C	8	43A	C		1	169
764	767	Budišov	Pod hradem Cave	UM1	R	Exact	27.48	27.51	27.51	Exact	17.44	17.5	17.51	III	C	8	43	E		1	171

Table. A1.1. (Continued.)

ID	Tooth ID	Storage ID	Site ID	Tooth Type ID	Left/Right	Measurement Type Length	Length1 (mm)	Length2 (mm)	Length3 (mm)	Measurement Type Breadth	Breadth1 (mm)	Breadth2 (mm)	Breadth3 (mm)	Wear Stage	Articulation ID	Layer	Box Number	Medium Box Number Suffix	Small Box Number Suffix	Log Book Number	Log Book Page Number
765	768	Budišov	Pod hradem Cave	UM1	L	Exact	30.92	30.94	30.91	Exact	19.38	19.37	19.41	IX	C	8	43	E		1	171
767	770	Budišov	Pod hradem Cave	UM1	L	Exact	27.66	27.55	27.68	Exact	17.09	17.11	17.12	IV	D	8	43	F		1	171
772	775	Budišov	Pod hradem Cave	UM1	R	Exact	28.64	28.65	28.55	Exact	18.54	18.53	18.52	II	C	8	42	C		1	172
775	778	Budišov	Pod hradem Cave	UM1	L	Exact	29.33	29.34	29.41	Exact	18.57	18.63	18.58	II	C	8	42	E		1	173
793	796	Budišov	Pod hradem Cave	UM1	R	Exact	28.51	28.52	28.53	Exact	18.41	18.42	18.44	II	D	8	40	A		1	176
796	799	Budišov	Pod hradem Cave	UM1	L	Close estimate	32.45	32.45	32.45	Close estimate	12.51	12.51	12.51	I	M	8	40	C		1	176
800	803	Budišov	Pod hradem Cave	UM1		No measurement				No measurement				II	D	8	39	B		1	177
747	750	Budišov	Pod hradem Cave	UM2	R	Exact	39.85	39.86	39.95	Exact	20.33	20.21	20.22	II	D	8	43A	B		1	169
753	756	Budišov	Pod hradem Cave	UM2	R	Exact	50.81	50.84	50.88	Exact	26.21	26.17	26.24	VI	C	8	43A	C		1	169
755	758	Budišov	Pod hradem Cave	UM2	L	Exact	50.54	50.68	50.65	Exact	24.65	24.44	24.4	VII	C	8	43A	C		1	169
759	762	Budišov	Pod hradem Cave	UM2	R	Exact	42.97	42.93	42.91	Exact	27.31	22.35	22.45	II	D	8	43	B		1	170
760	763	Budišov	Pod hradem Cave	UM2	L	Exact	46.7	46.95	46.9	Exact	23.33	23.22	23.34	II	D	8	43	C		1	171
763	766	Budišov	Pod hradem Cave	UM2	R	Exact	43.8	43.85	43.89	Exact	24.48	24.17	24.33	III	C	8	43	E		1	171
770	773	Budišov	Pod hradem Cave	UM2	R	Exact	49.82	49.83	49.75	Exact	24.73	24.73	24.72	II	D	8	42	C		1	172
780	783	Budišov	Pod hradem Cave	UM2	L	Exact	41.78	41.77	41.75	Exact	22.5	22.48	22.5	II	D	8	41	A		1	174
784	787	Budišov	Pod hradem Cave	UM2	R	Exact	46.25	46.14	46.19	Exact	23.5	23.48	23.51	III	D	8	41	B		1	174
799	802	Budišov	Pod hradem Cave	UM2	L	No measurement				Close estimate	24.81	24.81	24.81	III	D	8	39	B		1	177
741	744	Budišov	Pod hradem Cave	C		Exact	32.69	32.69	32.69	Exact	22.17	22.17	22.17		M	9	46	D		1	167
737	740	Budišov	Pod hradem Cave	LM1	L	Exact	31.98	31.94	31.97	Exact	14.19	14.31	14.22	II	M	9	46	B		1	167
738	741	Budišov	Pod hradem Cave	LM1	R	Exact	30.36	30.4	30.4	Exact	14.55	14.59	14.59	II	D	9	46	D		1	167
736	739	Budišov	Pod hradem Cave	UM1	R	Exact	26.44	26.55	26.5	Exact	19.03	19.22	19	III	D	9	46	A		1	167
739	742	Budišov	Pod hradem Cave	UM1	L	Exact	30.29	30.29	30.28	Exact	18.34	18.44	18.44	II	D	9	46	D		1	167
740	743	Budišov	Pod hradem Cave	UM1	R	Exact	30.1	30.1	30.09	Exact	19.21	19.21	19.18	III	D	9	46	D		1	167
23	23	Budišov	Pod hradem Cave	C	L	Exact	32.76	32.78	32.74	Exact	22.07	22.08	22.07		M	10	149			1	13
710	712	Budišov	Pod hradem Cave	C		Estimate	29.62	29.62	29.62	Close estimate	21.2	21.2	21.2		D	10	52	A		1	161
713	715	Budišov	Pod hradem Cave	C		Exact	29.01	29.26	29.11	Exact	21.33	21.36	21.2		D	10	52	B		1	162
714	716	Budišov	Pod hradem Cave	C		Close estimate	24.22	24.22	24.22	Exact	16.14	16.13	16.17		M	10	52	B		1	162
716	718	Budišov	Pod hradem Cave	C		No measurement				Close estimate	21.9	21.9	21.9		D	10	52	F		1	162
721	723	Budišov	Pod hradem Cave	C		Close estimate	32.6	32.6	32.6	Close estimate	20.69	20.69	20.69		D	10	51	D		1	163
729	731	Budišov	Pod hradem Cave	C		No measurement				No measurement					D	10	50		f	1	165
717	719	Budišov	Pod hradem Cave	C (J)		No measurement				Close estimate	17.79	17.79	17.79		D	10	52	F		1	162

Table. A1.1. (Continued.)

ID	Tooth ID	Storage ID	Site ID	Tooth Type ID	Left/Right	Measurement Type Length	Length1 (mm)	Length2 (mm)	Length3 (mm)	Measurement Type Breadth	Breadth1 (mm)	Breadth2 (mm)	Breadth3 (mm)	Wear Stage	Articulation ID	Layer	Box Number	Medium Box Number Suffix	Small Box Number Suffix	Log Book Number	Log Book Page Number
20	20	Budišov	Pod hradem Cave	LM1	L	Exact	31.21	31.22	31.22	Exact	15.17	15.13	15.17	IV	M	10	149			1	13
709	711	Budišov	Pod hradem Cave	LM1	R	Exact	32.13	32.01	32.09	Close estimate	12.45	12.45	12.45	I	D	10	53	C		1	161
718	720	Budišov	Pod hradem Cave	LM1	R	Close estimate	29.13	29.13	29.13	Close estimate	13.77	13.77	13.77	I	M	10	52	E		1	162
724	726	Budišov	Pod hradem Cave	LM1	L	Exact	29.03	29.05	29.14	Exact	13.37	13.41	13.47	II	D	10	51	D		1	163
725	727	Budišov	Pod hradem Cave	LM1	L	Exact	31.6	31.58	31.57	Exact	14.04	13.99	13.97	II	D	10	50	C		1	164
728	730	Budišov	Pod hradem Cave	LM1	R	Exact	30.67	30.41	30.44	Exact	13.3	13.36	13.39	III	D	10	50		f	1	165
19	19	Budišov	Pod hradem Cave	LM2	L	Exact	30.59	30.63	30.69	Exact	19.32	19.28	19.33	IV	M	10	149			1	13
707	709	Budišov	Pod hradem Cave	LM2	L	Exact	28.59	28.64	28.65	Exact	16.75	16.4	16.63	VII	D	10	53	B		1	160
711	713	Budišov	Pod hradem Cave	LM2	R	Exact	29.03	28.87	29.17	Exact	16.73	16.79	16.75	III	M	10	52	A		1	161
715	717	Budišov	Pod hradem Cave	LM2	R	Exact	33.24	33.2	33.27	Exact	19.14	19.14	19.11	II	D	10	52	D		1	162
720	722	Budišov	Pod hradem Cave	LM2	L	Exact	30.45	30.31	30.4	Exact	17.02	17.66	17.14	II	D	10	51	C		1	163
723	725	Budišov	Pod hradem Cave	LM2	L	Exact	31.19	31.2	31.21	Exact	18.82	18.84	18.9	VII	D	10	51	D		1	163
726	728	Budišov	Pod hradem Cave	LM2	L	Exact	30.46	30.33	30.41	Exact	17	17.01	17.04	I	D	10	50		c	1	164
730	732	Budišov	Pod hradem Cave	LM2	R	Exact	28.73	28.72	28.73	Exact	16.81	16.74	16.81	I	D	10	50		b	1	165
732	735	Budišov	Pod hradem Cave	LM2	R	Exact	30.2	30.3	30.25	Exact	18.53	18.61	18.48	III	D	10	50		a	1	165
733	736	Budišov	Pod hradem Cave	LM2		No measurement				No measurement				IX	D	10	50		b	1	165
18	18	Budišov	Pod hradem Cave	LM3	L	Exact	28.07	28.08	28.09	Exact	20.17	20.14	20.15	IV	M	10	149			1	13
712	714	Budišov	Pod hradem Cave	LM3	R	Exact	25.5	25.59	25.54	Exact	19.11	19.27	19.56	III	M	10	52	A		1	161
722	724	Budišov	Pod hradem Cave	LM3	L	Exact	26.5	26.51	26.53	Exact	21.02	21.08	21.21	VII	D	10	51	D		1	163
734	737	Budišov	Pod hradem Cave	LM3	L	Exact	31.17	31.18	31.19	Exact	20.93	20.96	20.95	III	D	10	50		a	1	165
708	710	Budišov	Pod hradem Cave	UM1	R	Exact	28.87	28.77	28.85	Exact	18.15	18.32	18.16	II	D	10	53	B		1	160
719	721	Budišov	Pod hradem Cave	UM1	R	Exact	28.81	28.88	28.93	Exact	18.34	18.11	18.64	II	D	10	51	B		1	163
727	729	Budišov	Pod hradem Cave	UM1	L	Exact	31.13	31.12	31.13	Exact	18.68	18.95	18.82	II	D	10	50		c	1	164
731	733	Budišov	Pod hradem Cave	UM1	L	Exact	27.27	27.21	27.25	Exact	18.33	18.34	18.34	II	D	10	50		b	1	165
735	738	Budišov	Pod hradem Cave	UM2	L	Exact	40.03	40.03	40.03	Exact	22.58	22.58	22.49	II	D	10	50		a	1	165
661	663	Budišov	Pod hradem Cave	C		Exact	24.88	24.85	24.79	Exact	17.64	17.73	17.63		D	11	56	D		1	154
669	671	Budišov	Pod hradem Cave	C		Exact	26.57	26.89	26.58	Exact	20.23	20.99	20.74		C	11	56	C		1	154
670	672	Budišov	Pod hradem Cave	C	L	Exact	23.32	23.33	23.35	Exact	15.48	15.47	15.51	V	M	11	56	C		1	155
674	676	Budišov	Pod hradem Cave	C		Close estimate	31.24	31.24	31.24	Close estimate	23.53	23.53	23.53		M	11	55A	A		1	155
692	694	Budišov	Pod hradem Cave	C		Estimate	31.35	31.35	31.35	Estimate	23.36	23.36	23.36		D	11	54	C		1	158
693	695	Budišov	Pod hradem Cave	C		No measurement				Close estimate	21.14	21.14	21.14		D	11	54	C		1	158

Table. A1.1. (Continued.)

ID	Tooth ID	Storage ID	Site ID	Tooth Type ID	Left/Right	Measurement Type Length	Length1 (mm)	Length2 (mm)	Length3 (mm)	Measurement Type Breadth	Breadth1 (mm)	Breadth2 (mm)	Breadth3 (mm)	Wear Stage	Articulation ID	Layer	Box Number	Medium Box Number Suffix	Small Box Number Suffix	Log Book Number	Log Book Page Number
697	699	Budišov	Pod hradem Cave	C	L	Exact	30.07	29.94	29.91	Exact	20.45	20.59	20.69		M	11	54	E		1	158
698	700	Budišov	Pod hradem Cave	C		No measurement				Close estimate	23.44	23.44	23.44		D	11	53	C		1	159
660	662	Budišov	Pod hradem Cave	C (J)		Exact	24.33	24.38	24.32	Exact	18.91	18.98	18.94		D	11	56	B		1	153
673	675	Budišov	Pod hradem Cave	LM1	L	Exact	29.51	29.08	29.08	Exact	13.54	13.44	13.66	V	M	11	56	C		1	155
675	677	Budišov	Pod hradem Cave	LM1	L	Exact	32.69	32.7	32.72	Exact	14.82	14.93	14.92	II	D	11	55A	A		1	155
679	681	Budišov	Pod hradem Cave	LM1	L	Exact	31.54	31.48	31.47	Exact	14.07	14.02	14.14	III	D	11	55A	D		1	156
680	682	Budišov	Pod hradem Cave	LM1	R	Exact	31.41	31.41	31.41	Exact	14.42	14.45	14.46	I	D	11	55A	E		1	156
681	683	Budišov	Pod hradem Cave	LM1	R	No measurement				No measurement				VI	D	11	55	A		1	156
685	687	Budišov	Pod hradem Cave	LM1	L	Estimate	28.48	28.48	28.48	Exact	15.27	15.21	15.23	VIII	D	11	55	D		1	157
688	690	Budišov	Pod hradem Cave	LM1	L	No measurement				No measurement				I	D	11	55		a	1	157
702	704	Budišov	Pod hradem Cave	LM1	L	Exact	31.3	31.24	31.28	Exact	15.06	15.16	15.51	II	D	11	53	C		1	159
703	705	Budišov	Pod hradem Cave	LM1	L	Exact	32.5	32.39	32.4	Exact	14.98	15.15	15.03	II	D	11	53	C		1	159
704	706	Budišov	Pod hradem Cave	LM1	L	Exact	31.69	31.61	31.69	Exact	13.9	13.9	13.9	II	D	11	53	C		1	159
653	655	Budišov	Pod hradem Cave	LM2	R	Exact	32.57	32.32	32.37	Exact	18.51	18.56	18.63	III	M	11	59	A		1	152
654	656	Budišov	Pod hradem Cave	LM2	R	Exact	32.32	32.35	32.29	Exact	18.83	18.61	18.51	II	D	11	59	A		1	152
655	657	Budišov	Pod hradem Cave	LM2	L	Exact	27.17	27.17	27.15	Exact	16.57	16.62	16.59	II	D	11	59	A		1	152
659	661	Budišov	Pod hradem Cave	LM2	R	Exact	33.26	33.24	33.3	Exact	18.81	18.8	18.84	II	D	11	56	B		1	153
662	664	Budišov	Pod hradem Cave	LM2	L	Exact	31.23	31.02	31.37	Exact	17.69	17.74	17.99	VI	M	11	56	C		1	154
663	665	Budišov	Pod hradem Cave	LM2	L	Close estimate	32.09	32.09	32.09	Exact	19.27	19.48	19.5	I	D	11	56	C		1	154
664	666	Budišov	Pod hradem Cave	LM2	R	Exact	31.17	31.17	31.03	Exact	18.6	18.43	18.43	I	D	11	56	C		1	154
665	667	Budišov	Pod hradem Cave	LM2	R	Exact	30.39	30.39	30.37	Exact	18.44	18.38	18.37	II	D	11	56	C		1	154
666	668	Budišov	Pod hradem Cave	LM2	L	Exact	30.39	30.39	30.37	Exact	18.44	18.38	18.37	II	D	11	56	C		1	154
672	674	Budišov	Pod hradem Cave	LM2	L	Exact	27.75	27.75	27.75	Exact	17.37	17.39	17.42	V	M	11	56	C		1	155
676	678	Budišov	Pod hradem Cave	LM2	L	Exact	32.69	32.7	32.72	Exact	17.97	17.98	18	I	D	11	55A	A		1	155
682	684	Budišov	Pod hradem Cave	LM2	R	Exact	32.3	32.28	32.29	Exact	17.97	17.97	17.92	II	D	11	55	A		1	157
684	686	Budišov	Pod hradem Cave	LM2	R	Exact	31.39	31.29	31.4	Exact	18.44	18.45	18.47	II	D	11	55	A		1	157
689	691	Budišov	Pod hradem Cave	LM2	L	Exact	32.85	32.83	32.85	Exact	19.06	19.04	19.08	II	D	11	55		a	1	157
694	696	Budišov	Pod hradem Cave	LM2	L	Exact	30.86	30.86	30.87	Exact	16.96	16.97	16.98	II	D	11	54	C		1	158
695	697	Budišov	Pod hradem Cave	LM2	L	Exact	31.34	31.35	31.51	Exact	19.23	19.23	19.25	VI	M	11	54	E		1	158
701	703	Budišov	Pod hradem Cave	LM2	R	Exact	31.07	31.02	30.92	Exact	18.1	18.43	18.65	I	D	11	53	C		1	159
705	707	Budišov	Pod hradem Cave	LM2	R	Close estimate	30.33	30.33	30.33	Exact	18.41	18.41	18.6	VI	M	11	53	E		1	160

Table. A1.1. (Continued.)

ID	Tooth ID	Storage ID	Site ID	Tooth Type ID	Left/Right	Measurement Type Length	Length1 (mm)	Length2 (mm)	Length3 (mm)	Measurement Type Breadth	Breadth1 (mm)	Breadth2 (mm)	Breadth3 (mm)	Wear Stage	Articulation ID	Layer	Box Number	Medium Box Number Suffix	Small Box Number Suffix	Log Book Number	Log Book Page Number
706	708	Budišov	Pod hradem Cave	LM2	R	Exact	31.97	32.06	31.91	Exact	20.61	20.86	20.64	IX	D	11	53	E	a	1	160
652	654	Budišov	Pod hradem Cave	LM3	R	Exact	27.96	28.2	28.16	Exact	20.76	20.64	20.82	III	M	11	59	A		1	152
668	670	Budišov	Pod hradem Cave	LM3	R	Exact	27.89	27.85	27.99	Exact	20	20	19.95	I	D	11	56	C		1	154
671	673	Budišov	Pod hradem Cave	LM3	L	Exact	26.08	26.14	26.12	Exact	18.43	18.2	18.33	V	M	11	56	C		1	155
677	679	Budišov	Pod hradem Cave	LM3	L	Exact	26.48	26.46	26.46	Exact	18.54	18.64	18.66	II	D	11	55A	A		1	155
687	689	Budišov	Pod hradem Cave	LM3	L	Exact	23.21	23.25	23.23	Exact	17.27	17.26	17.25	III	D	11	55	E		1	157
696	698	Budišov	Pod hradem Cave	LM3	L	Exact	29.68	29.75	29.63	Exact	20.59	20.9	20.87	VI	M	11	54	E		1	158
678	680	Budišov	Pod hradem Cave	UM1	L	Exact	29.96	30.02	30.02	Exact	18.18	18.21	18.3	II	D	11	55A	C		1	156
683	685	Budišov	Pod hradem Cave	UM1	R	Exact	30.35	30.33	30.23	Exact	17.92	17.97	17.72	II	D	11	55	A		1	157
691	693	Budišov	Pod hradem Cave	UM1	L	Exact	29.95	29.91	29.78	Exact	18.57	18.62	18.65	II	D	11	54	C		1	158
699	701	Budišov	Pod hradem Cave	UM1	R	Exact	31.04	31.09	31.15	Exact	19.8	20.06	19.95	II	D	11	53	C		1	159
656	658	Budišov	Pod hradem Cave	UM2	R	Exact	41.18	41.33	41.31	Exact	23.03	23.44	23.31	II	D	11	59	A		1	152
657	659	Budišov	Pod hradem Cave	UM2	L	Close estimate	41.22	41.22	41.22	Exact	23.53	23.6	23.58	VIII	C	11	57			1	153
658	660	Budišov	Pod hradem Cave	UM2	R	Exact	43.48	43.64	43.85	Exact	23.07	23.1	23.33	VIII	C	11	57			1	153
667	669	Budišov	Pod hradem Cave	UM2	L	Exact	49.81	49.69	49.56	Exact	23.97	24.19	24.14	II	D	11	56	C		1	154
686	688	Budišov	Pod hradem Cave	UM2	R	No measurement				No measurement				II	D	11	55	D		1	157
690	692	Budišov	Pod hradem Cave	UM2	L	Exact	37.62	37.54	37.72	Exact	19.64	19.06	19.85	IV	D	11	55		a	1	157
700	702	Budišov	Pod hradem Cave	UM2	R	Exact	45.9	45.9	45.87	Exact	23.18	23.15	23.29	IV	D	11	53	C		1	159
512	514	Budišov	Pod hradem Cave	C		Exact	22.07	21.99	22.03	Exact	16.23	16.3	16.21		D	12	77	E		1	129
572	574	Budišov	Pod hradem Cave	C		Exact	22.6	22.25	22	Exact	16.7	16.72	16.74		D	12	68	B		1	141
576	578	Budišov	Pod hradem Cave	C		No measurement				Exact	20.21	20.09	20.01			12	68	E		1	140
577	579	Budišov	Pod hradem Cave	C		Exact	31.4	31.45	31.41	Exact	21.62	21.61	21.55		D	12	68	F		1	140
580	582	Budišov	Pod hradem Cave	C		No measurement				No measurement					M	12	67	A		1	141
598	600	Budišov	Pod hradem Cave	C		Exact	24.76	24.86	24.96	Exact	19.36	19.04	19.18		D	12	66	B		1	143
601	603	Budišov	Pod hradem Cave	C		Exact	22.27	22.27	22.27	Exact	15.91	15.64	15.54		D	12	66	C		1	143
604	606	Budišov	Pod hradem Cave	C		No measurement				No measurement					D	12	66	F		1	143
612	614	Budišov	Pod hradem Cave	C		Close estimate	33.66	33.66	33.66	Close estimate	23.65	23.65	23.65		D	12	65	D		1	145
620	622	Budišov	Pod hradem Cave	C		Exact	23.29	23.36	23.36	Exact	16.26	16.19	16.2		M	12	63	A		1	146
626	628	Budišov	Pod hradem Cave	C		Exact	22.81	22.83	22.81	Close estimate	16	16	16		D	12	63	D		1	147
628	630	Budišov	Pod hradem Cave	C		No measurement				No measurement					D	12	62	B		1	148
640	642	Budišov	Pod hradem Cave	C		No measurement				Close estimate	17.82	17.82	17.82		D	12	61	A		1	150

Table. A1.1. (Continued.)

ID	Tooth ID	Storage ID	Site ID	Tooth Type ID	Left/Right	Measurement Type Length	Length1 (mm)	Length2 (mm)	Length3 (mm)	Measurement Type Breadth	Breadth1 (mm)	Breadth2 (mm)	Breadth3 (mm)	Wear Stage	Articulation ID	Layer	Box Number	Medium Box Number Suffix	Small Box Number Suffix	Log Book Number	Log Book Page Number
643	645	Budišov	Pod hradem Cave	C		Exact	23.79	23.75	23.76	Exact	18.23	18.19	18.18		D	12	61	C		1	150
645	647	Budišov	Pod hradem Cave	C	R	Exact	28.72	28.68	28.45	Exact	19.86	20.22	20.48		M	12	61	D		1	150
571	573	Budišov	Pod hradem Cave	C (J)		Exact	19.66	19.66	19.66	Exact	14.65	14.65	14.65		D	12	68	A		1	139
587	589	Budišov	Pod hradem Cave	C (J)		Exact	20.05	20.05	20.05	Exact	15.52	15.52	15.52		D	12	67	C		1	141
624	626	Budišov	Pod hradem Cave	C (J)		Exact	21	21	21	Exact	14.47	14.47	14.47		D	12	63	B		1	147
32	32	Budišov	Pod hradem Cave	LM1	L	Exact	31.74	31.8	31.8	Exact	14.51	14.57	14.55	III	M	12	149			1	15
511	513	Budišov	Pod hradem Cave	LM1	R	Exact	30.87	30.91	30.87	Exact	13.47	13.58	13.57	II	D	12	77	B		1	129
575	577	Budišov	Pod hradem Cave	LM1	L	Exact	33.45	33.46	33.46	Exact	17.3	17.26	17.31	IX	D	12	68	E		1	140
579	581	Budišov	Pod hradem Cave	LM1	L	Exact	31.8	31.77	31.79	Exact	14.91	14.9	14.91	III	D	12	67	A		1	141
581	583	Budišov	Pod hradem Cave	LM1	L	Exact	30.32	30.31	30.34	Exact	13.64	13.59	13.55	II	D	12	67	C		1	141
582	584	Budišov	Pod hradem Cave	LM1	R	Exact	32.01	32.02	32.01	Exact	14.03	14.08	14.07	IV	D	12	67	C		1	141
583	585	Budišov	Pod hradem Cave	LM1	L	Exact	29.8	29.85	29.65	Exact	13.86	13.86	13.66	II	D	12	67	C		1	141
588	590	Budišov	Pod hradem Cave	LM1	L	Exact	31.21	31.21	31.21	Exact	12.29	12.29	12.29	I	M	12	67	C		1	141
605	607	Budišov	Pod hradem Cave	LM1	L	No measurement				No measurement				I	D	12	66	F		1	143
608	610	Budišov	Pod hradem Cave	LM1	R	Exact	31.66	31.68	31.64	Exact	14.81	15.15	14.92	I	D	12	65	B		1	144
611	613	Budišov	Pod hradem Cave	LM1		Exact	29.09	29.09	29.09	Close estimate	12.46	12.46	12.46	I	M	12	65	C		1	144
617	619	Budišov	Pod hradem Cave	LM1	R	Exact	28.79	28.76	28.76	Exact	13.43	13.63	13.58	III	M	12	63	A		1	146
622	624	Budišov	Pod hradem Cave	LM1	L	Exact	32.31	32.31	32.28	Exact	14.86	14.73	14.85	I	D	12	63	B		1	147
623	625	Budišov	Pod hradem Cave	LM1	L	Exact	31.6	31.65	31.61	Exact	15.35	15.45	15.46	II	D	12	63	B		1	147
625	627	Budišov	Pod hradem Cave	LM1	L	Exact	31.3	31.3	31.25	Exact	14.15	14.21	14.04	I	M	12	63	C		1	147
627	629	Budišov	Pod hradem Cave	LM1	R	No measurement				Exact	14.8	14.84	14.85	II	D	12	62	B		1	148
636	638	Budišov	Pod hradem Cave	LM1	R	Exact	29.51	29.49	29.55	Exact	14.13	14.21	14.17		M	12	61A	D		1	149
637	639	Budišov	Pod hradem Cave	LM1	R	Exact	29.87	29.87	29.89	Exact	14.18	14.2	14.18	II	D	12	61A	D		1	149
641	643	Budišov	Pod hradem Cave	LM1	L	Close estimate	30.3	30.3	30.3	Exact	14.06	14.09	14.05	VIII	D	12	61	A		1	150
642	644	Budišov	Pod hradem Cave	LM1	L	Exact	30.85	30.84	30.82	Exact	15.24	15.23	15.24	VIII	D	12	61	A		1	150
644	646	Budišov	Pod hradem Cave	LM1	R	Exact	30.17	30.17	30.24	Exact	13.63	13.69	13.63	III	M	12	61	D		1	150
31	31	Budišov	Pod hradem Cave	LM2	L	Exact	31.22	31.17	31.15	Exact	17.96	17.95	17.94	III	M	12	149			1	15
569	571	Budišov	Pod hradem Cave	LM2	L	Exact	34.05	34.05	34.05	Close estimate	17.31	17.31	17.31	II	M	12	69	B		1	139
574	576	Budišov	Pod hradem Cave	LM2	L	Exact	30.43	30.29	30.31	Exact	17.9	17.88	17.86	III	D	12	68	C		1	140
585	587	Budišov	Pod hradem Cave	LM2	L	Exact	30.57	30.55	30.55	Exact	18.41	18.33	18.32	II	D	12	67	C		1	141
589	591	Budišov	Pod hradem Cave	LM2	L	Exact	32.5	32.49	32.25	Exact	18.42	18.44	18.42	III	D	12	67	E		1	142

Table. A1.1. (Continued.)

ID	Tooth ID	Storage ID	Site ID	Tooth Type ID	Left/Right	Measurement Type Length	Length1 (mm)	Length2 (mm)	Length3 (mm)	Measurement Type Breadth	Breadth1 (mm)	Breadth2 (mm)	Breadth3 (mm)	Wear Stage	Articulation ID	Layer	Box Number	Medium Box Number Suffix	Small Box Number Suffix	Log Book Number	Log Book Page Number
590	592	Budišov	Pod hradem Cave	LM2	R	Exact	29.74	29.52	29.57	Exact	17.13	17.13	17.13	I	D	12	67	E		1	142
591	593	Budišov	Pod hradem Cave	LM2	R	Exact	31.67	31.65	31.59	Exact	17.61	17.61	17.81	II	D	12	67	E		1	142
596	598	Budišov	Pod hradem Cave	LM2	R	Exact	32.17	32.02	32.21	Exact	19.62	19.21	19.32	III	D	12	66	B		1	143
600	602	Budišov	Pod hradem Cave	LM2	R	Exact	29.73	29.79	29.92	Exact	17.66	17.6	17.61	VI	D	12	66	C		1	143
603	605	Budišov	Pod hradem Cave	LM2	L	Exact	32.27	32.08	32.05	Exact	18.95	19	19	II	D	12	66	E		1	143
610	612	Budišov	Pod hradem Cave	LM2	R	Exact	29.7	29.81	29.57	Exact	18.1	18.37	18.53	III	M	12	65	C		1	144
616	618	Budišov	Pod hradem Cave	LM2	L	Exact	30.25	30.27	30.28	Exact	18.14	18.16	18.14	V	D	12	64	B		1	145
618	620	Budišov	Pod hradem Cave	LM2	R	Exact	30.39	30.43	30.38	Exact	18.24	18.19	18.19	III	M	12	63	A		1	146
621	623	Budišov	Pod hradem Cave	LM2	R	Exact	31.13	31.12	31.13	Exact	18.23	18.25	18.29	II	D	12	63	B		1	147
631	633	Budišov	Pod hradem Cave	LM2	R	Exact	30.43	30.44	30.47	Exact	18.2	18.21	18.21	II	D	12	62	D		1	148
632	634	Budišov	Pod hradem Cave	LM2	L	No measurement				Exact	18.02	17.97	17.92	III	D	12	62	B		1	148
634	636	Budišov	Pod hradem Cave	LM2	R	Exact	31.92	31.92	31.89	Exact	18.69	18.7	18.69	III	D	12	61A	B		1	149
635	637	Budišov	Pod hradem Cave	LM2	R	Exact	30.81	30.79	30.81	Exact	17.75	17.76	17.75		M	12	61A	D		1	149
638	640	Budišov	Pod hradem Cave	LM2	R	Exact	31.11	31.31	31.32	Exact	17.89	17.9	17.84	I	M	12	61	A		1	150
639	641	Budišov	Pod hradem Cave	LM2	L	Exact	30.82	30.83	30.79	Exact	17.53	17.65	17.55	I	D	12	61	A		1	150
646	648	Budišov	Pod hradem Cave	LM2	L	Exact	33.63	33.63	33.67	Exact	20.37	20.47	20.47	II	D	12	61	F		1	151
647	649	Budišov	Pod hradem Cave	LM2	R	Estimate	28.69	28.69	28.69	Exact	16.83	16.83	16.83	II	D	12	61	F		1	151
30	30	Budišov	Pod hradem Cave	LM3	L	Exact	24.82	25.29	25.06	Exact	19.6	19.49	19.55	III	M	12	149			1	14
568	570	Budišov	Pod hradem Cave	LM3	L	Exact	27.12	27.13	27.11	Exact	19.52	19.68	19.62	II	M	12	69	B		1	139
578	580	Budišov	Pod hradem Cave	LM3	R	Exact	28.15	28	27.99	Exact	18.95	18.95	18.98	III	D	12	67	A		1	141
584	586	Budišov	Pod hradem Cave	LM3	R	Exact	30.43	30.7	30.38	Exact	19.96	20.07	19.96	III	D	12	67	C		1	141
592	594	Budišov	Pod hradem Cave	LM3	L	Exact	28.79	28.79	28.79	Exact	19.11	19.11	19.11	III	D	12	67	E		1	142
599	601	Budišov	Pod hradem Cave	LM3	R	Exact	27.71	27.74	27.79	Exact	19.74	19.4	19.39	VIII	D	12	66	C		1	143
609	611	Budišov	Pod hradem Cave	LM3	R	Exact	27.56	27.56	27.59	Exact	19.24	19.2	19.11	III	M	12	65	C		1	144
614	616	Budišov	Pod hradem Cave	LM3	R	Exact	42.18	42.21	42.18	Exact	21.29	21.84	21.53	VII	D	12	65	E	a	1	145
615	617	Budišov	Pod hradem Cave	LM3	R	Exact	27.04	27.08	27.07	Exact	19.12	19.19	19.2	VIII	D	12	64	B		1	145
619	621	Budišov	Pod hradem Cave	LM3	R	Exact	25.91	25.91	25.85	Exact	18.35	18.38	18.13	III	M	12	63	A		1	146
629	631	Budišov	Pod hradem Cave	LM3	L	Exact	27.5	27.48	27.51	Exact	18.1	18.2	18.29	II	D	12	62	D		1	148
573	575	Budišov	Pod hradem Cave	UM1	L	Exact	29.66	29.51	29.85	Exact	18.81	18.77	18.78	II	D	12	68	B		1	140
593	595	Budišov	Pod hradem Cave	UM1	L	Exact	28.93	28.78	28.68	Exact	18.45	18.74	18.33	II	D	12	67		c	1	142
597	599	Budišov	Pod hradem Cave	UM1	R	No measurement				Exact	17.32	17.32	17.32	II	D	12	66	B		1	143

Table. A1.1. (Continued.)

ID	Tooth ID	Storage ID	Site ID	Tooth Type ID	Left/Right	Measurement Type Length	Length1 (mm)	Length2 (mm)	Length3 (mm)	Measurement Type Breadth	Breadth1 (mm)	Breadth2 (mm)	Breadth3 (mm)	Wear Stage	Articulation ID	Layer	Box Number	Medium Box Number Suffix	Small Box Number Suffix	Log Book Number	Log Book Page Number
606	608	Budišov	Pod hradem Cave	UM1	R	Exact	30.21	29.8	29.79	Exact	21.53	21.55	21.55	IX	D	12	65	A		1	144
607	609	Budišov	Pod hradem Cave	UM1	L	Exact	30.25	30.34	30.62	Exact	19.55	19.66	19.66	VII	D	12	65	B		1	144
630	632	Budišov	Pod hradem Cave	UM1	R	Exact	30.36	30.2	30	Estimate	17.35	17.35	17.35	II	D	12	62	D		1	148
567	569	Budišov	Pod hradem Cave	UM2	L	Exact	51.15	51.33	51.22	Exact	22.29	22.73	22.72	II	D	12	69	B		1	139
570	572	Budišov	Pod hradem Cave	UM2	L	Exact	51.42	51.54	51.52	Exact	26.41	26.84	26.98	II	D	12	68	A		1	139
586	588	Budišov	Pod hradem Cave	UM2	R	Exact	47.58	47.58	47.54	Exact	23.23	23.77	23.51	II	D	12	67	C		1	141
594	596	Budišov	Pod hradem Cave	UM2	L	Exact	46.29	46.15	46.29	Exact	22.8	22.94	22.93	II	D	12	67		c	1	142
595	597	Budišov	Pod hradem Cave	UM2	L	Exact	43.07	43.08	43.02	Close estimate	23.16	23.16	23.16		D	12	67		a	1	142
602	604	Budišov	Pod hradem Cave	UM2	R	Exact	47.92	47.66	47.99	Exact	23	23.16	23.41	III	D	12	66	D		1	143
613	615	Budišov	Pod hradem Cave	UM2	R	Exact	48.73	48.73	48.41	Exact	23.26	23.26	23.14	II	D	12	65	E		1	145
633	635	Budišov	Pod hradem Cave	UM2	L	Exact	49.71	49.73	49.7	Exact	26.09	25.8	25.8	III	D	12	61A	A		1	149
517	519	Budišov	Pod hradem Cave	C		Exact	32.39	32.63	32.61	Exact	23.96	24.4	24.13		D	13	76	C		1	130
520	522	Budišov	Pod hradem Cave	C		Exact	23.64	23.63	23.62	Exact	14.97	15.03	15.13		D	13	76	E		1	131
521	523	Budišov	Pod hradem Cave	C		Exact	23.96	23.94	23.79	Exact	16.96	17.02	17.09		M	13	75	B		1	131
528	530	Budišov	Pod hradem Cave	C		Exact	22.7	22.1	22.16	Exact	18.18	18.22	18.04		D	13	74	B		1	132
529	531	Budišov	Pod hradem Cave	C		Exact	27.42	27.2	27.33	Exact	20.92	21.2	20.5		D	13	74	E		1	132
530	532	Budišov	Pod hradem Cave	C		Exact	26.48	27.48	27.08	Exact	20.39	20.48	20.2		D	13	74	E		1	132
531	533	Budišov	Pod hradem Cave	C		No measurement				Close estimate	16.6	16.6	16.6		D	13	74	E		1	133
537	539	Budišov	Pod hradem Cave	C		Exact	22.55	22.53	22.29	Exact	19.94	20.02	20.04		D	13	73	C		1	134
544	546	Budišov	Pod hradem Cave	C		Exact	20.9	20.87	20.34	Exact	15.61	15.62	15.71		D	13	73		a	1	135
546	548	Budišov	Pod hradem Cave	C		Exact	24.67	24.13	24.09	Exact	21.62	21.62	21.19		D	13	73		b	1	135
549	551	Budišov	Pod hradem Cave	C		Exact	24.63	24.83	24.57	Exact	19.8	19.82	19.85		D	13	72	A		1	136
550	552	Budišov	Pod hradem Cave	C		No measurement				Exact	22.79	23.52	22.95		D	13	72	C	a	1	136
561	563	Budišov	Pod hradem Cave	C		Exact	34.07	34.07	34.05	Exact	24.87	24.63	24.85		M	13	70	A		1	138
566	568	Budišov	Pod hradem Cave	C		Close estimate	26.21	26.21	26.21	Close estimate	16.27	16.27	16.27		D	13	70		a	1	138
516	518	Budišov	Pod hradem Cave	C (J)		Exact	21.13	21.13	21.13	Exact	15.59	15.59	15.59		D	13	76	C		1	130
519	521	Budišov	Pod hradem Cave	C (J)		Exact	20.7	20.89	20.39	Exact	15.31	15.4	15.41		D	13	76	C		1	130
522	524	Budišov	Pod hradem Cave	C (J)		Exact	22.37	21.95	21.51	Exact	15.71	15.74	15.76		D	13	75	B		1	131
532	534	Budišov	Pod hradem Cave	C (J)		Exact	20.1	20.1	20.1	Exact	15	15	15		D	13	74	E		1	133
551	553	Budišov	Pod hradem Cave	C (J)		No measurement				No measurement					M	13	72	E		1	136
555	557	Budišov	Pod hradem Cave	C (J)		Exact	20.21	20.21	20.21	Exact	15.14	15.14	15.14		D	13	71	A		1	137

Table. A1.1. (Continued.)

ID	Tooth ID	Storage ID	Site ID	Tooth Type ID	Left/Right	Measurement Type Length	Length1 (mm)	Length2 (mm)	Length3 (mm)	Measurement Type Breadth	Breadth1 (mm)	Breadth2 (mm)	Breadth3 (mm)	Wear Stage	Articulation ID	Layer	Box Number	Medium Box Number Suffix	Small Box Number Suffix	Log Book Number	Log Book Page Number
525	527	Budišov	Pod hradem Cave	LM1	L	Exact	31.25	31.41	31.27	Exact	13.49	13.94	13.93	I	D	13	75	C		1	131
543	545	Budišov	Pod hradem Cave	LM1	R	No measurement				Exact	14.11	14.27	14.15	II	D	13	73	D		1	134
563	565	Budišov	Pod hradem Cave	LM1	R	Exact	29.88	29.87	29.88	Exact	12.71	12.74	12.74	I	D	13	70	B		1	138
518	520	Budišov	Pod hradem Cave	LM2	R	Exact	31.38	31.57	31.41	Exact	17.84	17.18	17.05	VIII	D	13	76	C		1	130
527	529	Budišov	Pod hradem Cave	LM2	R	Exact	29.39	29.3	29.38	Exact	19.47	19.48	19.48	IV	M	13	74	B		1	132
534	536	Budišov	Pod hradem Cave	LM2	R	Exact	31.19	31.09	31.15	Exact	18.64	18.53	18.64	II	D	13	74		b	1	133
538	540	Budišov	Pod hradem Cave	LM2	L	Exact	32.7	32.66	32.72	Exact	17.82	17.84	17.83	II	D	13	73	C		1	134
539	541	Budišov	Pod hradem Cave	LM2	L	Exact	33.08	33.03	33.08	Exact	19.2	19.29	19.18	II	D	13	73	A		1	134
547	549	Budišov	Pod hradem Cave	LM2	L	Exact	30.66	30.79	30.82	Exact	17.85	17.8	17.8	II	D	13	73		b	1	135
548	550	Budišov	Pod hradem Cave	LM2	R	No measurement				Exact	18.53	18.54	18.54	II	D	13	73		d	1	135
554	556	Budišov	Pod hradem Cave	LM2	L	Exact	31.19	31.27	31.25	Exact	18.6	18.59	18.6	I	D	13	71	A		1	137
557	559	Budišov	Pod hradem Cave	LM2	L	Exact	32.09	32.16	32.2	Exact	18.57	18.5	18.47	II	D	13	71	C		1	137
559	561	Budišov	Pod hradem Cave	LM2	R	Exact	29.69	29.67	29.77	Exact	16.39	16.33	16.34	II	D	13	71		b	1	137
564	566	Budišov	Pod hradem Cave	LM2	R	Exact	30.28	30.3	30.29	Exact	17.36	17.38	17.45	III	D	13	70	C		1	138
524	526	Budišov	Pod hradem Cave	LM3	L	Exact	27.92	27.93	27.92	Exact	19.36	19.34	19.35	IV	D	13	75	D		1	131
526	528	Budišov	Pod hradem Cave	LM3	R	Exact	30.27	30.24	30.24	Exact	18.81	18.74	18.79	IV	M	13	74	B		1	132
545	547	Budišov	Pod hradem Cave	LM3	R	Exact	27.52	27.53	27.53	Exact	20.32	20.62	20.59	VII	D	13	73		c	1	135
552	554	Budišov	Pod hradem Cave	LM3	R	Exact	28.81	28.65	28.67	Exact	19.98	20.04	20.05	II	D	13	72	E		1	136
562	564	Budišov	Pod hradem Cave	LM3	L	Exact	31.03	31.02	31.03	Exact	19.63	19.57	19.58	VII	D	13	70	A		1	138
649	651	Budišov	Pod hradem Cave	LM3	R	Exact	30.25	30.26	30.24	Exact	20.91	20.63	20.94	III	D	13	60	B		1	151
651	653	Budišov	Pod hradem Cave	LM3	L	Exact	30.3	30.23	30.29	Exact	22.02	21.88	21.98	II	D	13	60	E		1	152
514	516	Budišov	Pod hradem Cave	UM1	R	Exact	27.53	27.63	27.5	Exact	18.86	18.66	18.67	VIII	C	13	76	D		1	130
553	555	Budišov	Pod hradem Cave	UM1	R	Exact	30.07	30.28	30.14	Exact	18.68	18.65	18.84	II	D	13	71	A		1	137
558	560	Budišov	Pod hradem Cave	UM1	L	Exact	29.91	29.99	29.91	Exact	18.04	18.14	18.11	II	D	13	71	C		1	137
560	562	Budišov	Pod hradem Cave	UM1	R	No measurement				Exact	20.16	19.72	19.93	II	D	13	71		b	1	137
648	650	Budišov	Pod hradem Cave	UM1	R	Exact	30.12	30.16	30.19	Exact	18.12	18.15	18.13	II	D	13	60	A		1	151
513	515	Budišov	Pod hradem Cave	UM2	L	Exact	43.96	43.94	43.95	Exact	22.49	22.5	22.42	VII	D	13	76	D		1	130
515	517	Budišov	Pod hradem Cave	UM2	R	Exact	41.56	41.46	41.46	Exact	23.29	23.14	23.24	VIII	C	13	76	D		1	130
523	525	Budišov	Pod hradem Cave	UM2	R	Exact	43.68	43.7	43.69	Exact	24.11	24.09	24.42	II	D	13	75	C		1	131
533	535	Budišov	Pod hradem Cave	UM2	R	Exact	42.8	42.81	42.82	Exact	20.71	20.41	20.15	II	D	13	74	E		1	133
535	537	Budišov	Pod hradem Cave	UM2	R	Exact	49.27	49.19	49.18	Exact	23.62	23.73	23.12	II	D	13	74		a	1	133

Table. A1.1. (Continued.)

ID	Tooth ID	Storage ID	Site ID	Tooth Type ID	Left/Right	Measurement Type Length	Length1 (mm)	Length2 (mm)	Length3 (mm)	Measurement Type Breadth	Breadth1 (mm)	Breadth2 (mm)	Breadth3 (mm)	Wear Stage	Articulation ID	Layer	Box Number	Medium Box Number Suffix	Small Box Number Suffix	Log Book Number	Log Book Page Number
536	538	Budišov	Pod hradem Cave	UM2	L	No measurement				No measurement				II	D	13	73	A		1	134
540	542	Budišov	Pod hradem Cave	UM2	R	Exact	44.52	44.51	44.51	Exact	22.84	23.09	22.75	II	D	13	73	E		1	134
541	543	Budišov	Pod hradem Cave	UM2	L	Exact	42.79	42.9	42.9	Exact	20.9	21.03	20.93	II	D	13	73	D		1	134
542	544	Budišov	Pod hradem Cave	UM2	L	Exact	47.72	47.69	47.59	Exact	23.75	23.18	23.76	III	D	13	73	D		11	134
556	558	Budišov	Pod hradem Cave	UM2	R	Exact	52.01	51.89	51.99	Exact	23.94	23.99	23.76	VIII	C	13	71	B		1	137
565	567	Budišov	Pod hradem Cave	UM2	L	Exact	47.06	47.06	47.12	Exact	24.72	24.86	24.66	II	D	13	70	D		1	138
650	652	Budišov	Pod hradem Cave	UM2	R	Exact	46.01	45.98	45.95	Exact	22.75	22.43	22.39	II	D	13	60	D		1	152
507	509	Budišov	Pod hradem Cave	C		Exact	26.51	27.1	27.04	Exact	19.78	19.77	19.79		D	13A	78	A		1	128
508	510	Budišov	Pod hradem Cave	C		Exact	26.88	26.89	26.88	Exact	20.17	19.96	19.96		D	13A	78	E		1	128
510	512	Budišov	Pod hradem Cave	LM2	R	Exact	28.7	28.65	28.66	Exact	17.2	17.2	17.16	III	M	13A	78	E		1	129
509	511	Budišov	Pod hradem Cave	LM3	R	Exact	26.55	26.1	26.14	Exact	19.18	19.37	19.29	III	M	13A	78	E		1	129
506	508	Budišov	Pod hradem Cave	UM2	R	Exact	49.15	48.79	49.18	Exact	23.54	23.32	23.39	II	D	13A	78		b	1	128
496	498	Budišov	Pod hradem Cave	C		Exact	34.06	34.56	34.63	Exact	23.06	23.33	23.4		D	13B	79	C		1	127
497	499	Budišov	Pod hradem Cave	C		Exact	20.31	20.46	20.53	Exact	15.32	15.41	15.42		D	13B	79	C		1	127
498	500	Budišov	Pod hradem Cave	C		Exact	27.44	27.3	27.44	Exact	21.18	21.35	21.41		D	13B	79	C		1	127
501	503	Budišov	Pod hradem Cave	C		Exact	20.73	20.8	20.81	Exact	16.38	16.4	16.42		D	13B	79	E		1	127
502	504	Budišov	Pod hradem Cave	C		Exact	21.96	21.97	22.17	Exact	16.58	16.68	16.59		D	13B	79	E		1	127
500	502	Budišov	Pod hradem Cave	C (J)		Exact	23.56	23.7	23.8	Exact	18.93	18.98	18.95		D	13B	79	C		1	127
503	505	Budišov	Pod hradem Cave	LM2	R	Exact	29.94	29.96	29.98	Exact	19	19.03	19.06	IV	D	13B	79	E		1	127
495	497	Budišov	Pod hradem Cave	UM1	L	Exact	29.83	29.79	29.88	Exact	18.25	18.75	18.55	III	D	13B	79	B		1	127
499	501	Budišov	Pod hradem Cave	UM1	R	Exact	30.35	30.56	30.6	Exact	18.21	18.78	18.26	II	D	13B	79	C		1	127
479	481	Budišov	Pod hradem Cave	C		Exact	28.96	29.47	29.48	Exact	19.74	20.29	19.88		D	13C	80	B		1	124
481	483	Budišov	Pod hradem Cave	C		Exact	21.29	20.81	20.73	Exact	16.58	16.42	16.83		D	13C	80	C		1	125
483	485	Budišov	Pod hradem Cave	LM1	L	Exact	30.47	30.49	30.5	Exact	15.47	15.09	15.74	IX	D	13C	80	C		1	125
485	487	Budišov	Pod hradem Cave	LM1	L	Exact	29.1	29.11	29.1	Exact	13.84	13.82	14.04	II	D	13C	80	A		1	125
504	506	Budišov	Pod hradem Cave	LM1	L	Exact	32.08	32.19	32.09	Exact	15.75	15.26	15.27	II	D	13C	79	E	b	1	128
486	488	Budišov	Pod hradem Cave	LM3	L	Exact	27.71	27.73	27.69	Exact	20.4	20.36	20.29	II	D	13C	80	A		1	125
480	482	Budišov	Pod hradem Cave	UM1	L	Exact	30.07	30.03	30.05	Exact	17.87	17.89	18.1	II	D	13C	80	B		1	124
482	484	Budišov	Pod hradem Cave	UM1	R	Exact	26.4	26.39	26.38	Exact	18	18.05	17.98	IX	D	13C	80	C		1	125
487	489	Budišov	Pod hradem Cave	UM1	R	Exact	24.96	25.07	25.01	Exact	15.15	15.17	15.19	II	D	13C	80	A		1	125
484	486	Budišov	Pod hradem Cave	UM2	R	Exact	48.64	48.6	48.62	Exact	24.39	24.24	24.3	II	D	13C	80	A		1	125

Table. A1.1. (Continued.)

ID	Tooth ID	Storage ID	Site ID	Tooth Type ID	Left/Right	Measurement Type Length	Length1 (mm)	Length2 (mm)	Length3 (mm)	Measurement Type Breadth	Breadth1 (mm)	Breadth2 (mm)	Breadth3 (mm)	Wear Stage	Articulation ID	Layer	Box Number	Medium Box Number Suffix	Small Box Number Suffix	Log Book Number	Log Book Page Number
505	507	Budišov	Pod hradem Cave	UM2	R	Exact	45.76	45.8	45.76	Exact	23.62	23.62	23.59	II	D	13C	79	E	b	1	128
452	454	Budišov	Pod hradem Cave	C		Exact	31.2	31.94	31.52	Exact	23.57	23.4	23.08		D	14	84	B		1	119
454	456	Budišov	Pod hradem Cave	C		Exact	33.64	33.21	33.51	Exact	24.71	24.84	24.36		M	14	84	C		1	120
455	457	Budišov	Pod hradem Cave	C		Exact	21.32	21.73	21.23	Exact	16.33	16.32	16.28		D	14	84	D		1	120
456	458	Budišov	Pod hradem Cave	C		Exact	31.09	30.41	30.45	Exact	22.43	23.16	22.48		D	14	84	D		1	120
463	465	Budišov	Pod hradem Cave	C		Exact	32.11	32.7	32.63	Exact	23.24	23.04	23.21		D	14	83		d	1	121
470	472	Budišov	Pod hradem Cave	C		Close estimate	27.55	27.55	27.55	Close estimate	21.34	21.99	21.81		D	14	81	A		1	123
471	473	Budišov	Pod hradem Cave	C		Exact	30.89	30.51	29.5	Exact	15.76	15.76	15.66		D	14	81	A		1	123
472	474	Budišov	Pod hradem Cave	C		Exact	22.62	20.75	20.86	Exact	15.14	15.09	15.01		D	14	81	A		1	123
473	475	Budišov	Pod hradem Cave	C		Exact	21.45	21.62	21.78	No measurement					D	14	81	A		1	123
474	476	Budišov	Pod hradem Cave	C		Exact	20.31	20.64	20.66	Exact	14.87	14.91	14.94		D	14	81	A		1	123
492	494	Budišov	Pod hradem Cave	C		Exact	20.44	20.05	20.33	Exact	15.06	15.1	15.02		D	14	80		b	1	126
494	496	Budišov	Pod hradem Cave	C		Exact	30.85	30.4	30.41	Exact	22.08	22.41	22.31		D	14	80		a	1	126
462	464	Budišov	Pod hradem Cave	C (J)		Exact	27.28	27.21	27.38	Exact	21.85	21.68	21.99		D	14	83	D	a	1	121
469	471	Budišov	Pod hradem Cave	C (J)		Exact	18.9	18.9	18.9	Exact	14.93	14.93	14.93		D	14	81	A		1	123
453	455	Budišov	Pod hradem Cave	LM1	R	Exact	32.15	32.15	32.15	Exact	14.85	14.9	14.9	I	D	14	84	B		1	119
458	460	Budišov	Pod hradem Cave	LM1	R	Exact	27.39	27.4	27.38	Exact	12.95	12.81	12.76	II	D	14	84	D		1	120
459	461	Budišov	Pod hradem Cave	LM2	R	Exact	30.34	30.39	30.26	Exact	17.06	17.35	17.13	II	D	14	84	D		1	120
460	462	Budišov	Pod hradem Cave	LM2	R	Close estimate	24.97	24.97	24.97	Close estimate	14.98	14.98	14.98	I	M	14	84	D		1	120
461	463	Budišov	Pod hradem Cave	LM2	L	Exact	32.02	32.42	32.6	Exact	20.08	20.81	20.85	IV	D	14	83	A		1	121
477	479	Budišov	Pod hradem Cave	LM2	R	Exact	33.45	33.5	33.49	Exact	19.32	19.43	19.46	II	D	14	81	B		1	124
478	480	Budišov	Pod hradem Cave	LM2	R	Exact	34.18	34.02	34.3	Exact	19.97	19.98	19.94	II	D	14	81	B		1	124
493	495	Budišov	Pod hradem Cave	LM3	R	Exact	28.26	28.28	28.28	Exact	20.38	20.4	20.43	II	D	14	80		b	1	126
457	459	Budišov	Pod hradem Cave	UM1	R	Exact	29.2	29.39	29.09	Exact	18.75	18.83	18.81	III	D	14	84	D		1	120
466	468	Budišov	Pod hradem Cave	UM1	L	Exact	27.88	27.62	27.68	Exact	20.08	19.5	19.7	IX	C	14	82			1	122
468	470	Budišov	Pod hradem Cave	UM1	R	Exact	29.86	29.55	29.69	Exact	20.09	20.11	20.53	IX	C	14	82			1	122
476	478	Budišov	Pod hradem Cave	UM1	L	Exact	26.32	26.77	26.87	Exact	18.9	19.02	19.15	IX	C	14	81	A		1	123
488	490	Budišov	Pod hradem Cave	UM1	R	Exact	28.02	28.11	28.15	Exact	17.19	17.13	17.33	III	D	14	80	E		1	125
464	466	Budišov	Pod hradem Cave	UM2	L	Exact	43.43	43.42	43.36	Exact	21.51	21.44	21.01	II	D	14	83		d	1	121
465	467	Budišov	Pod hradem Cave	UM2	L	Exact	48.78	48.67	48.5	Exact	21.04	21.88	21.11	IX	C	14	82			1	122
467	469	Budišov	Pod hradem Cave	UM2	R	Exact	46.95	47.99	48.01	Exact	22.35	22.29	22.33	IX	C	14	82			1	122

Table. A1.1. (Continued.)

ID	Tooth ID	Storage ID	Site ID	Tooth Type ID	Left/Right	Measurement Type Length	Length1 (mm)	Length2 (mm)	Length3 (mm)	Measurement Type Breadth	Breadth1 (mm)	Breadth2 (mm)	Breadth3 (mm)	Wear Stage	Articulation ID	Layer	Box Number	Medium Box Number Suffix	Small Box Number Suffix	Log Book Number	Log Book Page Number
475	477	Budišov	Pod hradem Cave	UM2	L	Exact	42.82	43.15	43.7	Exact	23.92	23.27	23.15	IX	C	14	81	A		1	123
489	491	Budišov	Pod hradem Cave	UM2	R	Exact	48.07	48.08	48.08	Exact	23.36	23.68	23.54	III	D	14	80		c	1	126
490	492	Budišov	Pod hradem Cave	UM2	L	Exact	45.92	45.98	46	Exact	24.57	24.01	24.4	II	D	14	80		c	1	126
491	493	Budišov	Pod hradem Cave	UM2	R	Exact	48.1	48.13	48.14	Exact	24.08	24.11	24.12	II	D	14	80		c	1	126
314	316	Budišov	Pod hradem Cave	C		No measurement				No measurement					D	15	99	B		1	96
317	319	Budišov	Pod hradem Cave	C		No measurement				No measurement					D	15	99	C		1	97
324	326	Budišov	Pod hradem Cave	C		Exact	27.97	28.95	28.91	Exact	20.77	20.45	20.58		D	15	98	C		1	98
327	329	Budišov	Pod hradem Cave	C		Exact	26.02	26.36	26.23	Exact	20.79	20.43	20.51		D	15	98	E		1	98
328	330	Budišov	Pod hradem Cave	C		Exact	28.87	28.63	28.53	Exact	20.95	20.91	20.97		D	15	98	F		1	99
340	342	Budišov	Pod hradem Cave	C		Exact	28.41	28.87	28.82	Exact	20.88	20.74	20.77		M	15	96	A		1	101
341	343	Budišov	Pod hradem Cave	C		Close estimate	29.79	29.79	29.79	Close estimate	22.08	22.08	22.08		D	15	96	A		1	101
345	347	Budišov	Pod hradem Cave	C		Close estimate	47.49	47.52	47.49	No measurement					D	15	96	E		1	102
347	349	Budišov	Pod hradem Cave	C		Close estimate	23.49	23.49	23.49	No measurement					D	15	96	E		1	102
350	352	Budišov	Pod hradem Cave	C		Exact	21.08	20.81	21.08	Exact	15.62	15.7	15.87		D	15	95	B		1	103
353	355	Budišov	Pod hradem Cave	C		No measurement				No measurement					D	15	95	B		1	103
357	359	Budišov	Pod hradem Cave	C		Exact	28.18	28.14	28.15	Exact	21.38	21.63	21.87		D	15	95	D		1	104
358	360	Budišov	Pod hradem Cave	C		Exact	26.7	26.54	26.57	Exact	18.8	18.76	18.84		D	15	95	D		1	104
359	361	Budišov	Pod hradem Cave	C		Exact	30.92	32.01	31.82	Exact	21.88	21.83	21.96		D	15	95	E		1	104
364	366	Budišov	Pod hradem Cave	C		No measurement				No measurement					D	15	95		a	1	105
368	370	Budišov	Pod hradem Cave	C		Exact	26.8	26.93	26.84	Exact	20.45	20.84	20.54		M	15	94	C		1	106
369	371	Budišov	Pod hradem Cave	C		No measurement				Close estimate	21.1	21.1	21.1		D	15	94	C		1	106
377	379	Budišov	Pod hradem Cave	C		No measurement				Close estimate	20.19	20.19	20.19		M	15	94		b	1	106
379	381	Budišov	Pod hradem Cave	C		Exact	23.31	23.38	23.3	Exact	20.45	20.28	19.99		D	15	93	A		1	107
380	382	Budišov	Pod hradem Cave	C		Exact	24.35	24.45	24.23	Exact	20.5	20.31	20.24		D	15	93	A		1	107
384	386	Budišov	Pod hradem Cave	C		No measurement				No measurement					D	15	93	E		1	108
385	387	Budišov	Pod hradem Cave	C		Exact	20.85	21.19	21.55	Exact	16.12	16.18	16.17		D	15	93	E		1	108
389	391	Budišov	Pod hradem Cave	C		Exact	27.47	27.77	27.84	Exact	19.63	20.49	19.98		D	15	93		b	1	108
390	392	Budišov	Pod hradem Cave	C		Exact	28.11	27.9	28.24	Exact	21.6	21.58	21.77		D	15	93	E		1	108
391	393	Budišov	Pod hradem Cave	C		Exact	22.72	22.42	22.87	Exact	15.43	15.98	15.56		D	15	92	E		1	109
402	404	Budišov	Pod hradem Cave	C		Close estimate	28.2	28.2	28.2	Close estimate	23.62	23.62	23.62		D	15	91	E		1	110
412	414	Budišov	Pod hradem Cave	C		Exact	30.27	30.74	30.14	Exact	23.46	23.04	22.89		D	15	89	E		1	112

Table. A1.1. (Continued.)

ID	Tooth ID	Storage ID	Site ID	Tooth Type ID	Left/Right	Measurement Type Length	Length1 (mm)	Length2 (mm)	Length3 (mm)	Measurement Type Breadth	Breadth1 (mm)	Breadth2 (mm)	Breadth3 (mm)	Wear Stage	Articulation ID	Layer	Box Number	Medium Box Number Suffix	Small Box Number Suffix	Log Book Number	Log Book Page Number
413	415	Budišov	Pod hradem Cave	C		Exact	21.74	22.57	22.3	Exact	16.54	16.26	16.44		D	15	90		d	1	112
415	417	Budišov	Pod hradem Cave	C		Exact	30.61	30.61	30.61	Exact	21.65	21.48	21.59		D	15	89	E	a	1	113
416	418	Budišov	Pod hradem Cave	C		Exact	30.45	30.21	29.92	Exact	22.14	22.14	22.19		D	15	89	E	a	1	113
420	422	Budišov	Pod hradem Cave	C		No measurement				No measurement					D	15	89	F	a	1	113
424	426	Budišov	Pod hradem Cave	C		Exact	27.36	27.25	27.42	Exact	21.15	20.03	20.87		D	15	87	B		1	115
425	427	Budišov	Pod hradem Cave	C		Exact	28.42	28.82	28.78	Exact	20.58	20.35	20.48		D	15	87	C		1	115
427	429	Budišov	Pod hradem Cave	C		Close estimate	22.6	22.6	22.6	Close estimate	15.77	15.77	15.77		D	15	87	E		1	115
429	431	Budišov	Pod hradem Cave	C		Exact	26.69	27.17	27.05	Exact	19.98	20.1	20.81		D	15	86	B		1	116
435	437	Budišov	Pod hradem Cave	C		Exact	29.02	29.36	29.36	Exact	22.18	21.95	22.29		D	15	86		c	1	117
436	438	Budišov	Pod hradem Cave	C		Exact	20.37	20.23	20.63	Exact	14.82	14.94	14.87		D	15	86		c	1	117
441	443	Budišov	Pod hradem Cave	C		Exact	24.21	23.82	24.58	Exact	17.53	17.56	17.61		C	15	86		b	1	117
442	444	Budišov	Pod hradem Cave	C		No measurement				No measurement					M	15	86		e	1	117
342	344	Budišov	Pod hradem Cave	C (J)		Exact	24.96	24.99	24.96	Exact	21.47	21.31	21.31		D	15	96	A		1	101
400	402	Budišov	Pod hradem Cave	C (J)		Exact	28.27	27.84	28.22	Exact	20.93	20.63	20.17		D	15	91	D		1	110
319	321	Budišov	Pod hradem Cave	LM1	R	No measurement				Exact	14.07	14.07	14.07	II	D	15	99	D		1	97
320	322	Budišov	Pod hradem Cave	LM1	L	Exact	31.45	31.46	31.44	Exact	13.67	13.64	13.65	I	D	15	99	D		1	97
321	323	Budišov	Pod hradem Cave	LM1	L	Exact	29	29.12	29.11	Exact	14.23	14.04	13.92	V	D	15	99	D		1	97
331	333	Budišov	Pod hradem Cave	LM1	L	Exact	31.43	31.37	31.34	Exact	15.3	15.47	15.51	III	D	15	98	F		1	99
337	339	Budišov	Pod hradem Cave	LM1	L	Exact	27.74	27.68	27.69	Exact	13.11	13.18	13.06	II	D	15	97	D		1	100
338	340	Budišov	Pod hradem Cave	LM1	L	Exact	30.37	30.32	30.35	Exact	14.37	14.4	14.5	III	D	15	97	E		1	101
348	350	Budišov	Pod hradem Cave	LM1	R	Exact	22.12	22.42	22.35	Exact	12.9	13.06	12.99	I	D	15	96		a	1	102
354	356	Budišov	Pod hradem Cave	LM1	L	Exact	32.01	32.02	32.04	Close estimate	13.7	13.7	13.7	I	M	15	95	D		1	104
355	357	Budišov	Pod hradem Cave	LM1	L	Exact	31.09	31.07	31.11	Exact	13.16	13.09	13.18	I	D	15	95	D		1	104
356	358	Budišov	Pod hradem Cave	LM1	R	Exact	30.2	30.14	30.18	Exact	13.89	13.84	13.88	IX	D	15	95	D		1	104
365	367	Budišov	Pod hradem Cave	LM1	L	Exact	30.56	30.67	30.59	Exact	14.46	14.64	14.34	IX	D	15	95		c	1	105
367	369	Budišov	Pod hradem Cave	LM1	L	Exact	32.12	32.12	32.1	Exact	14.51	14.91	14.85	IX	D	15	94	D		1	105
374	376	Budišov	Pod hradem Cave	LM1	L	Exact	32.53	32.46	32.5	Exact	14.21	14.22	14.35	VII	D	15	94	C		1	106
383	385	Budišov	Pod hradem Cave	LM1	R	Exact	30.6	30.67	30.47	Exact	14.63	14.62	14.53	V	M	15	93	C		1	107
387	389	Budišov	Pod hradem Cave	LM1	L	Exact	33.58	33.39	33.41	Exact	15.31	15.63	15.48	II	D	15	93	E		1	108
388	390	Budišov	Pod hradem Cave	LM1	L	Exact	29.16	29.14	29.13	Exact	14.46	14.7	14.9	III	D	15	93	E		1	108
398	400	Budišov	Pod hradem Cave	LM1	R	Exact	31.83	31.82	31.83	Exact	14.29	14.25	14.3	I	D	15	91	C		1	110

Table. A1.1. (Continued.)

ID	Tooth ID	Storage ID	Site ID	Tooth Type ID	Left/Right	Measurement Type Length	Length1 (mm)	Length2 (mm)	Length3 (mm)	Measurement Type Breadth	Breadth1 (mm)	Breadth2 (mm)	Breadth3 (mm)	Wear Stage	Articulation ID	Layer	Box Number	Medium Box Number Suffix	Small Box Number Suffix	Log Book Number	Log Book Page Number
405	407	Budišov	Pod hradem Cave	LM1	L	Close estimate	28.75	28.75	28.75	Exact	20.4	20.84	20.99	II	D	15	90	A		1	111
409	411	Budišov	Pod hradem Cave	LM1	L	Exact	29.85	29.86	29.76	Exact	14.44	14.44	14.44	II	D	15	90		d	1	112
411	413	Budišov	Pod hradem Cave	LM1	R	Exact	29.83	29.83	29.83	Exact	12.54	12.54	12.54	I	M	15	89	D		1	112
414	416	Budišov	Pod hradem Cave	LM1	L	Exact	32.16	32.02	32.12	Exact	14.93	15.13	15.07	III	D	15	89	E	a	1	113
419	421	Budišov	Pod hradem Cave	LM1	R	Exact	28.83	28.81	28.81	Exact	13.19	13.48	13.44	II	D	15	89	F	a	1	113
421	423	Budišov	Pod hradem Cave	LM1	R	Exact	29.59	29.59	29.59	Exact	13.65	13.66	13.58	III	D	15	88		a	1	114
423	425	Budišov	Pod hradem Cave	LM1	R	Exact	30.1	30.1	30.07	Exact	13.35	13.76	13.66	II	D	15	87	A		1	114
428	430	Budišov	Pod hradem Cave	LM1	L	Exact	32.86	32.86	32.86	Close estimate	14.01	14.01	14.01	I	M	15	87	E		1	115
431	433	Budišov	Pod hradem Cave	LM1	R	Exact	30.19	30.29	30.24	Exact	13.69	13.66	13.68	II	D	15	86	C		1	116
444	446	Budišov	Pod hradem Cave	LM1	R	Exact	31.84	32.05	31.93	Exact	15.27	15.6	15.45	II	D	15	86		e	1	117
445	447	Budišov	Pod hradem Cave	LM1	L	Exact	29.09	29.87	29.98	Exact	12.88	13.43	12.53	IX	D	15	85	B		1	118
313	315	Budišov	Pod hradem Cave	LM2	R	Exact	28.7	28.89	28.96	Exact	16.66	16.52	16.57	III	D	15	99	B		1	96
316	318	Budišov	Pod hradem Cave	LM2	L	Exact	33.03	32.86	32.87	Exact	20.19	20.15	20.22	VIII	D	15	99	C		1	97
323	325	Budišov	Pod hradem Cave	LM2	L	Exact	30.7	31	31.08	Exact	18.61	18.32	18.41	IX	D	15	98	B		1	98
326	328	Budišov	Pod hradem Cave	LM2	L	Exact	30.78	30.84	30.78	Exact	17.26	17.41	17.38	IX	D	15	98	D		1	98
330	332	Budišov	Pod hradem Cave	LM2	L	Exact	29.59	29.59	29.59	Exact	17.26	17.43	17.48	I	D	15	98	F		1	99
332	334	Budišov	Pod hradem Cave	LM2	R	Exact	31.85	31.63	31.88	Exact	19.02	19.05	19.03	I	D	15	97	A		1	100
335	337	Budišov	Pod hradem Cave	LM2	R	Exact	33.47	33.53	33.59	Exact	18.73	18.83	18.76	II	D	15	97	B		1	100
339	341	Budišov	Pod hradem Cave	LM2	R	No measurement				Exact	20.77	20.65	20.66	IX	D	15	97	E		1	101
343	345	Budišov	Pod hradem Cave	LM2	R	Exact	29.29	29.26	29.29	Exact	18.13	17.92	17.67	II	D	15	96	B		1	102
344	346	Budišov	Pod hradem Cave	LM2	R	Exact	32.37	32.31	32.26	Exact	19.27	19.05	19.15	IV	D	15	96	D		1	102
352	354	Budišov	Pod hradem Cave	LM2	L	Exact	32.18	32.19	32.22	Exact	18.33	18.04	18.44	VII	D	15	95	B		1	103
362	364	Budišov	Pod hradem Cave	LM2	R	Exact	30.72	30.54	30.72	Exact	20.65	20.98	20.7	IX	M	15	95	E		1	104
366	368	Budišov	Pod hradem Cave	LM2	R	Exact	33.41	33.25	33.22	Exact	18.69	18.68	18.67	I	D	15	95		c	1	105
371	373	Budišov	Pod hradem Cave	LM2	R	Exact	30.67	31.16	30.92	Exact	18.39	18.45	18.42	V	D	15	94	C		1	106
373	375	Budišov	Pod hradem Cave	LM2	L	Exact	31.76	31.81	31.23	Exact	18.33	18.31	18.32	IV	D	15	94	C		1	106
376	378	Budišov	Pod hradem Cave	LM2	L	Close estimate	29.57	29.57	29.57	Exact	18.79	18.7	18.84	IX	M	15	94	E		1	106
382	384	Budišov	Pod hradem Cave	LM2	R	Exact	30.17	30.24	30.13	Exact	18.66	18.74	18.54	V	M	15	93	C		1	107
386	388	Budišov	Pod hradem Cave	LM2	R	Exact	32.03	31.74	31.75	Exact	18.56	18.6	18.75	II	D	15	93	E		1	108
392	394	Budišov	Pod hradem Cave	LM2	L	Close estimate	30.05	30.05	30.05	Exact	19.06	19.75	20.01	VIII	D	15	92	E		1	109
399	401	Budišov	Pod hradem Cave	LM2	L	No measurement				No measurement					D	15	91	C		1	110

Table. A1.1. (Continued.)

ID	Tooth ID	Storage ID	Site ID	Tooth Type ID	Left/Right	Measurement Type Length	Length1 (mm)	Length2 (mm)	Length3 (mm)	Measurement Type Breadth	Breadth1 (mm)	Breadth2 (mm)	Breadth3 (mm)	Wear Stage	Articulation ID	Layer	Box Number	Medium Box Number Suffix	Small Box Number Suffix	Log Book Number	Log Book Page Number
404	406	Budišov	Pod hradem Cave	LM2	L	Exact	31.95	32.03	32.03	Exact	18	18.22	18.18	II	D	15	91	E	a	1	110
410	412	Budišov	Pod hradem Cave	LM2	L	No measurement				Exact	18.81	18.81	18.81	II	D	15	90		d	1	112
426	428	Budišov	Pod hradem Cave	LM2	L	Exact	32.07	31.97	32.05	Exact	18.73	18.74	18.71	VI	D	15	87	E		1	115
432	434	Budišov	Pod hradem Cave	LM2	L	Exact	31.43	31.44	31.47	Exact	18.19	17.75	17.74	II	D	15	86	C		1	116
437	439	Budišov	Pod hradem Cave	LM2	L	Exact	32.61	32.48	32.8	Exact	19.15	19.06	19	II	D	15	86		d	1	117
443	445	Budišov	Pod hradem Cave	LM2	L	Exact	29.12	28.99	29.02	Exact	16.67	16.63	16.67	I	D	15	86		e	1	117
315	317	Budišov	Pod hradem Cave	LM3	R	Exact	32.87	32.8	32.86	Exact	22.07	22.07	22.07	III	M	15	99	C		1	97
322	324	Budišov	Pod hradem Cave	LM3	L	Exact	26.97	27	26.97	Exact	19.22	19.23	19.23	II	D	15	98	B		1	98
349	351	Budišov	Pod hradem Cave	LM3	R	Exact	26.72	26.47	26.73	Exact	20.99	20.74	20.88	II	D	15	95	A		1	103
363	365	Budišov	Pod hradem Cave	LM3	R	Exact	27.35	27.4	27.31	Exact	18.73	18.53	18.64	III	D	15	95		a	1	105
370	372	Budišov	Pod hradem Cave	LM3	R	Exact	27.32	27.67	27.67	Exact	20.91	21.03	20.97	V	D	15	94	C		1	106
375	377	Budišov	Pod hradem Cave	LM3	L	Exact	30.32	30.51	30.31	Exact	21.53	21.64	21.53	IX	M	15	94	E		1	106
381	383	Budišov	Pod hradem Cave	LM3	R	Exact	28.58	28.59	28.73	Exact	19.67	19.95	19.59	V	M	15	93	C		1	107
394	396	Budišov	Pod hradem Cave	LM3	R	Exact	30.83	30.71	30.65	Exact	21.88	21.72	21.81	VI	D	15	92	E		1	109
401	403	Budišov	Pod hradem Cave	LM3	L	Exact	25.41	25.4	25.4	Exact	18.53	18.77	18.93		M	15	91	E		1	110
407	409	Budišov	Pod hradem Cave	LM3	R	Exact	25.82	26.22	26.08	Exact	17.63	17.86	17.8	VII	D	15	90	C		1	111
417	419	Budišov	Pod hradem Cave	LM3	R	Exact	27.86	27.82	27.87	Exact	18.06	18	18.03	IX	D	15	89	B		1	113
418	420	Budišov	Pod hradem Cave	LM3	L	Exact	26.06	26.01	25.99	Exact	18.39	18.21	18.3	IX	D	15	89	B		1	113
439	441	Budišov	Pod hradem Cave	LM3	L	Exact	28.46	28.39	28.44	Exact	20.89	20.34	20.87	II	D	15	86		d	1	117
446	448	Budišov	Pod hradem Cave	LM3	L	Exact	28.99	29.02	29.02	Exact	18.69	18.84	18.85	II	D	15	85	B		1	118
448	450	Budišov	Pod hradem Cave	LM3	L	No measurement				No measurement				II	D	15	85	B		1	118
336	338	Budišov	Pod hradem Cave	UM1	R	Exact	29.7	29.6	29.65	Exact	18.58	18.8	18.58	II	D	15	97	D		1	100
361	363	Budišov	Pod hradem Cave	UM1	R	Exact	29.88	29.93	29.58	Exact	18.22	18.62	18.07	II	D	15	95	E		1	104
372	374	Budišov	Pod hradem Cave	UM1	L	Exact	29.43	29.53	29.43	Exact	17.21	17.17	17.38	I	D	15	94	C		1	106
378	380	Budišov	Pod hradem Cave	UM1	L	Exact	29.67	29.85	30.05	Exact	19.03	19.64	19.41	II	D	15	94		b	1	106
397	399	Budišov	Pod hradem Cave	UM1	L	Exact	29.39	29.46	29.49	Exact	18.77	18.94	18.63	II	D	15	92	B		1	109
422	424	Budišov	Pod hradem Cave	UM1	R	Exact	28.86	28.9	28.9	Exact	18.4	18.6	18.64	V	D	15	87	A		1	114
450	452	Budišov	Pod hradem Cave	UM1	R	Exact	28.75	28.93	29.02	Exact	18.91	18.79	18.87	II	D	15	85		a	1	119
318	320	Budišov	Pod hradem Cave	UM2	L	Exact	49.46	49.37	49.33	Exact	22.54	22.38	22.64	II	D	15	99	D		1	97
325	327	Budišov	Pod hradem Cave	UM2	R	Exact	47.71	47.59	47.65	Exact	23.97	23.97	23.97	II	D	15	98	C		1	98
329	331	Budišov	Pod hradem Cave	UM2	L	Close estimate	44.63	44.63	44.63	Exact	23.88	23.98	23.87	IX	D	15	98	F		1	99

Table. A1.1. (Continued.)

ID	Tooth ID	Storage ID	Site ID	Tooth Type ID	Left/Right	Measurement Type Length	Length1 (mm)	Length2 (mm)	Length3 (mm)	Measurement Type Breadth	Breadth1 (mm)	Breadth2 (mm)	Breadth3 (mm)	Wear Stage	Articulation ID	Layer	Box Number	Medium Box Number Suffix	Small Box Number Suffix	Log Book Number	Log Book Page Number
333	335	Budišov	Pod hradem Cave	UM2	L	Exact	46.23	46.82	46.35	Exact	22.59	22.6	22.64	II	D	15	97	B		1	100
334	336	Budišov	Pod hradem Cave	UM2	R	Exact	49.53	49.34	49.54	Exact	23.73	23.13	22.9	II	D	15	97	B		1	100
346	348	Budišov	Pod hradem Cave	UM2	L	Exact	47.49	47.52	47.49	Exact	21.99	21.84	22.03	II	D	15	96	E		1	102
351	353	Budišov	Pod hradem Cave	UM2	L	Exact	42.66	42.39	42.41	Exact	22.98	23	23.18	VI	D	15	95	B		1	103
360	362	Budišov	Pod hradem Cave	UM2	L	Exact	36.19	35.96	36.22	Exact	18.25	19.02	19	II	D	15	95	E		1	104
393	395	Budišov	Pod hradem Cave	UM2	R	Exact	47.11	47.24	47.07	Exact	21.95	22.51	22.82	II	D	15	92	E		1	109
395	397	Budišov	Pod hradem Cave	UM2	R	Estimate	46.27	46.27	46.27	Close estimate	23.77	23.77	23.77		D	15	92		b	1	109
396	398	Budišov	Pod hradem Cave	UM2	L	Exact	45.87	45.9	45.89	Exact	24.82	24.5	24.4	III	D	15	91	B		1	109
403	405	Budišov	Pod hradem Cave	UM2	R	Exact	45.8	45.77	45.85	Exact	21.86	21.47	21.94	III	D	15	91	E	a	1	110
406	408	Budišov	Pod hradem Cave	UM2	R	Exact	42.12	42.11	42.12	Exact	20.4	20.84	20.99	II	D	15	90	B		1	111
408	410	Budišov	Pod hradem Cave	UM2	L	Exact	48.91	48.67	48.95	Exact	23.88	23.7	23.97	II	D	15	90	D		1	111
430	432	Budišov	Pod hradem Cave	UM2	R	Exact	44.03	44.01	44.06	Exact	20.97	21.55	21.66	II	D	15	86	B		1	116
433	435	Budišov	Pod hradem Cave	UM2	L	Exact	43.89	43.75	43.81	Exact	23	22.99	23	II	D	15	86	E		1	116
434	436	Budišov	Pod hradem Cave	UM2	L	Exact	41.53	41.39	41.55	Exact	20.14	20.07	20.08	II	D	15	86	E		1	116
438	440	Budišov	Pod hradem Cave	UM2	R	Close estimate	46.04	46.04	46.04	Exact	23.34	23.26	23.38	II	D	15	86		d	1	117
440	442	Budišov	Pod hradem Cave	UM2	L	No measurement				No measurement					D	15	86		d	1	117
447	449	Budišov	Pod hradem Cave	UM2	R	Exact	46.57	46.78	46.82	Exact	25.53	26.04	25.79	V	D	15	85	B		1	118
449	451	Budišov	Pod hradem Cave	UM2	R	Exact	45.9	45.87	45.95	Exact	22.27	22.19	23.32	II	D	15	85		a	1	119
451	453	Budišov	Pod hradem Cave	UM2	R	No measurement				No measurement				II	D	15	85		a	1	119
255	257	Budišov	Pod hradem Cave	C		Exact	32.34	32.58	32.49	Exact	24.65	23.96	23.82		D	16	104	B		1	82
261	263	Budišov	Pod hradem Cave	C		Exact	24.56	24.31	24.18	Exact	17.03	16.8	16.84		D	16	104	D		1	83
262	264	Budišov	Pod hradem Cave	C		Exact	23.37	23.34	23.24	Exact	16.31	16.34	16.3		D	16	104	E		1	84
270	272	Budišov	Pod hradem Cave	C		Close estimate	28.87	28.87	28.87	Exact	20.79	20.84	20.94		D	16	103	D		1	86
271	273	Budišov	Pod hradem Cave	C		Estimate	26.89	26.89	26.89	Close estimate	18.6	18.6	18.6		D	16	103	E		1	86
275	277	Budišov	Pod hradem Cave	C		Close estimate	29.13	29.13	29.13	Exact	22.3	22.89	22.34		D	16	103	E		1	87
276	278	Budišov	Pod hradem Cave	C		Close estimate	29.96	29.96	29.96	Exact	23.06	22.99	22.94		D	16	102	A		1	88
284	286	Budišov	Pod hradem Cave	C		Exact	26.05	26.41	26.4	Exact	21.22	21.81	21.8		D	16	101	E		1	90
287	289	Budišov	Pod hradem Cave	C		Exact	16.48	16.41	16.47	Exact	13.91	13.81	13.99		D	16	101		c	1	91
294	296	Budišov	Pod hradem Cave	C		Exact	28.91	29.03	29.43	Exact	20.98	20.68	20.92		D	16	1		b	1	92
296	298	Budišov	Pod hradem Cave	C	R	Exact	26.02	26.67	26.84	Exact	20.61	20.22	20.24		M	16	100	A		1	93
303	305	Budišov	Pod hradem Cave	C		Exact	20.66	20.92	21.83	Exact	14.25	14.14	14.45		D	16	100	B		1	94

Table. A1.1. (Continued.)

ID	Tooth ID	Storage ID	Site ID	Tooth Type ID	Left/Right	Measurement Type Length	Length1 (mm)	Length2 (mm)	Length3 (mm)	Measurement Type Breadth	Breadth1 (mm)	Breadth2 (mm)	Breadth3 (mm)	Wear Stage	Articulation ID	Layer	Box Number	Medium Box Number Suffix	Small Box Number Suffix	Log Book Number	Log Book Page Number
306	308	Budišov	Pod hradem Cave	C		Exact	27.2	27.76	27.24	Exact	21.64	21.51	21.6		D	16	100	D		1	94
311	313	Budišov	Pod hradem Cave	C		Exact	26.9	26.53	26.97	Exact	21.54	22.04	21.55		D	16	100		e	1	96
256	258	Budišov	Pod hradem Cave	C (J)		Exact	21.66	21.33	21.62	Exact	16.42	16.5	16.44		D	16	104	C		1	83
264	266	Budišov	Pod hradem Cave	C (J)		Exact	25.07	25.84	26.16	Exact	20.51	20.55	20.46		D	16	104	F		1	84
272	274	Budišov	Pod hradem Cave	C (J)		Exact	22.74	22.85	22.9	Exact	15.3	15.44	15.71		D	16	103	E		1	86
280	282	Budišov	Pod hradem Cave	C (J)		Exact	22.13	22.04	21.96	Exact	18.26	18.22	18.24		D	16	101	B		1	89
289	291	Budišov	Pod hradem Cave	C (J)		No measurement				No measurement					D	16	101		c	1	91
257	259	Budišov	Pod hradem Cave	LM1	L	Exact	31.92	32.12	31.9	Exact	15.11	15.15	15.27	II	D	16	104	D		1	83
258	260	Budišov	Pod hradem Cave	LM1	R	Exact	27.72	27.69	27.79	Exact	14.28	14.11	14.13	V	D	16	104	D		1	83
274	276	Budišov	Pod hradem Cave	LM1	L	Exact	29.6	29.7	29.7	Exact	13.7	13.55	13.76	II	D	16	103	E		1	87
281	283	Budišov	Pod hradem Cave	LM1	L	Exact	30.94	30.94	31.03	Exact	14.25	13.67	13.66	II	D	16	101	C		1	89
283	285	Budišov	Pod hradem Cave	LM1	R	Exact	30.31	30.31	30.31	Exact	14.46	14.8	14.89	VIII	D	16	101	D		1	90
288	290	Budišov	Pod hradem Cave	LM1	L	Exact	30.59	30.69	30.57	Exact	14.58	14.59	14.66	II	D	16	101		c	1	91
291	293	Budišov	Pod hradem Cave	LM1	R	Exact	32.06	32.11	32.07	Exact	15.08	15.43	15.71	IV	D	16	101		b	1	91
302	304	Budišov	Pod hradem Cave	LM1	R	Exact	27.72	27.79	27.73	Exact	13.25	13.25	13.25	I	D	16	100	B		1	94
307	309	Budišov	Pod hradem Cave	LM1	R	Close estimate	27.8	27.8	27.8	Close estimate	12.74	12.74	12.74	I	M	16	100		a	1	95
252	254	Budišov	Pod hradem Cave	LM2	L	Exact	30.34	30.31	30.1	Exact	17.88	18.09	18.21	II	D	16	104	A		1	82
253	255	Budišov	Pod hradem Cave	LM2	L	Exact	30.4	30.43	30.5	Exact	18.09	18.13	18.08	VI	D	16	104	A		1	82
259	261	Budišov	Pod hradem Cave	LM2	L	Exact	28.18	28.5	28.52	Exact	17.88	17.69	17.83	III	D	16	104	D		1	83
269	271	Budišov	Pod hradem Cave	LM2	L	Exact	32.57	32.51	32.53	Exact	18.81	18.91	18.94	II	D	16	103	C		1	86
277	279	Budišov	Pod hradem Cave	LM2	L	Exact	31.5	31.48	31.48	Exact	18.47	18.15	18.31	II	D	16	101	A		1	89
290	292	Budišov	Pod hradem Cave	LM2	R	Close estimate	31.98	31.98	31.98	Exact	19.56	19.72	19.67	VII	D	16	101		c	1	91
292	294	Budišov	Pod hradem Cave	LM2	L	Exact	30.32	30.38	30.31	Exact	16.85	16.96	16.97	II	D	16	101		b	1	91
300	302	Budišov	Pod hradem Cave	LM2	L	Exact	31.15	31.31	31.47	Exact	18.21	18.73	18.59	III	D	16	100	A		1	93
305	307	Budišov	Pod hradem Cave	LM2	R	Exact	32.95	33.35	33.61	Exact	20.23	20.55	20.57	IX	M	16	100	D		1	94
308	310	Budišov	Pod hradem Cave	LM2	R	Exact	30.74	30.71	30.79	Exact	18	17.93	17.84	I	D	16	100		b	1	95
266	268	Budišov	Pod hradem Cave	LM3	R	Exact	27.55	27.55	27.55	Exact	20.7	20.49	20.66	II	D	16	103	B		1	85
295	297	Budišov	Pod hradem Cave	LM3	R	Exact	28.24	28.43	28.44	Exact	21	20.87	20.84	IV	M	16	100	A		1	93
298	300	Budišov	Pod hradem Cave	LM3	L	Close estimate	27.84	27.84	27.84	Exact	20.29	20.26	20.24	III	D	16	100	A		1	93
304	306	Budišov	Pod hradem Cave	LM3	R	Exact	28.89	28.75	28.93	Exact	19.5	19.48	19.71	IX	M	16	100	D		1	94
263	265	Budišov	Pod hradem Cave	UM1	L	Exact	30.26	30.4	30.47	Exact	17.73	17.88	17.87	I	D	16	104	E		104	84

Table. A1.1. (Continued.)

ID	Tooth ID	Storage ID	Site ID	Tooth Type ID	Left/Right	Measurement Type Length	Length1 (mm)	Length2 (mm)	Length3 (mm)	Measurement Type Breadth	Breadth1 (mm)	Breadth2 (mm)	Breadth3 (mm)	Wear Stage	Articulation ID	Layer	Box Number	Medium Box Number Suffix	Small Box Number Suffix	Log Book Number	Log Book Page Number
265	267	Budišov	Pod hradem Cave	UM1	R	Exact	30.15	30.16	30.15	Exact	20.05	19.62	19.67	II	D	16	103	A		1	85
267	269	Budišov	Pod hradem Cave	UM1	L	Exact	29.38	29.51	29.66	Exact	19.88	20.13	19.88	II	D	16	103	C		1	85
268	270	Budišov	Pod hradem Cave	UM1	R	Exact	30.87	31.03	30.96	Exact	14.1	14.11	14.1	II	D	16	103	C		1	85
282	284	Budišov	Pod hradem Cave	UM1	L	Exact	29.39	29.47	29.61	Exact	18.75	18.83	18.73	II	D	16	101	D		1	90
297	299	Budišov	Pod hradem Cave	UM1	L	Exact	30.32	30.4	30.38	Exact	18.74	18.43	18.44	II	D	16	100	A		1	93
254	256	Budišov	Pod hradem Cave	UM2	R	Exact	40.92	41.06	41.02	Exact	19.44	19.32	19.2	IX	D	16	104	A		1	82
260	262	Budišov	Pod hradem Cave	UM2	L	Exact				Exact					D	16	104	D		1	83
273	275	Budišov	Pod hradem Cave	UM2	R	Exact	44.74	44.85	44.78	Exact	24.26	24.05	24.99	VI	D	16	103	E		1	87
278	280	Budišov	Pod hradem Cave	UM2	R	Exact	46.53	46.4	46.86	Exact	22.9	22.83	22.79	II	D	16	101	B		1	89
279	281	Budišov	Pod hradem Cave	UM2	L	No measurement				No measurement				III	D	16	101	B		1	89
285	287	Budišov	Pod hradem Cave	UM2	L	Exact	46.49	46.75	46.67	Exact	24.18	24.39	24.17	II	D	16	101	E		1	90
286	288	Budišov	Pod hradem Cave	UM2	L	Exact	45.51	45.46	45.43	Exact	22.86	22.76	22.96	III	D	16	101	E		1	90
293	295	Budišov	Pod hradem Cave	UM2	L	Exact	48.72	49.07	49.59	Exact	26.06	26.11	26.14	IX	D	16	101		b	1	92
299	301	Budišov	Pod hradem Cave	UM2	R	Exact	45.19	45.53	45.59	Exact	25.25	25.72	25.76	IV	D	16	100	A		1	93
301	303	Budišov	Pod hradem Cave	UM2	L	Exact	44.43	44.31	44.31	Exact	22.97	22.6	22.66	II	D	16	100	A		1	93
309	311	Budišov	Pod hradem Cave	UM2	R	Exact	40.83	41.03	41.08	Exact	23.74	23.62	23.54	IX	D	16	100		c	1	95
310	312	Budišov	Pod hradem Cave	UM2	L	Exact	42.62	42.49	42.55	Exact	19.16	19.52	19.91	II	D	16	100		d	1	95
312	314	Budišov	Pod hradem Cave	UM2	R	Exact	47.5	47.73	47.77	Exact	22.99	23.54	23.24	II	D	16	100		e	1	96
26	26	Budišov	Pod hradem Cave	C		Exact	29.49	30.12	29.76	Exact	19.69	19.68	19.75			17	149			1	14
117	117	Budišov	Pod hradem Cave	C		Exact	20.57	20.45	20.47	Exact	14.75	14.74	14.6		D	17	132	B		1	40
126	126	Budišov	Pod hradem Cave	C		No measurement				Close estimate	19.29	19.34	19.14			17	129	A		1	42
128	128	Budišov	Pod hradem Cave	C		Exact	31.85	31.88	31.9	Exact	21.62	21.64	21.69			17	129	A	a	1	43
129	129	Budišov	Pod hradem Cave	C		Exact	23.87	23.5	23.63	Exact	17.96	17.62	17.81			17	128	A		1	44
131	131	Budišov	Pod hradem Cave	C		Exact	26.47	26.46	26.8	Exact	18.02	17.79	17.25			17	1	B		1	44
135	135	Budišov	Pod hradem Cave	C		Exact	22.76	22.87	22.58	Exact	15.86	16.1	15.74		D	17	1	B		1	45
139	139	Budišov	Pod hradem Cave	C		Exact	22.65	22.39	21.81	Exact	17.57	17.56	17.69		D	17	1	D		1	46
143	143	Budišov	Pod hradem Cave	C	L	Close estimate	30.46	29.73	29.91	Close estimate	22.39	22.18	22.25		M	17	126	B		1	47
150	150	Budišov	Pod hradem Cave	C		Exact	27.9	27.76	28.06	Exact	21.34	21.53	21.56		D	17	126	D		1	48
151	151	Budišov	Pod hradem Cave	C		Exact	27.01	26.3	25.85	Close estimate	22.28	21.55	21.95		D	17	126	D		1	49
157	158	Budišov	Pod hradem Cave	C		Exact	21.92	21.19	21.58	Exact	19.98	19.09	19.56			17	125	E	a	1	50
161	162	Budišov	Pod hradem Cave	C		Exact	22.77	22.93	23.43	Exact	18.92	18.41	18.76		D	17	124	B		1	52

Table. A1.1. (Continued.)

ID	Tooth ID	Storage ID	Site ID	Tooth Type ID	Left/Right	Measurement Type Length	Length1 (mm)	Length2 (mm)	Length3 (mm)	Measurement Type Breadth	Breadth1 (mm)	Breadth2 (mm)	Breadth3 (mm)	Wear Stage	Articulation ID	Layer	Box Number	Medium Box Number Suffix	Small Box Number Suffix	Log Book Number	Log Book Page Number
162	163	Budišov	Pod hradem Cave	C		Exact	22.3	22.13	22.62	Exact	17.26	17.38	17.46		D	17	123	D		1	52
163	164	Budišov	Pod hradem Cave	C		Exact	23.83	23.8	24.12	Exact	16.85	16.89	16.8		D	17	122	A		1	53
166	167	Budišov	Pod hradem Cave	C		Exact	31.58	31.9	31.74	Exact	22.28	22.26	22.08		D	17	122	D		1	54
173	174	Budišov	Pod hradem Cave	C		Exact	27.48	27.33	27.64	Exact	20.58	20.4	20.38		D	17	120	C		1	57
183	184	Budišov	Pod hradem Cave	C		Estimate	32.72	32.72	32.72	Exact	23.99	24.24	24.42		D	17	119	D		1	60
190	191	Budišov	Pod hradem Cave	C		Exact	27.44	27.44	27.44	Exact	20.28	20.19	20.09		D	17	118	C		1	62
193	194	Budišov	Pod hradem Cave	C		Exact	30.11	30.45	30.71	Exact	21.18	21.26	21.26		D	17	118		a	1	63
194	195	Budišov	Pod hradem Cave	C		Exact	28.12	28.56	28.86	Exact	20.96	20.92	20.85		D	17	115	B		1	64
203	205	Budišov	Pod hradem Cave	C		Exact	23.01	23.15	22.7	Exact	16.18	16.22	16.22		D	17	113		a	1	67
205	207	Budišov	Pod hradem Cave	C		Exact	27.92	28.28	28.46	Exact	20.6	20.43	20.39		D	17	112	C		1	68
213	215	Budišov	Pod hradem Cave	C		Estimate	31.2	31.2	31.2	Close estimate	22.2	22.2	22.2		D	17	111	A		1	71
217	219	Budišov	Pod hradem Cave	C		Close estimate	31.2	31.2	31.2	Close estimate	22.89	22.89	22.89		D	17	110	A		1	72
218	220	Budišov	Pod hradem Cave	C		Exact	29.61	29.46	29.51	Exact	20.5	20.34	20.58		D	17	110	C		1	72
227	229	Budišov	Pod hradem Cave	C		Exact	27.71	27.95	28.01	Exact	22.43	22.3	22.28		D	17	109		h	1	75
229	231	Budišov	Pod hradem Cave	C		Exact	30.73	31	30.73	Exact	19.24	19.61	19.58		D	17	107	A		1	77
230	232	Budišov	Pod hradem Cave	C		Exact	26.75	26.04	26.34	Exact	18.85	18.36	18.44		D	17	107	A		1	77
232	234	Budišov	Pod hradem Cave	C		Close estimate	28.9	29.02	29.1	Close estimate	20.46	20.46	20.46		D	17	107	B		1	77
239	241	Budišov	Pod hradem Cave	C		Close estimate	31.46	31.46	31.46	Exact	23.09	23.26	23.15		M	17	107	E		1	78
248	250	Budišov	Pod hradem Cave	C		Exact	20.9	21.09	21.47	Exact	15.37	15.09	15.03		D	17	106		e	1	80
250	252	Budišov	Pod hradem Cave	C		Exact	28.62	27.56	27.69	Exact	19.67	19.21	19.81		M	17	105			1	81
123	123	Budišov	Pod hradem Cave	C (J)		Estimate	19.18	19.18	19.18	Estimate	15.08	15.08	15.08		D	17	129	C		1	42
130	130	Budišov	Pod hradem Cave	C (J)		Exact	19.2	19.44	19.47	Exact	16.15	16.11	16.16		D	17	128	B		1	44
144	144	Budišov	Pod hradem Cave	C (J)		Exact	20.56	20.5	20.49	Exact	17.67	17.65	17.75		D	17	126	B		1	48
145	145	Budišov	Pod hradem Cave	C (J)		Exact	20.59	21.06	20.82	Exact	15.9	16	16.03		D	17	126	B		1	48
149	149	Budišov	Pod hradem Cave	C (J)		No measurement				No measurement					D	17	126	D		1	48
158	159	Budišov	Pod hradem Cave	C (J)		Exact	19.87	19.83	19.73	Exact	18.25	17.92	17.93		D	17	125	E	a	1	51
160	161	Budišov	Pod hradem Cave	C (J)		Exact	46.25	46.14	46.25	Exact	17.23	17.32	17.41		D	17	124	A		1	51
171	172	Budišov	Pod hradem Cave	C (J)		No measurement				No measurement					D	17	122	E		1	55
221	223	Budišov	Pod hradem Cave	C (J)		Exact	19.12	19.14	19.4	Exact	14.34	14.35	14.09		D	17	109	D		1	74
223	225	Budišov	Pod hradem Cave	C (J)		No measurement				No measurement					D	17	109		a	1	74
228	230	Budišov	Pod hradem Cave	C (J)		Exact	19.26	19.07	19.1	Exact	16.29	16.47	16.41		D	17	108	E		1	76

Table. A1.1. (Continued.)

ID	Tooth ID	Storage ID	Site ID	Tooth Type ID	Left/Right	Measurement Type Length	Length1 (mm)	Length2 (mm)	Length3 (mm)	Measurement Type Breadth	Breadth1 (mm)	Breadth2 (mm)	Breadth3 (mm)	Wear Stage	Articulation ID	Layer	Box Number	Medium Box Number Suffix	Small Box Number Suffix	Log Book Number	Log Book Page Number
240	242	Budišov	Pod hradem Cave	C (J)		No measurement				No measurement					D	17	106	A		1	79
243	245	Budišov	Pod hradem Cave	C (J)		No measurement				No measurement					D	17	106	C		1	79
244	246	Budišov	Pod hradem Cave	C (J)		Exact	18.35	18.27	18.44	Exact	15.05	15.15	15.03		D	17	106		c	1	79
245	247	Budišov	Pod hradem Cave	C (J)		No measurement				No measurement					D	17	106		c	1	79
116	116	Budišov	Pod hradem Cave	LM1	R	Exact	29.53	29.6	29.62	Exact	13.96	14.17	14.18	II	D	17	131	B		1	40
120	120	Budišov	Pod hradem Cave	LM1	R	Exact	31.07	30.93	30.93	Exact	16.09	16.03	15.89	IX	D	17	131	C		1	41
121	121	Budišov	Pod hradem Cave	LM1	R	Exact	31.67	31.64	31.68	Exact	14.94	14.87	14.9	II	D	17	131	C		1	41
125	125	Budišov	Pod hradem Cave	LM1	L	Exact	34.49	34.61	34.56	Exact	15.18	15.36	15.3	II	D	17	129	A		1	42
138	138	Budišov	Pod hradem Cave	LM1	L	No measurement				No measurement				II	D	17	127	D		1	46
168	169	Budišov	Pod hradem Cave	LM1	R	Exact	31.74	31.73	31.77	Exact	16	16.16	15.9	VI	D	17	122	D		1	55
172	173	Budišov	Pod hradem Cave	LM1	L	Exact	31.08	31.05	31.09	Exact	13.79	13.58	14.05	II	D	17	121	D		1	56
182	183	Budišov	Pod hradem Cave	LM1	L	Exact	32.9	32.88	32.87	Close estimate	16.43	16.43	16.43	IX	D	17	119	D		1	59
199	200	Budišov	Pod hradem Cave	LM1	R	Exact	29.39	29.44	29.41	Exact	14.37	14.46	14.42	VIII	D	17	114		a	1	65
231	233	Budišov	Pod hradem Cave	LM1	R	Estimate	31	31	31	Exact	15.04	15.11	15.1	II	D	17	107	A		1	77
235	237	Budišov	Pod hradem Cave	LM1	L	Exact	30.9	30.96	30.88	Exact	14.61	14.24	14.31	IX	M	17	107	B		1	77
247	249	Budišov	Pod hradem Cave	LM1	R	Exact	31.06	30.9	30.98	Exact	14.9	14.63	14.57	III	D	17	106		b	1	80
25	25	Budišov	Pod hradem Cave	LM2	L	Exact	29.75	29.75	29.72	Exact	17.72	17.76	17.74	III	M	17	149			1	14
136	136	Budišov	Pod hradem Cave	LM2	L	Exact	31.63	31.59	31.54	Exact	18.44	18.44	18.46	II	M	17	127	C		1	46
148	148	Budišov	Pod hradem Cave	LM2	L	Exact	33.14	33.14	33.12	Exact	18.77	18.83	18.88	II	D	17	126	B		1	48
164	165	Budišov	Pod hradem Cave	LM2	R	Exact	33.79	33.82	33.73	Exact	18.52	18.37	18.69	II	D	17	122	A		1	53
165	166	Budišov	Pod hradem Cave	LM2	R	Exact	32.16	32.2	31.9	Exact	20.24	20.06	20.21	IX	D	17	122	A		1	53
167	168	Budišov	Pod hradem Cave	LM2	R	Exact	31.34	31.26	31.33	Exact	19.46	19.45	19.47	IV	M	17	122	D		1	54
170	171	Budišov	Pod hradem Cave	LM2	R	Exact	40.14	40.02	40.14	Exact	19.87	19.8	19.79	III	M	17	122	E		1	55
178	179	Budišov	Pod hradem Cave	LM2	L	Close estimate	31.12	31.12	31.12	Exact	19.51	19.35	19.32	IX	D	17	120	E		1	58
179	180	Budišov	Pod hradem Cave	LM2	R	Exact	30.66	30.57	30.45	Exact	17.89	17.93	17.85	II	D	17	119	A		1	59
180	181	Budišov	Pod hradem Cave	LM2	L	Exact	31.82	31.84	31.88	Exact	17.69	17.65	17.75	II	D	17	119	A		1	59
184	185	Budišov	Pod hradem Cave	LM2	L	Exact	29.06	29.11	29.17	Exact	16.94	17.05	17.02	V	D	17	119		a	1	60
186	187	Budišov	Pod hradem Cave	LM2	R	Exact	29.71	29.72	29.89	Exact	18.74	18.24	18.12	VII	M	17	118	A		1	61
202	204	Budišov	Pod hradem Cave	LM2	R	Exact	31.57	32.1	31.94	Exact	18.19	18.1	18.13	III	M	17	113	B		1	67
207	209	Budišov	Pod hradem Cave	LM2	R	Exact	29.4	29.4	29.57	Exact	18.61	18.33	18.36	VIII	D	17	112	C		1	69
209	211	Budišov	Pod hradem Cave	LM2	R	Exact	29.45	29.06	29	Exact	17.72	17.93	17.42	IX	M	17	112	F		1	69

Table. A1.1. (Continued.)

ID	Tooth ID	Storage ID	Site ID	Tooth Type ID	Left/Right	Measurement Type Length	Length1 (mm)	Length2 (mm)	Length3 (mm)	Measurement Type Breadth	Breadth1 (mm)	Breadth2 (mm)	Breadth3 (mm)	Wear Stage	Articulation ID	Layer	Box Number	Medium Box Number Suffix	Small Box Number Suffix	Log Book Number	Log Book Page Number
219	221	Budišov	Pod hradem Cave	LM2	R	Exact	30.65	30.67	30.75	Exact	17.54	17.5	17.71	II	D	17	110	E		1	73
222	224	Budišov	Pod hradem Cave	LM2	R	Exact	30.16	30.39	30.21	Exact	17.56	17.6	17.6	III	D	17	109		a	1	74
225	227	Budišov	Pod hradem Cave	LM2	L	No measurement				No measurement				VII	D	17	109		c	1	75
234	236	Budišov	Pod hradem Cave	LM2	L	Exact	29.25	29.35	29.27	Exact	18.09	18.03	18.39	IX	M	17	107	B		1	77
238	240	Budišov	Pod hradem Cave	LM2	R	Exact	31.73	31.6	31.63	Exact	20.36	20.19	20.14	VIII	M	17	107	E		1	78
24	24	Budišov	Pod hradem Cave	LM3	L	Exact	26.21	26.22	26.21	Exact	18.69	18.76	18.74	III	M	17	149			1	14
127	127	Budišov	Pod hradem Cave	LM3	R	Exact	31.07	31.04	31.03	Exact	22.33	22.36	22.39	II	D	17	129	A	a	1	42
133	133	Budišov	Pod hradem Cave	LM3	R	Exact	26.72	26.88	26.88	Exact	19.55	19.55	19.56	V	D	17	127	B		1	45
137	137	Budišov	Pod hradem Cave	LM3	R	Exact	25.06	25.04	25.34	Exact	15.46	15.8	15.4	II	M	17	127	D		1	46
147	147	Budišov	Pod hradem Cave	LM3	L	Exact	26.13	26.21	26.22	Exact	18.81	18.79	18.78	V	M	17	126	B		1	48
152	152	Budišov	Pod hradem Cave	LM3	R	Exact	25.09	25.02	25.02	Exact	19.85	19.77	19.76	IX	D	17	126		e	1	49
156	157	Budišov	Pod hradem Cave	LM3	L	Exact	28.15	28.1	28.15	Exact	19.57	19.48	19.38	V	D	17	125	E	a	1	50
174	175	Budišov	Pod hradem Cave	LM3	L	Exact	27.75	27.87	27.85	Exact	19.29	19.19	19.4	IV	D	17	120	C		1	57
181	182	Budišov	Pod hradem Cave	LM3	L	Exact	26.58	26.62	26.62	Exact	20.68	20.83	20.85	IV	M	17	119	D		1	59
185	186	Budišov	Pod hradem Cave	LM3	R	Exact	27.07	27.19	27.01	Exact	19.1	19.01	19.63	VII	M	17	118	A		1	61
188	189	Budišov	Pod hradem Cave	LM3	R	Exact	29.46	29.68	29.62	Exact	19.16	19.12	19.23	IV	D	17	118	A		1	61
201	203	Budišov	Pod hradem Cave	LM3	R	Exact	27.55	27.4	27.4	Exact	20.26	20.13	20.12	III	M	17	113	B		1	67
206	208	Budišov	Pod hradem Cave	LM3	R	Exact	25.53	25.39	25.43	Exact	18.97	18.9	18.97	VIII	M	17	112	C		1	69
208	210	Budišov	Pod hradem Cave	LM3	R	Exact	25.95	26.22	26.21	Exact	19.31	19.26	19.44	IX	M	17	112	F		1	69
210	212	Budišov	Pod hradem Cave	LM3	R	Exact	27.03	27.43	27.35	Exact	20.02	20.16	20.07	IX	M	17	112	F		1	69
233	235	Budišov	Pod hradem Cave	LM3	L	Exact	26.19	26.65	26.19	Exact	20.08	20.48	20.69	IX	M	17	107	B		1	77
237	239	Budišov	Pod hradem Cave	LM3	R	Exact	26.56	26.04	26.44	Exact	20.27	20.14	20.29	VIII	M	17	107	E		1	78
249	251	Budišov	Pod hradem Cave	LM3	R	Exact	24.12	24	23.94	Exact	20.19	20.35	20.21	III	M	17	105			1	81
119	119	Budišov	Pod hradem Cave	UM1	L	Exact	29.65	29.61	29.62	Exact	18.14	17.96	18	II	D	17	131	A		1	41
134	134	Budišov	Pod hradem Cave	UM1	L	Exact	29.78	29.97	29.85	Exact	18.63	18.57	18.56	II	D	17	127	B		1	45
141	141	Budišov	Pod hradem Cave	UM1	R	Exact	24.6	24.88	24.88	Exact	18.45	18.45	18.47	II	D	17	127		b	1	46
142	142	Budišov	Pod hradem Cave	UM1	R	Exact	27.48	27.61	27.57	Exact	19.95	19.95	19.95	IX	D	17	127		b	1	47
154	154	Budišov	Pod hradem Cave	UM1	R	Exact	27.45	27.35	27.44	Exact	17.38	17.41	18.14	II	D	17	126		d	1	49
155	155	Budišov	Pod hradem Cave	UM1	L	Exact	28.47	28.64	28.62	Exact	17.24	17.09	17.33	II	D	17	126		d	1	49
169	170	Budišov	Pod hradem Cave	UM1	R	Exact	25.92	25.85	25.88	Exact	15.83	15.64	15.66	III	M	17	122	E		1	55
187	188	Budišov	Pod hradem Cave	UM1	R	Exact	28.95	29.06	29.18	Exact	18.53	18.24	18.53	II	D	17	118	A		1	61

Table. A1.1. (Continued.)

ID	Tooth ID	Storage ID	Site ID	Tooth Type ID	Left/Right	Measurement Type Length	Length1 (mm)	Length2 (mm)	Length3 (mm)	Measurement Type Breadth	Breadth1 (mm)	Breadth2 (mm)	Breadth3 (mm)	Wear Stage	Articulation ID	Layer	Box Number	Medium Box Number Suffix	Small Box Number Suffix	Log Book Number	Log Book Page Number
196	197	Budišov	Pod hradem Cave	UM1	R	Exact	29.54	29.56	29.61	Exact	18.7	18.49	18.99	II	D	17	115	F		1	64
211	213	Budišov	Pod hradem Cave	UM1		Exact	25.71	25.85	25.92	Exact	21.52	21.44	21.41	IX	C	17	111			1	70
215	217	Budišov	Pod hradem Cave	UM1	R	Exact	29.72	29.96	29.98	Exact	19.33	19.21	19.03	II	D	17	111	A		1	71
220	222	Budišov	Pod hradem Cave	UM1	L	Exact	29.91	29	29.14	Exact	19.65	19.33	19.46	II	D	17	109	C	a	1	74
242	244	Budišov	Pod hradem Cave	UM1	R	Exact	27.34	27.28	27.38	Exact	18.09	18	18.11	II	D	17	106	C		1	79
122	122	Budišov	Pod hradem Cave	UM2	L	Exact	45.06	44.97	44.98	Exact	22.63	22.71	22.55	III	D	17	129	C		1	42
124	124	Budišov	Pod hradem Cave	UM2	R	Exact	45.02	45.1	45.13	Exact	23.17	23.23	23.05	III	D	17	129	B		1	42
132	132	Budišov	Pod hradem Cave	UM2	R	Exact	29.88	29.71	29.89	Exact	21.86	21.17	21.92	I	D	17	127	B		1	45
140	140	Budišov	Pod hradem Cave	UM2	R	Exact	45.67	45.58	45.63	Exact	22.26	22.58	22.36	II	D	17	127		b	1	46
146	146	Budišov	Pod hradem Cave	UM2	R	Exact	43.2	43.2	43.21	Exact	18.9	18.97	19.3	II	D	17	126	B		1	48
153	153	Budišov	Pod hradem Cave	UM2	R	Exact	46.43	46.42	46.44	Exact	23.02	23.1	23.16	II	D	17	126		d	1	49
159	160	Budišov	Pod hradem Cave	UM2	L	Exact	46.25	46.14	46.25	Exact	22.24	22.23	22.51	II	D	17	1	A		1	51
176	177	Budišov	Pod hradem Cave	UM2	L	Exact	45.33	45.19	45.62	Exact	23.7	22.53	23.6	II	D	17	120		a	1	58
189	190	Budišov	Pod hradem Cave	UM2	L	Exact	41.25	41.64	41.83	Exact	21.24	21.14	21.36	II	D	17	118	B		1	62
191	192	Budišov	Pod hradem Cave	UM2	R	Exact	46.29	46.71	46.69	Exact	24.45	24.59	24.07	VII	C	17	118	D		1	63
192	193	Budišov	Pod hradem Cave	UM2	L	Exact	46.3	46.23	46.17	Exact	23.75	23.84	2384	VIII	D	17	118	D		1	63
197	198	Budišov	Pod hradem Cave	UM2	L	Estimate	40.8	40.8	40.8	Exact	22.32	22.49	22.52	VIII	D	17	114	B		1	65
198	199	Budišov	Pod hradem Cave	UM2	R	Exact	47.12	46.82	46.85	Exact	22.8	23.19	23.09	IX	D	17	114		a	1	65
200	201	Budišov	Pod hradem Cave	UM2	R	Exact	44.92	44.9	44.9	Exact	20.48	21.53	21.96	II	D	17	114	C		1	66
212	214	Budišov	Pod hradem Cave	UM2		Exact	43.61	43.49	43.55	Exact	23.45	23.74	23.96	IX	C	17	111			1	70
214	216	Budišov	Pod hradem Cave	UM2		Estimate	41	41	41	Estimate	22.07	22.07	22.07	IX	D	17	111	A		1	71
216	218	Budišov	Pod hradem Cave	UM2	R	Exact	42.42	42.54	42.48	Exact	20.43	21.04	20.47	II	D	17	111	A		1	71
224	226	Budišov	Pod hradem Cave	UM2	R	No measurement				No measurement				IX	D	17	109		c	1	75
226	228	Budišov	Pod hradem Cave	UM2	L	No measurement				No measurement				II	D	17	109		h	1	75
236	238	Budišov	Pod hradem Cave	UM2	R	Exact	44.04	44	44.1	Exact	20.71	20.91	20.82	II	D	17	107	B		1	77
241	243	Budišov	Pod hradem Cave	UM2	R	Exact	42.63	42.63	42.67	Exact	21.85	21.55	21.94	VI	D	17	106	C		1	79
246	248	Budišov	Pod hradem Cave	UM2	L	Exact	47.74	47.59	47.92	Exact	22.51	22.35	22.74	II	D	17	106		b	1	80
78	78	Budišov	Pod hradem Cave	C		Exact	20.4	20.07	20.08	Exact	17.08	16.79	16.59		D	18	136		b	1	30
79	79	Budišov	Pod hradem Cave	C		Exact	25.54	26.16	25.23	Exact	17.82	18.47	17.4		D	18	136	C		1	31
80	80	Budišov	Pod hradem Cave	C		Estimate	25.54	26.16	25.23	Exact	20.65	20.64	20.63		D	18	136	C		1	31
85	85	Budišov	Pod hradem Cave	C		Estimate	29.62	29.62	29.62	Estimate	18.72	18.72	18.72		D	18	134	E		1	33

Table. A1.1. (Continued.)

ID	Tooth ID	Storage ID	Site ID	Tooth Type ID	Left/Right	Measurement Type Length	Length1 (mm)	Length2 (mm)	Length3 (mm)	Measurement Type Breadth	Breadth1 (mm)	Breadth2 (mm)	Breadth3 (mm)	Wear Stage	Articulation ID	Layer	Box Number	Medium Box Number Suffix	Small Box Number Suffix	Log Book Number	Log Book Page Number
89	89	Budišov	Pod hradem Cave	C	R	Exact	30.47	30.16	30.21	Exact	21.78	22.44	22.32		M	18	134	E		1	33
91	91	Budišov	Pod hradem Cave	C		No measurement				Estimate	20.31	20.31	20.31		M	18	134	A		1	34
96	96	Budišov	Pod hradem Cave	C	R	Exact	28.35	28.34	28.27	Exact	21.61	21.51	21.62		M	18	133	A		1	35
99	99	Budišov	Pod hradem Cave	C		Exact	26.89	26.99	27.11	Exact	21.14	21.27	21.29		D	18	133	B		1	36
100	100	Budišov	Pod hradem Cave	C		Exact	30.65	31.25	31.58	Exact	21.3	21.28	21.24		D	18	133	B		1	36
102	102	Budišov	Pod hradem Cave	C		Exact	28.63	28.63	28.63	Exact	11.79	12.06	12.36		D	18	133	D		1	37
103	103	Budišov	Pod hradem Cave	C		Exact	22.79	22.46	23.15	Exact	17.28	17.65	17.08		D	18	133	D		1	37
107	107	Budišov	Pod hradem Cave	C	L	Exact	24.6	24.2	24.35	Exact	17.88	16.92	17.31		M	18	132	A		1	38
111	111	Budišov	Pod hradem Cave	C	L	Exact	27.02	26.5	27.8	Exact	20.05	19.19	19.21		M	18	132	C		1	39
84	84	Budišov	Pod hradem Cave	LM1	R	Exact	31.78	31.74	31.6	Exact	14.95	15.04	15	VI	M	18	134	E		1	33
88	88	Budišov	Pod hradem Cave	LM1	R	Exact	32.19	32.33	32.18	Exact	16.67	16.44	16.27	IX	M	18	134	E		1	33
95	95	Budišov	Pod hradem Cave	LM1	R	Exact	31	31	30.87	Exact	14	13.99	13.96	IV	M	18	133	A		1	35
106	106	Budišov	Pod hradem Cave	LM1	L	Exact				Exact	13.89	13.6	13.57	IX	M	18	132	A		1	38
108	108	Budišov	Pod hradem Cave	LM1	R	Exact	28.26	28.16	28.57	Exact	11.64	11.7	11.67	IX	D	18	1	B		1	38
110	110	Budišov	Pod hradem Cave	LM1	L	Exact	29.32	29.68	29.52	Exact	16.71	17	17.01	IV	M	18	132	C		1	38
83	83	Budišov	Pod hradem Cave	LM2	R	Exact	31.24	31.17	31.17	Exact	18.28	18.23	18.22	VI	M	18	134	E		1	32
87	87	Budišov	Pod hradem Cave	LM2	R	Exact	33.97	33.38	33.29	Exact	20.11	20.1	20.05	IX	M	18	134	E		1	33
90	90	Budišov	Pod hradem Cave	LM2		Exact	30.91	30.81	30.63	Exact	19.66	19.74	19.79	VIII	M	18	134	A		1	34
93	93	Budišov	Pod hradem Cave	LM2	R	Exact	33.52	33.61	33.57	Exact	18.09	18.25	18.31	II	D	18	134	C		1	35
94	94	Budišov	Pod hradem Cave	LM2	R	Exact	31.13	31.28	31.35	Exact	18.05	18.05	18.05	IV	M	18	133	A		1	35
105	105	Budišov	Pod hradem Cave	LM2	L	Exact	27.37	27.64	27.66	Exact	17.51	17.71	17.54	IX	M	18	132	A		1	38
109	109	Budišov	Pod hradem Cave	LM2	L	Exact	29.32	29.68	29.52	Exact	16.71	17	17.01	IV	M	18	132	C		1	38
113	113	Budišov	Pod hradem Cave	LM2	R	Exact	29.08	29.06	29.02	Exact	17.44	17.36	17.37	III	D	18	132		f	1	39
82	82	Budišov	Pod hradem Cave	LM3	R	Exact	27.79	27.72	27.62	Exact	19.32	19.24	19.24	VI	M	18	134	E		1	32
86	86	Budišov	Pod hradem Cave	LM3	R	Exact	28.17	27.67	27.97	Exact	21.51	21.12	21.21	IX	M	18	134	E		1	33
92	92	Budišov	Pod hradem Cave	LM3	L	Exact	29.55	29.74	29.83	Exact	20.92	20.99	20.86	IV	D	18	134	C		1	34
97	97	Budišov	Pod hradem Cave	LM3	L	Exact	27.31	27.28	27.23	Exact	18.78	18.78	18.76	IV	D	18	133	B		1	36
98	98	Budišov	Pod hradem Cave	LM3	R	Exact	25.99	25.87	25.9	Exact	18.91	18.96	18.91	IX	D	18	133	B		1	36
104	104	Budišov	Pod hradem Cave	LM3	L	Exact	23.49	23.91	23.61	Exact	17.93	17.44	17.73	IX	M	18	132	A		1	38
76	76	Budišov	Pod hradem Cave	UM2	R	Exact	47.95	47.83	47.85	Exact	25.59	25.55	25.53	III	D	18	136	B		1	30
81	81	Budišov	Pod hradem Cave	UM2	L	Exact	47.58	47.65	47.66	Exact	24.59	24.76	24.55	III	D	18	134		c	1	32

Table. A1.1. (Continued.)

ID	Tooth ID	Storage ID	Site ID	Tooth Type ID	Left/Right	Measurement Type Length	Length1 (mm)	Length2 (mm)	Length3 (mm)	Measurement Type Breadth	Breadth1 (mm)	Breadth2 (mm)	Breadth3 (mm)	Wear Stage	Articulation ID	Layer	Box Number	Medium Box Number Suffix	Small Box Number Suffix	Log Book Number	Log Book Page Number
112	112	Budišov	Pod hradem Cave	UM2	L	Exact	43.25	43.33	43.2	Exact	19.84	19.74	19.7	III		18	132		d	1	39
115	115	Budišov	Pod hradem Cave	UM2	L	Exact	43.87	43.78	43.97	Exact	21.74	21.98	21.85	III	D	18	132		e	1	40
72	72	Budišov	Pod hradem Cave	LM1	R	Exact	30.58	30.65	30.65	Exact	14.19	14.27	14.15	VIII	M	18a	138	A		1	28
74	74	Budišov	Pod hradem Cave	LM1	L	Close estimate	31.02	31.02	31.02	Close estimate	12.4	12.4	12.4	I	M	18a	1	D		1	29
75	75	Budišov	Pod hradem Cave	LM1	R	Exact	31.48	31.49	31.47	Exact	13.63	13.6	13.59	II	D	18a	137		e	1	29
71	71	Budišov	Pod hradem Cave	LM2	R	Exact	32.35	32.14	32.03	Exact	19.24	19.63	19.78	VIII	M	18a	138	A		1	28
70	70	Budišov	Pod hradem Cave	LM3	R	Exact	29.38	29.51	29.54	Exact	20.03	20.25	20.2	VIII	M	18a	138	A		1	28
73	73	Budišov	Pod hradem Cave	UM2	R	Exact	40.56	40.58	40.61	Exact	21.94	22.18	22.17	II	D	18a	137	D		1	29
21	21	Budišov	Pod hradem Cave	C	L	Exact	26.72	26.85	26.72	Exact	19.82	19.67	19.91		M	18b	149			1	13
22	22	Budišov	Pod hradem Cave	C	R	Exact	26.81	26.44	26.93	Exact	20.55	20.73	20.83		M	18b	149			1	13
68	68	Budišov	Pod hradem Cave	C		Exact	21.54	21.96	21.9	Exact	16.78	16.92	16.79		D	18b	139	D		1	27
17	17	Budišov	Pod hradem Cave	LM2	L	Exact	31.3	31.2	31.6	Exact	19.4	19.44	19.45	VI	M	18b	149			1	12
69	69	Budišov	Pod hradem Cave	LM2	R	Exact	30.43	30.38	30.4	Exact	17.61	17.53	17.65	II	M	18b	139	D		1	27
16	16	Budišov	Pod hradem Cave	LM3	L	Exact	29.5	29.64	29.6	Exact	20.2	20.2	20.13	VI	M	18b	149			1	12
57	57	Budišov	Pod hradem Cave	C		Estimate	19.54	19.54	19.54	Exact	18.77	18.79	18.72		D	18c	143	B		1	23
58	58	Budišov	Pod hradem Cave	C		Estimate	19.76	19.76	19.76	Estimate	16	16	16		D	18c	143	A		1	24
59	59	Budišov	Pod hradem Cave	C		Exact	27.2	26.29	27.28	Exact	20.51	20.27	20.22		D	18c	143	A		1	24
60	60	Budišov	Pod hradem Cave	C		Exact	27.37	26.9	26.78	Exact	19.85	19.98	19.96		D	18c	143	E		1	24
62	62	Budišov	Pod hradem Cave	C		Estimate	21.53	21.08	21.32	Exact	14.57	14.34	14.33		D	18c	141	A		1	25
65	65	Budišov	Pod hradem Cave	C		Estimate	25.69	25.69	25.69	Exact	19.59	19.22	19.13		D	18c	140	A		1	26
67	67	Budišov	Pod hradem Cave	C		Exact	22.09	22.5	22.38	Exact	16.27	15.94	16		D	18c	140		a	1	27
66	66	Budišov	Pod hradem Cave	C (J)		Estimate	25.69	25.69	25.69	Exact	19.59	19.22	19.13		D	18c	140	D		1	26
61	61	Budišov	Pod hradem Cave	LM1	L	Exact	27.8	27.81	27.79	Exact	14.59	14.58	14.59	VIII	D	18c	143		d	1	24
64	64	Budišov	Pod hradem Cave	LM1	R	Exact	30.01	30.06	30.04	Exact	15.49	15.48	15.54	VIII	D	18c	140	B		1	26
50	50	Budišov	Pod hradem Cave	LM2	L	Exact	27.36	27.31	27.41	Exact	15.3	15.35	15.37	II	M	18c	145		f	1	21
56	56	Budišov	Pod hradem Cave	LM2	R	Exact	30.52	30.53	30.51	Exact	18.03	18.16	18.14	II	D	18c	143	B		1	23
63	63	Budišov	Pod hradem Cave	LM2	R	Exact	32.67	32.62	32.62	Exact	18.9	18.91	18.9	II	D	18c	141	B		1	25
49	49	Budišov	Pod hradem Cave	C	L	Close estimate	26.28	25.5	25.53	Exact	19.48	19.5	19.46		M	18f	145	A		1	21
54	54	Budišov	Pod hradem Cave	C		Estimate	33.67	33.67	33.67	Estimate	23.5	23.5	23.5		M	18f	144			1	22
55	55	Budišov	Pod hradem Cave	C		Estimate	32.84	32.84	32.84	Estimate	24.7	24.7	24.7		D	18f	144			1	22
48	48	Budišov	Pod hradem Cave	LM1	L	Exact	30.08	30.05	30.14	Exact	13.73	13.37	13.32	VIII	M	18f	145	A		1	20

Table. A1.1. (Continued.)

ID	Tooth ID	Storage ID	Site ID	Tooth Type ID	Left/Right	Measurement Type Length	Length1 (mm)	Length2 (mm)	Length3 (mm)	Measurement Type Breadth	Breadth1 (mm)	Breadth2 (mm)	Breadth3 (mm)	Wear Stage	Articulation ID	Layer	Box Number	Medium Box Number Suffix	Small Box Number Suffix	Log Book Number	Log Book Page Number
51	51	Budišov	Pod hradem Cave	LM1	R	Exact	30.12	30.13	30.14	Exact	14.02	14.54	14.4	II	D	18f	145		d	1	21
47	47	Budišov	Pod hradem Cave	LM2	L	Exact	31.96	32.02	32.09	Exact	19.23	19.49	19.31	VIII	M	18f	145	A		1	20
52	52	Budišov	Pod hradem Cave	UM2	L	Exact	44.81	44.42	44.54	Exact	23.93	23.98	24.05	IX	D	18f	144			1	22
53	53	Budišov	Pod hradem Cave	UM2	R	Exact	43.28	43.06	43.17	Exact	24.11	24.1	24.07	IX	D	18f	144			1	22
37	37	Budišov	Pod hradem Cave	C		Exact	21.27	21.5	21.5	Close estimate	15.43	15.4	15.4		D	19	148		a	1	16
42	42	Budišov	Pod hradem Cave	C		Exact	26.96	27.6	27.02	Exact	19	19.09	19.14		M	19	146	C		1	19
44	44	Budišov	Pod hradem Cave	C		Estimate	21.85	21.59	21.52	Estimate	16.54	16.51	16.53		D	19	146	B		1	19
46	46	Budišov	Pod hradem Cave	C		No measurement				Close estimate	20.88	20.98	21.02		D	19	146		f	1	19
41	41	Budišov	Pod hradem Cave	LM1	R	Exact	29.53	29.11	29.09	Exact	13.14	13.09	13.1	VIII	M	19	146	C		1	18
43	43	Budišov	Pod hradem Cave	LM1	R	Exact	30.29	30.26	30.26	Exact	14.6	14.75	14.74	VIII	D	19	146	A		1	19
40	40	Budišov	Pod hradem Cave	LM2	R	Exact	29.92	30.25	29.99	Exact	18.08	18.15	17.97	VIII	M	19	146	C		1	18
45	45	Budišov	Pod hradem Cave	LM2	L	Exact	30.41	30.41	30.41	Exact	17.74	17.79	17.71	VI	D	19	146		g	1	19
38	38	Budišov	Pod hradem Cave	LM3	L	Exact	28.1	28.14	28.14	Exact	18.2	18.23	18.26	III	M	19	147	C		1	17
39	39	Budišov	Pod hradem Cave	LM3	R	Exact	26.67	26.96	26.95	Exact	20.07	20.08	20.13	VIII	M	19	146	C		1	18
29	29	Budišov	Pod hradem Cave	C	R	Exact	28.08	28.05	28.11	Exact	21.11	21.1	21.11		M		149			1	14
36	36	Budišov	Pod hradem Cave	C	L	Close estimate	32.38	32.89	32.88	Exact	20.14	20.15	20.23		M		149			1	15
101	101	Budišov	Pod hradem Cave	C		Estimate	29.7	29.86	29.9	Estimate	21.16	21.14	21.19		D		0			1	36
118	118	Budišov	Pod hradem Cave	C		No measurement				No measurement					D		0			1	40
177	178	Budišov	Pod hradem Cave	C		No measurement				No measurement					D		0			1	58
195	196	Budišov	Pod hradem Cave	C		No measurement				No measurement					D		0			1	64
833	836	Budišov	Pod hradem Cave	C		No measurement				Close estimate	19.3	19.3	19.3		D		31		f	1	184
114	114	Budišov	Pod hradem Cave	C (J)		Close estimate	19.58	19.2	19.2	Exact	14.46	14.6	14.85		D		0			1	39
175	176	Budišov	Pod hradem Cave	C (J)		No measurement				No measurement					D		120	C		1	57
6	6	Budišov	Pod hradem Cave	LM1		No measurement				No measurement				VII	D		161	A	a	1	3
7	7	Budišov	Pod hradem Cave	LM1	R	Exact	27.88	27.84	27.89	Exact	12.96	12.96	12.96	I	D		161	A	a	1	3
8	8	Budišov	Pod hradem Cave	LM1	L	Exact	31.4	31.46	31.48	Exact	14.04	13.97	13.98	II	D		161	A	a	1	3
9	9	Budišov	Pod hradem Cave	LM1	L	Exact	30.16	30.18	30.21	Exact	14.32	14.3	14.3	II	D		161	A	a	1	3
10	10	Budišov	Pod hradem Cave	LM1	R	Exact	29.94	30.05	30.04	Exact	14.1	14.08	14.08	II	D		161	A	a	1	3
35	35	Budišov	Pod hradem Cave	LM1	L	Exact	29.94	29.51	29.5	Exact	15.44	15.54	15.5	IX	M		149			1	15
1	1	Budišov	Pod hradem Cave	LM2	L	Exact	26.17	26.16	26.19	Exact	16.67	16.66	16.67	VI	M		161			1	2
4	4	Budišov	Pod hradem Cave	LM2	L	Exact	31.66	31.45	31.54	Exact	18.33	18.34	18.33	IX	D		161	A	a	1	3

Table. A1.1. (Continued.)

ID	Tooth ID	Storage ID	Site ID	Tooth Type ID	Left/Right	Measurement Type Length	Length1 (mm)	Length2 (mm)	Length3 (mm)	Measurement Type Breadth	Breadth1 (mm)	Breadth2 (mm)	Breadth3 (mm)	Wear Stage	Articulation ID	Layer	Box Number	Medium Box Number Suffix	Small Box Number Suffix	Log Book Number	Log Book Page Number
5	5	Budišov	Pod hradem Cave	LM2	L	Exact	29.55	29.58	29.54	Exact	17.63	17.65	17.58	II	D		161	A	a	1	3
13	13	Budišov	Pod hradem Cave	LM2	L	Exact	30.67	30.66	30.66	Exact	18.58	18.59	18.59	II	M		161	A	b	1	4
14	14	Budišov	Pod hradem Cave	LM2		Close estimate	29.78	29.77	29.77	Close estimate	18.95	18.92	18.95	II	M		161	A	b	1	5
15	15	Budišov	Pod hradem Cave	LM2	L	Close estimate	25.8	25.8	25.8	Close estimate	12.2	12.2	12.2	I	M		161	A	b	1	5
28	28	Budišov	Pod hradem Cave	LM2	R	Exact	31.21	31.18	31.17	Exact	19.86	20.1	20.06	VII	M		149			1	14
34	34	Budišov	Pod hradem Cave	LM2	L	Exact	29.32	29.26	29.3	Exact	18.23	18.33	18.33	IX	M		149			1	15
251	253	Budišov	Pod hradem Cave	LM2	L	Exact	30.67	30.55	30.77	Exact	19.12	19.2	19.43	II	D		105	A		1	81
2	2	Budišov	Pod hradem Cave	LM3	R	Exact	29.43	29.37	29.33	Exact	19.87	19.9	19.79	VIII	D		161	A	a	1	2
3	3	Budišov	Pod hradem Cave	LM3	R	Exact	29.43	27.97	27.97	Exact	18.22	18.27	18.29	VII	D		161	A	a	1	2
27	27	Budišov	Pod hradem Cave	LM3	R	Exact	24.57	24.53	24.53	Exact	20.28	20.36	20.38	VII	M		149			1	14
33	33	Budišov	Pod hradem Cave	LM3	L	Exact	27.38	27.36	27.39	Exact	20.26	20.25	20.27	IX	M		149			1	15
12	12	Budišov	Pod hradem Cave	UM1	R	Exact	27.5	27.5	27.52	Exact	17.3	17.31	17.29	II	D		161	A	a	1	4
11	11	Budišov	Pod hradem Cave	UM2	R	Exact	43.3	43.3	43.33	Exact	23.14	23.11	23.11	III	D		161	A	a	1	4
77	77	Budišov	Pod hradem Cave	UM2	R	No measurement				No measurement				II	D		0			1	0
204	206	Budišov	Pod hradem Cave	UM2	R	No measurement				Exact	22.56	22.11	22.56		D		0			1	67
879	4	Brno	Barová Cave	C		Exact	30.59	30.59	30.59	Exact	23.01	23.01	23.01								
880	5	Brno	Barová Cave	C		Exact	24.52	24.52	24.52	Exact	18.31	18.31	18.31								
883	8	Brno	Barová Cave	C		Exact	30.94	30.94	30.94	Exact	21.69	21.69	21.69								
884	9	Brno	Barová Cave	C		Exact	30.89	30.89	30.89	Exact	22.84	22.84	22.84								
890	15	Brno	Barová Cave	C		Exact	30	30	30	Exact	20.66	20.66	20.66								
902	27	Brno	Barová Cave	C		Exact	24.35	24.35	24.35	Exact	18.02	18.02	18.02								
906	31	Brno	Barová Cave	C		Exact	28.6	28.6	28.6	Exact	20.05	20.05	20.05								
953	78	Brno	Barová Cave	C		Exact	26.37	26.37	26.37	Exact	19.48	19.48	19.48								
954	79	Brno	Barová Cave	C		Exact	23.68	23.68	23.68	Exact	16.72	16.72	16.72								
955	80	Brno	Barová Cave	C		Exact	23.68	23.68	23.68	Exact	16.22	16.22	16.22								
956	81	Brno	Barová Cave	C		Exact	31.29	31.29	31.29	Exact	22.3	22.3	22.3								
957	82	Brno	Barová Cave	C		Exact	30.32	30.32	30.32	Exact	23.76	23.76	23.76								
958	83	Brno	Barová Cave	C		Exact	21.26	21.26	21.26	Exact	16.27	16.27	16.27								
959	84	Brno	Barová Cave	C		Exact	30.11	30.11	30.11	Exact	21	21	21								
960	85	Brno	Barová Cave	C		No measurement				Exact	23.44	23.44	23.44								
961	86	Brno	Barová Cave	C		Exact	20.89	20.89	20.89	Exact	16.44	16.44	16.44								

Table. A1.1. (Continued.)

ID	Tooth ID	Storage ID	Site ID	Tooth Type ID	Left/Right	Measurement Type Length	Length1 (mm)	Length2 (mm)	Length3 (mm)	Measurement Type Breadth	Breadth1 (mm)	Breadth2 (mm)	Breadth3 (mm)	Wear Stage	Articulation ID	Layer	Box Number	Medium Box Number Suffix	Small Box Number Suffix	Log Book Number	Log Book Page Number
962	87	Brno	Barová Cave	C		Exact	23.32	23.32	23.32	Exact	16.72	16.72	16.72								
963	88	Brno	Barová Cave	C		Exact	28.46	28.46	28.46	Exact	20.16	20.16	20.16								
964	89	Brno	Barová Cave	C		Exact	16.41	16.41	16.41	Exact	13.44	13.44	13.44								
965	90	Brno	Barová Cave	C		Exact	18.57	18.57	18.57	Exact	13.82	13.82	13.82								
966	91	Brno	Barová Cave	C		Exact	25.03	25.03	25.03	Exact	16.57	16.57	16.57								
967	92	Brno	Barová Cave	C		Exact	22.75	22.75	22.75	Exact	16.53	16.53	16.53								
968	93	Brno	Barová Cave	C		No measurement				Exact	20.82	20.82	20.82								
969	94	Brno	Barová Cave	C		Exact	21.84	21.84	21.84	Exact	15.96	15.96	15.96								
970	95	Brno	Barová Cave	C		Exact	22.61	22.61	22.61	Exact	15.93	15.93	15.93								
971	96	Brno	Barová Cave	C		Exact	21.54	21.54	21.54	Exact	15.93	15.93	15.93								
972	97	Brno	Barová Cave	C		Exact	21.91	21.91	21.91	Exact	14.99	14.99	14.99								
973	98	Brno	Barová Cave	C		Exact	18.92	18.92	18.92	Exact	13.86	13.86	13.86								
974	99	Brno	Barová Cave	C		Exact	25.33	25.33	25.33	Exact	17.41	17.41	17.41								
975	100	Brno	Barová Cave	C		Exact	23.02	23.02	23.02	Exact	17.5	17.5	17.5								
976	101	Brno	Barová Cave	C		Exact	21.66	21.66	21.66	Exact	16.87	16.87	16.87								
977	102	Brno	Barová Cave	C		Exact	28.47	28.47	28.47	Exact	18.94	18.94	18.94								
978	103	Brno	Barová Cave	C		Exact	20.39	20.39	20.39	Exact	14.53	14.53	14.53								
979	104	Brno	Barová Cave	C		Exact	22.17	22.17	22.17	Exact	16.87	16.87	16.87								
980	105	Brno	Barová Cave	C		Exact	22.58	22.58	22.58	Exact	16.32	16.32	16.32								
981	106	Brno	Barová Cave	C		Exact	24.33	24.33	24.33	Exact	17.38	17.38	17.38								
982	107	Brno	Barová Cave	C		Exact	28.03	28.03	28.03	No measurement											
983	108	Brno	Barová Cave	C		Exact	24.14	24.14	24.14	Exact	18.74	18.74	18.74								
984	109	Brno	Barová Cave	C		Exact	20	20	20	Exact	15.75	15.75	15.75								
985	110	Brno	Barová Cave	C		No measurement				Exact	21.5	21.5	21.5								
996	121	Brno	Barová Cave	C		Exact	20.19	20.19	20.19	Exact	14.83	14.83	14.83								
997	122	Brno	Barová Cave	C		Exact	21.72	21.72	21.72	Exact	16.23	16.23	16.23								
1012	137	Brno	Barová Cave	C	R	No measurement				Exact	13.99	13.99	13.99	VI							
1050	175	Brno	Barová Cave	C		Exact	21.85	21.85	21.85	Exact	15.76	15.76	15.76								
1051	176	Brno	Barová Cave	C		Exact	22.27	22.27	22.27	Exact	15.3	15.3	15.3								
1052	177	Brno	Barová Cave	C		Exact	22.24	22.24	22.24	Exact	16.64	16.64	16.64								
1053	178	Brno	Barová Cave	C		Exact	24.61	24.61	24.61	Exact	20.99	20.99	20.99								

Table. A1.1. (Continued.)

ID	Tooth ID	Storage ID	Site ID	Tooth Type ID	Left/Right	Measurement Type Length	Length1 (mm)	Length2 (mm)	Length3 (mm)	Measurement Type Breadth	Breadth1 (mm)	Breadth2 (mm)	Breadth3 (mm)	Wear Stage	Articulation ID	Layer	Box Number	Medium Box Number Suffix	Small Box Number Suffix	Log Book Number	Log Book Page Number
1054	179	Brno	Barová Cave	C		Exact	21.7	21.7	21.7	Exact	14.9	14.9	14.9								
1055	180	Brno	Barová Cave	C		Exact	20.32	20.32	20.32	Exact	16.05	16.05	16.05								
1056	181	Brno	Barová Cave	C		Exact	17.01	17.01	17.01	No measurement											
1057	182	Brno	Barová Cave	C		Exact	22.1	22.1	22.1	Exact	17.17	17.17	17.17								
1058	183	Brno	Barová Cave	C		Exact	26.38	26.38	26.38	Exact	17.68	17.68	17.68								
1059	184	Brno	Barová Cave	C		Exact	27.77	27.77	27.77	Exact	18.29	18.29	18.29								
1060	185	Brno	Barová Cave	C		No measurement				Exact	19.65	19.65	19.65								
1061	186	Brno	Barová Cave	C		Exact	26.22	26.22	26.22	Estimate	21.26	21.26	21.26								
1062	187	Brno	Barová Cave	C		Exact	28.95	28.95	28.95	Exact	20.96	20.96	20.96								
1063	188	Brno	Barová Cave	C		Exact	23.51	23.51	23.51	Exact	15.71	15.71	15.71								
1064	189	Brno	Barová Cave	C		No measurement				Exact	21.38	21.38	21.38								
1065	190	Brno	Barová Cave	C		Exact	29.42	29.42	29.42	Exact	24.14	24.14	24.14								
1066	191	Brno	Barová Cave	C		Exact	23.53	23.53	23.53	No measurement											
1067	192	Brno	Barová Cave	C		Exact	30.03	30.03	30.03	Exact	21.06	21.06	21.06								
1068	193	Brno	Barová Cave	C		Exact	24.03	24.03	24.03	Exact	16.52	16.52	16.52								
1069	194	Brno	Barová Cave	C		No measurement				Exact	16.54	16.54	16.54								
1070	195	Brno	Barová Cave	C		Exact	22.08	22.08	22.08	Exact	16.12	16.12	16.12								
1071	196	Brno	Barová Cave	C		Exact	23.99	23.99	23.99	Exact	16.51	16.51	16.51								
1072	197	Brno	Barová Cave	C		No measurement				No measurement											
1073	198	Brno	Barová Cave	C		Exact	23.21	23.21	23.21	Exact	16.92	16.92	16.92								
1074	199	Brno	Barová Cave	C		Exact	22.24	22.24	22.24	Exact	16.52	16.52	16.52								
1075	200	Brno	Barová Cave	C		Exact	24.32	24.32	24.32	Exact	17.01	17.01	17.01								
1101	226	Brno	Barová Cave	C		Exact	26.32	26.32	26.32	Exact	16.29	16.29	16.29								
1105	230	Brno	Barová Cave	C		Exact	29.44	29.44	29.44	Exact	21.51	21.51	21.51								
1108	233	Brno	Barová Cave	C		Exact	27.59	27.59	27.59	Exact	19.62	19.62	19.62								
1111	236	Brno	Barová Cave	C		Exact	26.33	26.33	26.33	Exact	19.33	19.33	19.33								
1164	299	Brno	Barová Cave	C		Exact	20.68	20.68	20.68	Exact	15.98	15.98	15.98								
1165	300	Brno	Barová Cave	C		Exact	22.75	22.75	22.75	Exact	16.11	16.11	16.11								
1166	302	Brno	Barová Cave	C		Exact	23.67	23.67	23.67	Exact	15.98	15.98	15.98								
1167	303	Brno	Barová Cave	C		Exact	26.78	26.78	26.78	Exact	17.67	17.67	17.67								
1168	304	Brno	Barová Cave	C		Exact	20.37	20.37	20.37	Exact	15.78	15.78	15.78								

Table. A1.1. (Continued.)

ID	Tooth ID	Storage ID	Site ID	Tooth Type ID	Left/Right	Measurement Type Length	Length1 (mm)	Length2 (mm)	Length3 (mm)	Measurement Type Breadth	Breadth1 (mm)	Breadth2 (mm)	Breadth3 (mm)	Wear Stage	Articulation ID	Layer	Box Number	Medium Box Number Suffix	Small Box Number Suffix	Log Book Number	Log Book Page Number
1169	305	Brno	Barová Cave	C		Exact	21.04	21.04	21.04	Exact	16.3	16.3	16.3								
1170	306	Brno	Barová Cave	C		Exact	21.26	21.26	21.26	Exact	15.8	15.8	15.8								
1171	307	Brno	Barová Cave	C		Exact	21.17	21.17	21.17	Exact	15.84	15.84	15.84								
1172	308	Brno	Barová Cave	C		Exact	20.34	20.34	20.34	Exact	16.44	16.44	16.44								
1173	309	Brno	Barová Cave	C		Exact	28.23	28.23	28.23	Exact	23.73	23.73	23.73								
1174	310	Brno	Barová Cave	C		Exact	19.94	19.94	19.94	Exact	16.21	16.21	16.21								
1175	311	Brno	Barová Cave	C		Exact	26.45	26.45	26.45	Exact	20.03	20.03	20.03								
1176	312	Brno	Barová Cave	C		Exact	24.35	24.35	24.35	Exact	17.48	17.48	17.48								
1177	313	Brno	Barová Cave	C		Exact	23.04	23.04	23.04	Exact	18.76	18.76	18.76								
1178	314	Brno	Barová Cave	C		Exact	23.93	23.93	23.93	Exact	18	18	18								
1179	315	Brno	Barová Cave	C		Exact	21.88	21.88	21.88	Exact	17.13	17.13	17.13								
1180	316	Brno	Barová Cave	C		Exact	21.49	21.49	21.49	Exact	17.8	17.8	17.8								
1181	317	Brno	Barová Cave	C		Exact	19.22	19.22	19.22	Exact	14.88	14.88	14.88								
1182	318	Brno	Barová Cave	C		Exact	31.11	31.11	31.11	Exact	23.77	23.77	23.77								
1183	319	Brno	Barová Cave	C		Exact	20.81	20.81	20.81	Exact	16.67	16.67	16.67								
1184	320	Brno	Barová Cave	C		Exact	24.72	24.72	24.72	Exact	16.65	16.65	16.65								
1185	321	Brno	Barová Cave	C		Exact	22.02	22.02	22.02	Exact	17.67	17.67	17.67								
1186	322	Brno	Barová Cave	C		Exact	22.86	22.86	22.86	Exact	16.31	16.31	16.31								
1187	323	Brno	Barová Cave	C		Exact	24.12	24.12	24.12	Exact	17.43	17.43	17.43								
1195	331	Brno	Barová Cave	C		Exact	31.16	31.16	31.16	Exact	23.06	23.06	23.06								
1235	382	Brno	Barová Cave	C		Exact	25.3	25.3	25.3	Exact	17.68	17.68	17.68								
1236	383	Brno	Barová Cave	C		Exact	28.08	28.08	28.08	Exact	20.21	20.21	20.21								
1239	386	Brno	Barová Cave	C		Exact	22.1	22.1	22.1	Exact	16.98	16.98	16.98								
1240	387	Brno	Barová Cave	C		No measurement				Exact	16.33	16.33	16.33								
1241	388	Brno	Barová Cave	C		Exact	22.13	22.13	22.13	Exact	16.8	16.8	16.8								
1242	389	Brno	Barová Cave	C		Exact	22.08	22.08	22.08	Exact	16.44	16.44	16.44								
1243	390	Brno	Barová Cave	C		Exact	22.71	22.71	22.71	Exact	16.59	16.59	16.59								
878	3	Brno	Barová Cave	LM1	L	Exact	30.23	30.23	30.23	Exact	15.5	15.5	15.5	IV							
881	6	Brno	Barová Cave	LM1	R	Exact	31.97	31.97	31.97	Exact	15.42	15.42	15.42	VII							
886	11	Brno	Barová Cave	LM1	L	Exact	28.97	28.97	28.97	Exact	13.97	13.97	13.97	III							
897	22	Brno	Barová Cave	LM1	L	Exact	31.36	31.36	31.36	Exact	14.27	14.27	14.27	III							

Table. A1.1. (Continued.)

ID	Tooth ID	Storage ID	Site ID	Tooth Type ID	Left/Right	Measurement Type Length	Length1 (mm)	Length2 (mm)	Length3 (mm)	Measurement Type Breadth	Breadth1 (mm)	Breadth2 (mm)	Breadth3 (mm)	Wear Stage	Articulation ID	Layer	Box Number	Medium Box Number Suffix	Small Box Number Suffix	Log Book Number	Log Book Page Number
905	30	Brno	Barová Cave	LM1	L	Exact	31.93	31.93	31.93	Exact	14.34	14.34	14.34	III							
941	66	Brno	Barová Cave	LM1	R	No measurement				No measurement				III							
944	69	Brno	Barová Cave	LM1	L	Exact	30.98	30.98	30.98	Exact	15.94	15.94	15.94	III							
945	70	Brno	Barová Cave	LM1	R	Exact	30.44	30.44	30.44	Exact	15.54	15.54	15.54	VI							
946	71	Brno	Barová Cave	LM1	R	Exact	31.65	31.65	31.65	Exact	14.14	14.14	14.14	IV							
947	72	Brno	Barová Cave	LM1	R	Exact	30.74	30.74	30.74	Exact	13.79	13.79	13.79	III							
948	73	Brno	Barová Cave	LM1	R	Exact	29.97	29.97	29.97	Exact	13.84	13.84	13.84	IV							
949	74	Brno	Barová Cave	LM1	R	Exact	28.77	28.77	28.77	Exact	13.16	13.16	13.16	II							
950	75	Brno	Barová Cave	LM1	R	Exact	29.03	29.03	29.03	Exact	13.23	13.23	13.23	II							
951	76	Brno	Barová Cave	LM1	R	Exact	26	26	26	Exact	12.5	12.5	12.5	III							
952	77	Brno	Barová Cave	LM1	L	Exact	26.97	26.97	26.97	Exact	13.11	13.11	13.11	IV							
1008	133	Brno	Barová Cave	LM1	L	Exact	32.33	32.33	32.33	Exact	14.82	14.82	14.82	III							
1009	134	Brno	Barová Cave	LM1	R	Exact	26.41	26.41	26.41	Exact	13.37	13.37	13.37	IX							
1010	135	Brno	Barová Cave	LM1	R	Exact	27.9	27.9	27.9	Exact	13.66	13.66	13.66	V							
1011	136	Brno	Barová Cave	LM1	L	Exact	28.19	28.19	28.19	Exact	12.1	12.1	12.1	VII							
1013	138	Brno	Barová Cave	LM1		No measurement				No measurement											
1014	139	Brno	Barová Cave	LM1	R	Exact	29.14	29.14	29.14	Exact	13.29	13.29	13.29	III							
1015	140	Brno	Barová Cave	LM1	R	No measurement				Exact	14.36	14.36	14.36	IX							
1016	141	Brno	Barová Cave	LM1	L	Exact	29.42	29.42	29.42	Exact	13.32	13.32	13.32	VII							
1017	142	Brno	Barová Cave	LM1	L	Exact	29.02	29.02	29.02	Exact	13.5	13.5	13.5	IX							
1018	143	Brno	Barová Cave	LM1	L	Exact	30.14	30.14	30.14	Exact	14.04	14.04	14.04	IV							
1021	146	Brno	Barová Cave	LM1	L	Exact	29.28	29.28	29.28	Exact	17.29	17.29	17.29	VI							
1022	147	Brno	Barová Cave	LM1	R	Exact	28.61	28.61	28.61	Exact	16.38	16.38	16.38	III							
1023	148	Brno	Barová Cave	LM1	L	Exact	30.12	30.12	30.12	Exact	17.95	17.95	17.95	IV							
1090	215	Brno	Barová Cave	LM1	R	Exact	29.26	29.26	29.26	Exact	13.52	13.52	13.52	II							
1098	223	Brno	Barová Cave	LM1	L	Exact	27.63	27.63	27.63	Exact	12.88	12.88	12.88	V							
1102	227	Brno	Barová Cave	LM1	L	Exact	29.49	29.49	29.49	Exact	13.74	13.74	13.74	V							
1120	245	Brno	Barová Cave	LM1	L	Exact	27.98	27.98	27.98	Exact	13.38	13.38	13.38	VIII							
1121	246	Brno	Barová Cave	LM1	L	Exact	30.92	30.92	30.92	Exact	16.11	16.11	16.11	III							
1150	285	Brno	Barová Cave	LM1	L	Exact	30.17	30.17	30.17	Exact	14.65	14.65	14.65	IV							
1151	286	Brno	Barová Cave	LM1	L	No measurement				Exact	14.49	14.49	14.49	V							

Table. A1.1. (Continued.)

ID	Tooth ID	Storage ID	Site ID	Tooth Type ID	Left/Right	Measurement Type Length	Length1 (mm)	Length2 (mm)	Length3 (mm)	Measurement Type Breadth	Breadth1 (mm)	Breadth2 (mm)	Breadth3 (mm)	Wear Stage	Articulation ID	Layer	Box Number	Medium Box Number Suffix	Small Box Number Suffix	Log Book Number	Log Book Page Number
1152	287	Brno	Barová Cave	LM1	R	Exact	31.31	31.31	31.31	Exact	15.27	15.27	15.27	IX							
1153	288	Brno	Barová Cave	LM1	L	Exact	29.54	29.54	29.54	Exact	13.61	13.61	13.61	II							
1154	289	Brno	Barová Cave	LM1	R	No measurement				No measurement				VI							
1155	290	Brno	Barová Cave	LM1	L	Exact	29.82	29.82	29.82	Exact	14.57	14.57	14.57	V							
1156	291	Brno	Barová Cave	LM1	R	Exact	29.93	29.93	29.93	Exact	14.7	14.7	14.7	VI							
1157	292	Brno	Barová Cave	LM1	R	Exact	28.73	28.73	28.73	Exact	13.99	13.99	13.99	V							
1188	324	Brno	Barová Cave	LM1	R	Exact	30.01	30.01	30.01	Exact	14.1	14.1	14.1	IV							
1191	327	Brno	Barová Cave	LM1	L	Exact	28.54	28.54	28.54	Exact	12.83	12.83	12.83	V							
1200	336	Brno	Barová Cave	LM1	R	Exact	29.78	29.78	29.78	Exact	14.83	14.83	14.83	IV							
1201	337	Brno	Barová Cave	LM1	R	Exact	29.32	29.32	29.32	Exact	14.77	14.77	14.77	III							
1205	341	Brno	Barová Cave	LM1	L	Exact	28.74	28.74	28.74	Exact	13.12	13.12	13.12	V							
1207	343	Brno	Barová Cave	LM1	R	Exact	29.13	29.13	29.13	Exact	14.72	14.72	14.72	III							
1213	349	Brno	Barová Cave	LM1		Exact	30.08	30.08	30.08	Exact	23.06	23.06	23.06								
1215	351	Brno	Barová Cave	LM1	R	Exact	27.92	27.92	27.92	Exact	19.87	19.87	19.87	V							
1219	355	Brno	Barová Cave	LM1	L	No measurement				No measurement				VIII							
1223	370	Brno	Barová Cave	LM1	R	No measurement				Exact	14.88	14.88	14.88	VIII							
1230	377	Brno	Barová Cave	LM1	R	Exact	28.77	28.77	28.77	Exact	14.06	14.06	14.06	VI							
1233	380	Brno	Barová Cave	LM1	R	Exact	29.82	29.82	29.82	Exact	14	14	14	IV							
877	2	Brno	Barová Cave	LM2	L	Exact	29.51	29.51	29.51	Exact	19.25	19.25	19.25	IV							
882	7	Brno	Barová Cave	LM2	R	Exact	33.46	33.46	33.46	Exact	21.62	21.62	21.62	VII							
887	12	Brno	Barová Cave	LM2	L	Exact	28.29	28.29	28.29	Exact	17.55	17.55	17.55	III							
889	14	Brno	Barová Cave	LM2	L	Exact	31.81	31.81	31.81	Exact	19.07	19.07	19.07	III							
894	19	Brno	Barová Cave	LM2	L	Exact	30.74	30.74	30.74	Exact	18.04	18.04	18.04	IV							
896	21	Brno	Barová Cave	LM2	L	Exact	30.58	30.58	30.58	Exact	17.88	17.88	17.88	III							
899	24	Brno	Barová Cave	LM2	L	Exact	28.46	28.46	28.46	Exact	17.06	17.06	17.06	III							
900	25	Brno	Barová Cave	LM2	R	Exact	29.4	29.4	29.4	Exact	16.97	16.97	16.97	III							
904	29	Brno	Barová Cave	LM2	L	Exact	30.92	30.92	30.92	Exact	17.85	17.85	17.85	III							
907	32	Brno	Barová Cave	LM2	L	Exact	27.76	27.76	27.76	Exact	17.45	17.45	17.45	VI							
908	33	Brno	Barová Cave	LM2	L	No measurement				Exact	18.47	18.47	18.47	VIII							
911	36	Brno	Barová Cave	LM2	R	Exact	33.31	33.31	33.31	Exact	19.8	19.8	19.8	II							
914	39	Brno	Barová Cave	LM2	R	Exact	31.08	31.08	31.08	Exact	17.71	17.71	17.71	VI							

Table. A1.1. (Continued.)

ID	Tooth ID	Storage ID	Site ID	Tooth Type ID	Left/Right	Measurement Type Length	Length1 (mm)	Length2 (mm)	Length3 (mm)	Measurement Type Breadth	Breadth1 (mm)	Breadth2 (mm)	Breadth3 (mm)	Wear Stage	Articulation ID	Layer	Box Number	Medium Box Number Suffix	Small Box Number Suffix	Log Book Number	Log Book Page Number
922	47	Brno	Barová Cave	LM2	L	Exact	29.56	29.56	29.56	Exact	17.23	17.23	17.23	II							
923	48	Brno	Barová Cave	LM2	L	Exact	28.81	28.81	28.81	Exact	17.74	17.74	17.74	III							
927	52	Brno	Barová Cave	LM2	R	No measurement				Exact	19.37	19.37	19.37	II							
928	53	Brno	Barová Cave	LM2	L	Exact	31	31	31	Exact	17.88	17.88	17.88	II							
930	55	Brno	Barová Cave	LM2	R	Exact	33.39	33.39	33.39	Exact	19	19	19	III							
933	58	Brno	Barová Cave	LM2	L	Exact	31.75	31.75	31.75	Exact	18.4	18.4	18.4	III							
1019	144	Brno	Barová Cave	LM2	L	Exact	32.82	32.82	32.82	Exact	15.23	15.23	15.23	III							
1020	145	Brno	Barová Cave	LM2	L	Exact	29.22	29.22	29.22	Exact	17.52	17.52	17.52	IV							
1024	149	Brno	Barová Cave	LM2	L	Exact	31.49	31.49	31.49	Exact	19.63	19.63	19.63	II							
1025	150	Brno	Barová Cave	LM2	L	Exact	28.3	28.3	28.3	Exact	16.4	16.4	16.4	III							
1026	151	Brno	Barová Cave	LM2	R	Exact	32.43	32.43	32.43	Exact	18.39	18.39	18.39	III							
1027	152	Brno	Barová Cave	LM2	R	Exact	28.98	28.98	28.98	Exact	17.57	17.57	17.57	V							
1028	153	Brno	Barová Cave	LM2	L	Exact	26.62	26.62	26.62	Exact	16.08	16.08	16.08	IV							
1029	154	Brno	Barová Cave	LM2	L	Exact	30.9	30.9	30.9	Exact	18.86	18.86	18.86	II							
1030	155	Brno	Barová Cave	LM2	R	Exact	28.62	28.62	28.62	Exact	17.28	17.28	17.28	VIII							
1031	156	Brno	Barová Cave	LM2	L	Exact	29.8	29.8	29.8	Exact	17.55	17.55	17.55	IV							
1032	157	Brno	Barová Cave	LM2	R	Exact	29.77	29.77	29.77	Exact	17.25	17.25	17.25	III							
1079	204	Brno	Barová Cave	LM2	R	Exact	29.2	29.2	29.2	Exact	18.3	18.3	18.3	VII							
1080	205	Brno	Barová Cave	LM2	R	Exact	29.37	29.37	29.37	Exact	17.05	17.05	17.05	III							
1091	216	Brno	Barová Cave	LM2	R	Exact	27.47	27.47	27.47	Exact	16.38	16.38	16.38	II							
1093	218	Brno	Barová Cave	LM2	R	Exact	27.59	27.59	27.59	Exact	17.15	17.15	17.15	V							
1096	221	Brno	Barová Cave	LM2	R	Exact	30.33	30.33	30.33	Exact	18.16	18.16	18.16	III							
1099	224	Brno	Barová Cave	LM2	L	Exact	26.93	26.93	26.93	Exact	17.25	17.25	17.25	V							
1104	229	Brno	Barová Cave	LM2	R	Exact	30.72	30.72	30.72	Exact	18.57	18.57	18.57	VI							
1122	247	Brno	Barová Cave	LM2	L	Exact	30.33	30.33	30.33	Exact	20.18	20.18	20.18	IV							
1132	257	Brno	Barová Cave	LM2	L	Exact	31.52	31.52	31.52	Exact	18.82	18.82	18.82	VI							
1133	258	Brno	Barová Cave	LM2	R	Exact	28.59	28.59	28.59	Exact	17.74	17.74	17.74	IV							
1134	259	Brno	Barová Cave	LM2	L	Exact	31.64	31.64	31.64	Exact	18.23	18.23	18.23	II							
1135	260	Brno	Barová Cave	LM2	R	Exact	32.62	32.62	32.62	Exact	19.81	19.81	19.81	VII							
1136	261	Brno	Barová Cave	LM2	L	Exact	29.6	29.6	29.6	Exact	17.6	17.6	17.6	III							
1137	262	Brno	Barová Cave	LM2	R	Exact	30.28	30.28	30.28	Exact	16.96	16.96	16.96	II							

Table. A1.1. (Continued.)

ID	Tooth ID	Storage ID	Site ID	Tooth Type ID	Left/Right	Measurement Type Length	Length1 (mm)	Length2 (mm)	Length3 (mm)	Measurement Type Breadth	Breadth1 (mm)	Breadth2 (mm)	Breadth3 (mm)	Wear Stage	Articulation ID	Layer	Box Number	Medium Box Number Suffix	Small Box Number Suffix	Log Book Number	Log Book Page Number
1138	263	Brno	Barová Cave	LM2	R	Exact	29.26	29.26	29.26	Exact	17.36	17.36	17.36	III							
1139	264	Brno	Barová Cave	LM2	L	Exact	29.31	29.31	29.31	Exact	17.33	17.33	17.33	III							
1140	265	Brno	Barová Cave	LM2	L	Exact	30.85	30.85	30.85	Exact	18.4	18.4	18.4	III							
1141	266	Brno	Barová Cave	LM2	L	Exact	29.69	29.69	29.69	Exact	17.79	17.79	17.79	IV							
1142	267	Brno	Barová Cave	LM2	R	Exact	34.2	34.2	34.2	Exact	19.41	19.41	19.41	II							
1143	268	Brno	Barová Cave	LM2	R	Exact	28.95	28.95	28.95	Exact	17.93	17.93	17.93	IV							
1192	328	Brno	Barová Cave	LM2	L	Exact	28.75	28.75	28.75	Exact	17.56	17.56	17.56	V							
1194	330	Brno	Barová Cave	LM2	L	Exact	25.74	25.74	25.74	Exact	19.12	19.12	19.12	VI							
1198	334	Brno	Barová Cave	LM2	L	Exact	29.29	29.29	29.29	Exact	17.52	17.52	17.52	II							
1199	335	Brno	Barová Cave	LM2	L	Exact	28.51	28.51	28.51	Exact	17.56	17.56	17.56	VIII							
1202	338	Brno	Barová Cave	LM2	R	Exact	30.32	30.32	30.32	Exact	20.03	20.03	20.03	III							
1203	339	Brno	Barová Cave	LM2	L	Exact	28.34	28.34	28.34	Exact	17.14	17.14	17.14	VII							
1204	340	Brno	Barová Cave	LM2	L	Exact	29.95	29.95	29.95	Exact	17.49	17.49	17.49	II							
1206	342	Brno	Barová Cave	LM2	L	Exact	27.68	27.68	27.68	Exact	17.38	17.38	17.38	V							
1208	344	Brno	Barová Cave	LM2	R	Exact	31.13	31.13	31.13	Exact	19.34	19.34	19.34	III							
1211	347	Brno	Barová Cave	LM2	L	Exact	29.56	29.56	29.56	Exact	19.42	19.42	19.42	VI							
1216	352	Brno	Barová Cave	LM2	R	Exact	29.86	29.86	29.86	Exact	14.86	14.86	14.86	IV							
1217	353	Brno	Barová Cave	LM2	R	Exact	29.2	29.2	29.2	Exact	18.18	18.18	18.18	IV							
1220	366	Brno	Barová Cave	LM2	L	Exact	29.91	29.91	29.91	Exact	18.19	18.19	18.19	VIII							
1221	368	Brno	Barová Cave	LM2	R	Exact	29.68	29.68	29.68	Exact	18.39	18.39	18.39	VI							
1224	371	Brno	Barová Cave	LM2	R	Exact	30.44	30.44	30.44	Exact	19.5	19.5	19.5	VIII							
1229	376	Brno	Barová Cave	LM2	R	Exact	26.24	26.24	26.24	Exact	17.97	17.97	17.97	III							
1231	378	Brno	Barová Cave	LM2	R	Exact	30.47	30.47	30.47	Exact	19.77	19.77	19.77	VI							
1234	381	Brno	Barová Cave	LM2	R	Exact	29.58	29.58	29.58	Exact	18.73	18.73	18.73	IV							
1237	384	Brno	Barová Cave	LM2	L	Exact	27.59	27.59	27.59	Exact	17.01	17.01	17.01	IV							
876	1	Brno	Barová Cave	LM3	L	Exact	29.16	29.16	29.16	Exact	20.12	20.12	20.12	IV							
885	10	Brno	Barová Cave	LM3	L	Exact	26.29	26.29	26.29	Exact	18.89	18.89	18.89	III							
888	13	Brno	Barová Cave	LM3	L	Exact	28.01	28.01	28.01	Exact	18.74	18.74	18.74	III							
891	16	Brno	Barová Cave	LM3	R	Exact	28.91	28.91	28.91	Exact	19.31	19.31	19.31	V							
892	17	Brno	Barová Cave	LM3	L	Exact	28.06	28.06	28.06	Exact	18.91	18.91	18.91	III							
893	18	Brno	Barová Cave	LM3	L	Exact	30.53	30.53	30.53	Exact	22.32	22.32	22.32	IV							

Table. A1.1. (Continued.)

ID	Tooth ID	Storage ID	Site ID	Tooth Type ID	Left/Right	Measurement Type Length	Length1 (mm)	Length2 (mm)	Length3 (mm)	Measurement Type Breadth	Breadth1 (mm)	Breadth2 (mm)	Breadth3 (mm)	Wear Stage	Articulation ID	Layer	Box Number	Medium Box Number Suffix	Small Box Number Suffix	Log Book Number	Log Book Page Number
895	20	Brno	Barová Cave	LM3	L	Exact	26.06	26.06	26.06	Exact	18.94	18.94	18.94	III							
898	23	Brno	Barová Cave	LM3	L	Exact	25.02	25.02	25.02	Close estimate	16.23	16.23	16.23	III							
901	26	Brno	Barová Cave	LM3	R	Exact	26.15	26.15	26.15	Exact	18.09	18.09	18.09	III							
903	28	Brno	Barová Cave	LM3	L	Exact	30.72	30.72	30.72	Exact	19.8	19.8	19.8	III							
916	41	Brno	Barová Cave	LM3	R	Exact	30.22	30.22	30.22	Exact	21.47	21.47	21.47	III							
917	42	Brno	Barová Cave	LM3	L	Exact	27.78	27.78	27.78	Exact	18.88	18.88	18.88	III							
919	44	Brno	Barová Cave	LM3	R	Exact	28.27	28.27	28.27	Exact	21	21	21	VII							
924	49	Brno	Barová Cave	LM3	R	Exact	26.86	26.86	26.86	Exact	18.84	18.84	18.84	VII							
925	50	Brno	Barová Cave	LM3	L	Exact	29.14	29.14	29.14	Exact	20.95	20.95	20.95	III							
926	51	Brno	Barová Cave	LM3	R	Exact	30.8	30.8	30.8	Exact	20.73	20.73	20.73	III							
929	54	Brno	Barová Cave	LM3	R	Exact	27	27	27	Exact	20.13	20.13	20.13	III							
932	57	Brno	Barová Cave	LM3	L	Exact	25.3	25.3	25.3	Exact	17.83	17.83	17.83	III							
940	65	Brno	Barová Cave	LM3	R	Exact	28.06	28.06	28.06	Exact	20.85	20.85	20.85	VIII							
943	68	Brno	Barová Cave	LM3	L	Exact	22.95	22.95	22.95	Exact	17.18	17.18	17.18	V							
995	120	Brno	Barová Cave	LM3	R	Exact	29.42	29.42	29.42	Exact	20.79	20.79	20.79	IV							
1033	158	Brno	Barová Cave	LM3	L	Exact	27.02	27.02	27.02	Exact	19.25	19.25	19.25	V							
1034	159	Brno	Barová Cave	LM3	R	Exact	30.45	30.45	30.45	Exact	20.81	20.81	20.81	III							
1035	160	Brno	Barová Cave	LM3	L	Exact	27.32	27.32	27.32	Exact	19.84	19.84	19.84	III							
1036	161	Brno	Barová Cave	LM3	R	Exact	27.62	27.62	27.62	Exact	20.04	20.04	20.04	III							
1037	162	Brno	Barová Cave	LM3	L	Exact	26.35	26.35	26.35	Exact	18.76	18.76	18.76	IV							
1038	163	Brno	Barová Cave	LM3	R	Exact	26.64	26.64	26.64	Exact	18.46	18.46	18.46	VII							
1039	164	Brno	Barová Cave	LM3	R	Exact	27.07	27.07	27.07	Exact	18.88	18.88	18.88	IV							
1040	165	Brno	Barová Cave	LM3	R	Exact	25.68	25.68	25.68	Exact	18.6	18.6	18.6	V							
1041	166	Brno	Barová Cave	LM3	R	Exact	27.18	27.18	27.18	Exact	18.71	18.71	18.71	III							
1078	203	Brno	Barová Cave	LM3	R	Exact	27.15	27.15	27.15	Exact	18.91	18.91	18.91	III							
1081	206	Brno	Barová Cave	LM3	L	No measurement				Exact	20.47	20.47	20.47	IV							
1092	217	Brno	Barová Cave	LM3	R	Exact	23.94	23.94	23.94	Exact	17.35	17.35	17.35	II							
1094	219	Brno	Barová Cave	LM3	R	Exact	27.52	27.52	27.52	Exact	17.75	17.75	17.75	V							
1095	220	Brno	Barová Cave	LM3	R	Exact	28.21	28.21	28.21	Exact	19.53	19.53	19.53	IV							
1097	222	Brno	Barová Cave	LM3	R	Exact	27.03	27.03	27.03	Exact	19.9	19.9	19.9	III							
1100	225	Brno	Barová Cave	LM3	L	Exact	27.76	27.76	27.76	Exact	17.92	17.92	17.92	V							

Table. A1.1. (Continued.)

ID	Tooth ID	Storage ID	Site ID	Tooth Type ID	Left/Right	Measurement Type Length	Length1 (mm)	Length2 (mm)	Length3 (mm)	Measurement Type Breadth	Breadth1 (mm)	Breadth2 (mm)	Breadth3 (mm)	Wear Stage	Articulation ID	Layer	Box Number	Medium Box Number Suffix	Small Box Number Suffix	Log Book Number	Log Book Page Number
1103	228	Brno	Barová Cave	LM3	L	Exact	27.97	27.97	27.97	Exact	19.55	19.55	19.55	IX							
1119	244	Brno	Barová Cave	LM3	L	Exact	28.05	28.05	28.05	Exact	19.46	19.46	19.46	III							
1123	248	Brno	Barová Cave	LM3	L	Exact	29.44	29.44	29.44	Exact	19.73	19.73	19.73	IV							
1144	269	Brno	Barová Cave	LM3	L	Exact	25.35	25.35	25.35	Exact	18.47	18.47	18.47	IV							
1145	270	Brno	Barová Cave	LM3	L	Exact	28.24	28.24	28.24	Exact	17.8	17.8	17.8	III							
1146	271	Brno	Barová Cave	LM3	R	Exact	26.08	26.08	26.08	Exact	19.21	19.21	19.21	III							
1147	272	Brno	Barová Cave	LM3	R	Exact	29.02	29.02	29.02	Exact	19.52	19.52	19.52	III							
1148	273	Brno	Barová Cave	LM3	R	Exact	26.93	26.93	26.93	Exact	19.82	19.82	19.82	III							
1149	274	Brno	Barová Cave	LM3	L	Exact	28.38	28.38	28.38	Exact	19.18	19.18	19.18	III							
1193	329	Brno	Barová Cave	LM3	L	Exact	29.73	29.73	29.73	Exact	19.61	19.61	19.61	VI							
1196	332	Brno	Barová Cave	LM3	R	Exact	25.07	25.07	25.07	Exact	19.07	19.07	19.07	IV							
1197	333	Brno	Barová Cave	LM3	R	Exact	27.93	27.93	27.93	Exact	20.02	20.02	20.02	IV							
1212	348	Brno	Barová Cave	LM3	L	Exact	25.31	25.31	25.31	Exact	19.38	19.38	19.38	VI							
1214	350	Brno	Barová Cave	LM3	R	Exact	30.02	30.02	30.02	Exact	14.23	14.23	14.23	V							
1218	354	Brno	Barová Cave	LM3	R	Exact	25.49	25.49	25.49	Exact	19.03	19.03	19.03	V							
1222	369	Brno	Barová Cave	LM3	R	Exact	30.13	30.13	30.13	Exact	19.45	19.45	19.45	VI							
1225	372	Brno	Barová Cave	LM3	R	Exact	19.37	19.37	19.37	Exact	27.39	27.39	27.39	VIII							
1226	373	Brno	Barová Cave	LM3	L	Exact	29.12	29.12	29.12	Exact	20.34	20.34	20.34	VII							
1227	374	Brno	Barová Cave	LM3	L	Exact	25.98	25.98	25.98	Exact	18.68	18.68	18.68	IV							
1228	375	Brno	Barová Cave	LM3	R	Exact	30.89	30.89	30.89	Exact	18.68	18.68	18.68	III							
1232	379	Brno	Barová Cave	LM3	R	Exact	29.42	29.42	29.42	Exact	20.24	20.24	20.24	VI							
1238	385	Brno	Barová Cave	LM3	L	Exact	27.01	27.01	27.01	Exact	18.15	18.15	18.15	IV							
909	34	Brno	Barová Cave	UM1	R	Exact	24.97	24.97	24.97	Exact	19.06	19.06	19.06	IX							
910	35	Brno	Barová Cave	UM1	L	Exact	27.57	27.57	27.57	Exact	17.52	17.52	17.52	IV							
918	43	Brno	Barová Cave	UM1	R	Exact	30.24	30.24	30.24	Exact	18.87	18.87	18.87	II							
920	45	Brno	Barová Cave	UM1	L	Exact	32.15	32.15	32.15	Exact	19.91	19.91	19.91	III							
921	46	Brno	Barová Cave	UM1	L	Exact	31.06	31.06	31.06	Exact	20.87	20.87	20.87	II							
931	56	Brno	Barová Cave	UM1	R	Exact	28	28	28	Exact	17.39	17.39	17.39	II							
935	60	Brno	Barová Cave	UM1	R	Exact	32.6	32.6	32.6	Exact	21.79	21.79	21.79	VIII							
936	61	Brno	Barová Cave	UM1	L	Exact	28.98	28.98	28.98	Exact	19.19	19.19	19.19	VIII							
986	111	Brno	Barová Cave	UM1	R	Exact	30.18	30.18	30.18	Exact	18.44	18.44	18.44	IV							

Table. A1.1. (Continued.)

ID	Tooth ID	Storage ID	Site ID	Tooth Type ID	Left/Right	Measurement Type Length	Length1 (mm)	Length2 (mm)	Length3 (mm)	Measurement Type Breadth	Breadth1 (mm)	Breadth2 (mm)	Breadth3 (mm)	Wear Stage	Articulation ID	Layer	Box Number	Medium Box Number Suffix	Small Box Number Suffix	Log Book Number	Log Book Page Number
987	112	Brno	Barová Cave	UM1	R	Exact	28.62	28.62	28.62	Exact	20.36	20.36	20.36	VI							
989	114	Brno	Barová Cave	UM1	L	Exact	29.88	29.88	29.88	Exact	18.49	18.49	18.49	IV							
991	116	Brno	Barová Cave	UM1	R	Exact	28.86	28.86	28.86	Exact	18.73	18.73	18.73	VIII							
994	119	Brno	Barová Cave	UM1	L	Exact	30.06	30.06	30.06	Exact	18.55	18.55	18.55	IV							
999	124	Brno	Barová Cave	UM1	R	Exact	29.05	29.05	29.05	Exact	18.72	18.72	18.72	III							
1000	125	Brno	Barová Cave	UM1	L	Exact	27.82	27.82	27.82	Exact	18.02	18.02	18.02	V							
1001	126	Brno	Barová Cave	UM1	R	Exact	29.66	29.66	29.66	Exact	18.85	18.85	18.85	V							
1002	127	Brno	Barová Cave	UM1	R	Exact	31.14	31.14	31.14	Exact	19.53	19.53	19.53	III							
1003	128	Brno	Barová Cave	UM1	L	Exact	31.08	31.08	31.08	Exact	21.08	21.08	21.08	III							
1004	129	Brno	Barová Cave	UM1	R	Exact	29.31	29.31	29.31	Exact	18.37	18.37	18.37	II							
1005	130	Brno	Barová Cave	UM1	R	Exact	28.72	28.72	28.72	Exact	18.28	18.28	18.28	V							
1006	131	Brno	Barová Cave	UM1	R	No measurement				Exact	19.54	19.54	19.54	III							
1007	132	Brno	Barová Cave	UM1	L	Exact	28.29	28.29	28.29	Exact	17.87	17.87	17.87	II							
1083	208	Brno	Barová Cave	UM1	R	Exact	25.79	25.79	25.79	Exact	17.89	17.89	17.89	III							
1085	210	Brno	Barová Cave	UM1	R	Exact	28.1	28.1	28.1	Exact	17.37	17.37	17.37	III							
1089	214	Brno	Barová Cave	UM1	L	Exact	30.16	30.16	30.16	Exact	19	19	19	II							
1107	232	Brno	Barová Cave	UM1	R	Exact	26.65	26.65	26.65	Exact	20.1	20.1	20.1	VII							
1110	235	Brno	Barová Cave	UM1	L	Exact	25.58	25.58	25.58	Exact	19.57	19.57	19.57	VII							
1113	238	Brno	Barová Cave	UM1	R	Exact	29.66	29.66	29.66	Exact	19.98	19.98	19.98	V							
1115	240	Brno	Barová Cave	UM1	R	Exact	45.32	45.32	45.32	Exact	17.08	17.08	17.08	IV							
1117	242	Brno	Barová Cave	UM1	L	Exact	28.92	28.92	28.92	Exact	17.25	17.25	17.25	IV							
1124	249	Brno	Barová Cave	UM1	R	Exact	28.89	28.89	28.89	Exact	18.26	18.26	18.26	II							
1125	250	Brno	Barová Cave	UM1	R	Exact	29.62	29.62	29.62	Exact	19.07	19.07	19.07	II							
1126	251	Brno	Barová Cave	UM1	R	Exact	27.01	27.01	27.01	Exact	19.13	19.13	19.13	VII							
1127	252	Brno	Barová Cave	UM1	R	Exact	29.33	29.33	29.33	Exact	18.82	18.82	18.82	IV							
1128	253	Brno	Barová Cave	UM1	L	Exact	25.47	25.47	25.47	Exact	17.68	17.68	17.68	VI							
1129	254	Brno	Barová Cave	UM1	L	Exact	27.19	27.19	27.19	Exact	16.64	16.64	16.64	III							
1130	255	Brno	Barová Cave	UM1	L	Exact	28.46	28.46	28.46	Exact	17.79	17.79	17.79	II							
1131	256	Brno	Barová Cave	UM1	L	No measurement				Exact	18.51	18.51	18.51	II							
1190	326	Brno	Barová Cave	UM1	L	Exact	28.25	28.25	28.25	Exact	18.17	18.17	18.17	IV							
1209	345	Brno	Barová Cave	UM1	L	Exact	28.9	28.9	28.9	Exact	18.37	18.37	18.37	IV							

Table. A1.1. (Continued.)

ID	Tooth ID	Storage ID	Site ID	Tooth Type ID	Left/Right	Measurement Type Length	Length1 (mm)	Length2 (mm)	Length3 (mm)	Measurement Type Breadth	Breadth1 (mm)	Breadth2 (mm)	Breadth3 (mm)	Wear Stage	Articulation ID	Layer	Box Number	Medium Box Number Suffix	Small Box Number Suffix	Log Book Number	Log Book Page Number
912	37	Brno	Barová Cave	UM2	R	Exact	42.49	42.49	42.49	Exact	21.49	21.49	21.49	III							
913	38	Brno	Barová Cave	UM2	R	Exact	43.02	43.02	43.02	Exact	22.77	22.77	22.77	III							
915	40	Brno	Barová Cave	UM2	R	Exact	41.27	41.27	41.27	Exact	22.61	22.61	22.61	III							
934	59	Brno	Barová Cave	UM2	R	Exact	45.26	45.26	45.26	Exact	22.32	22.32	22.32	IX							
937	62	Brno	Barová Cave	UM2	L	Exact	46.85	46.85	46.85	Exact	26.1	26.1	26.1	IV							
938	63	Brno	Barová Cave	UM2	R	Exact	44.94	44.94	44.94	Exact	24.67	24.67	24.67	III							
939	64	Brno	Barová Cave	UM2	L	Exact	47.39	47.39	47.39	Exact	23.1	23.1	23.1	III							
942	67	Brno	Barová Cave	UM2	L	No measurement				Exact	22.88	22.88	22.88	II							
988	113	Brno	Barová Cave	UM2	R	Exact	43.18	43.18	43.18	Exact	23.63	23.63	23.63	VI							
990	115	Brno	Barová Cave	UM2	L	Exact	44.09	44.09	44.09	Exact	21.64	21.64	21.64	IV							
992	117	Brno	Barová Cave	UM2	R	Exact	42.92	42.92	42.92	Exact	21.74	21.74	21.74	VIII							
993	118	Brno	Barová Cave	UM2	L	Exact	45.43	45.43	45.43	Exact	23.91	23.91	23.91	IV							
998	123	Brno	Barová Cave	UM2	L	No measurement				Exact	20.59	20.59	20.59	II							
1042	167	Brno	Barová Cave	UM2	L	Exact	46.88	46.88	46.88	Exact	24.58	24.58	24.58	III							
1043	168	Brno	Barová Cave	UM2	R	Exact	43.86	43.86	43.86	Exact	27.14	27.14	27.14	IV							
1044	169	Brno	Barová Cave	UM2	L	Exact	42.46	42.46	42.46	Exact	21.86	21.86	21.86	III							
1045	170	Brno	Barová Cave	UM2	R	Exact	49.28	49.28	49.28	Exact	23.57	23.57	23.57	III							
1046	171	Brno	Barová Cave	UM2	R	Exact	45.15	45.15	45.15	Exact	23.64	23.64	23.64	IV							
1047	172	Brno	Barová Cave	UM2	LR	Exact	44.09	44.09	44.09	Exact	22.89	22.89	22.89	II							
1048	173	Brno	Barová Cave	UM2	R	Exact	36.63	36.63	36.63	Exact	21.99	21.99	21.99	III							
1049	174	Brno	Barová Cave	UM2		Exact	35.37	35.37	35.37	Exact	20.24	20.24	20.24								
1076	201	Brno	Barová Cave	UM2	R	Exact	44.61	44.61	44.61	Exact	23.39	23.39	23.39	II							
1077	202	Brno	Barová Cave	UM2	L	Exact	48.75	48.75	48.75	Exact	23.46	23.46	23.46	IV							
1082	207	Brno	Barová Cave	UM2	L	Exact	43.05	43.05	43.05	Exact	23.44	23.44	23.44	IV							
1084	209	Brno	Barová Cave	UM2	R	Exact	43.03	43.03	43.03	Exact	23.58	23.58	23.58	III							
1086	211	Brno	Barová Cave	UM2	R	Exact	44.17	44.17	44.17	Exact	24.64	24.64	24.64	III							
1087	212	Brno	Barová Cave	UM2	R	Exact	42.4	42.4	42.4	Exact	22.96	22.96	22.96	IV							
1088	213	Brno	Barová Cave	UM2	R	Exact	45.52	45.52	45.52	Exact	23.37	23.37	23.37	III							
1106	231	Brno	Barová Cave	UM2	R	Exact	39.98	39.98	39.98	Exact	20.69	20.69	20.69	VII							
1109	234	Brno	Barová Cave	UM2	L	Exact	40.52	40.52	40.52	Exact	21.07	21.07	21.07	VII							
1112	237	Brno	Barová Cave	UM2	R	Exact	47.78	47.78	47.78	Exact	23.56	23.56	23.56	V							

Table. A1.1. (Continued.)

ID	Tooth ID	Storage ID	Site ID	Tooth Type ID	Left/Right	Measurement Type Length	Length1 (mm)	Length2 (mm)	Length3 (mm)	Measurement Type Breadth	Breadth1 (mm)	Breadth2 (mm)	Breadth3 (mm)	Wear Stage	Articulation ID	Layer	Box Number	Medium Box Number Suffix	Small Box Number Suffix	Log Book Number	Log Book Page Number
1114	239	Brno	Barová Cave	UM2	R	Exact	44.64	44.64	44.64	Exact	24.5	24.5	24.5	IV							
1116	241	Brno	Barová Cave	UM2	L	Exact	27.96	27.96	27.96	Exact	25.71	25.71	25.71	IV							
1118	243	Brno	Barová Cave	UM2	L	Exact	41.77	41.77	41.77	Exact	23.1	23.1	23.1	IX							
1158	293	Brno	Barová Cave	UM2	L	Exact	39.3	39.3	39.3	Exact	20.82	20.82	20.82	VIII							
1159	294	Brno	Barová Cave	UM2	L	Exact	44.61	44.61	44.61	Exact	23.73	23.73	23.73	III							
1160	295	Brno	Barová Cave	UM2	L	Exact	45	45	45	Exact	23.85	23.85	23.85	III							
1161	296	Brno	Barová Cave	UM2	L	No measurement				Exact	22.97	22.97	22.97	VII							
1162	297	Brno	Barová Cave	UM2	R	Exact	41.66	41.66	41.66	Exact	22.25	22.25	22.25	III							
1163	298	Brno	Barová Cave	UM2	R	Exact	48.24	48.24	48.24	Exact	23.45	23.45	23.45	VIII							
1189	325	Brno	Barová Cave	UM2	L	Exact	44.37	44.37	44.37	Exact	22.63	22.63	22.63	IV							
1210	346	Brno	Barová Cave	UM2	L	Exact	43.87	43.87	43.87	Exact	22.5	22.5	22.5	IV							
1244	391	Brno	Šipka Cave	C		Exact	29.67	29.54	29.53	Exact	23.14	23.14	23.09								
1245	392	Brno	Šipka Cave	C		Exact	27.04	27.37	27.14	Exact	21.11	21.14	21.13								
1246	393	Brno	Šipka Cave	C		Exact	25	25.19	25.26	Exact	19.65	19.6	19.6								
1247	394	Brno	Šipka Cave	C		Exact	20.67	20.77	20.83	Exact	16.82	16.82	16.81								
1248	395	Brno	Šipka Cave	C		Exact	20.21	19.87	20.08	Exact	13.88	13.97	18.53								
1249	396	Brno	Šipka Cave	C		No measurement				Exact	19.41	19.35	19.38								
1250	397	Brno	Šipka Cave	C		Exact	22.15	22.15	22.11	Exact	15.66	15.75	15.63								
1251	398	Brno	Šipka Cave	C		Exact	23.12	23.09	22.85	Exact	16.04	16.03	16.01								
1252	399	Brno	Šipka Cave	C		Exact	21	21.08	20.95	Exact	16.44	16.42	16.41								
1253	400	Brno	Šipka Cave	C		Exact	18.34	18.17	18.04	Exact	13.12	13.2	13.21								
1254	401	Brno	Šipka Cave	C		Close estimate	20.34	20.34	20.34	Close estimate	15.26	15.26	15.26								
1255	402	Brno	Šipka Cave	C		Exact	18.78	18.6	18.59	Exact	13.99	14.01	14.04								
1256	403	Brno	Šipka Cave	C		Exact	22.83	22.99	22.99	Exact	16.3	16.3	16.3								
1257	404	Brno	Šipka Cave	C		Exact	21.86	21.94	21.83	Exact	15.38	15.36	15.31								
1258	405	Brno	Šipka Cave	C		Exact	27.56	27.66	27.39	Exact	22.5	22.43	22.45								
1259	406	Brno	Šipka Cave	C		Exact	19.87	19.91	19.92	Exact	16.14	16.17	16.18								
1260	407	Brno	Šipka Cave	C		Exact	18.82	18.68	18.73	Exact	13.82	13.76	13.81								
1261	408	Brno	Šipka Cave	C		Exact	20.13	20.42	20.29	Exact	14.96	14.8	14.73								
1262	409	Brno	Šipka Cave	C		Close estimate	20.44	20.44	20.44	Exact	16.31	16.24	16.2								
1263	410	Brno	Šipka Cave	C		Exact	24.11	24.11	24.11	Exact	18.88	18.63	18.54								

Table. A1.1. (Continued.)

ID	Tooth ID	Storage ID	Site ID	Tooth Type ID	Left/Right	Measurement Type Length	Length1 (mm)	Length2 (mm)	Length3 (mm)	Measurement Type Breadth	Breadth1 (mm)	Breadth2 (mm)	Breadth3 (mm)	Wear Stage	Articulation ID	Layer	Box Number	Medium Box Number Suffix	Small Box Number Suffix	Log Book Number	Log Book Page Number
1264	411	Brno	Šipka Cave	C		No measurement				No measurement											
1265	412	Brno	Šipka Cave	C		Exact	20.15	20.2	20.01	Exact	16.34	16.35	16.34								
1266	413	Brno	Šipka Cave	C		No measurement				Close estimate	13.34	13.34	13.34								
1267	414	Brno	Šipka Cave	C		Exact	27.04	27	27.01	Exact	19.56	19.56	19.54								
1268	415	Brno	Šipka Cave	C		Close estimate	21.18	21.18	21.18	Exact	16.2	15.9	15.93								
1269	416	Brno	Šipka Cave	C		Exact	21.25	20.93	21.24	Exact	14.65	14.68	14.69								
1270	417	Brno	Šipka Cave	C		Exact	22	22	22.05	Exact	15.89	15.87	15.91								
1271	418	Brno	Šipka Cave	C		Exact	23.94	23.84	23.64	Exact	17.51	17.52	17.91								
1272	419	Brno	Šipka Cave	C		Exact	20.78	20.21	20.25	Exact	14.46	14.48	14.51								
1273	420	Brno	Šipka Cave	C		Exact	22.32	22.92	22.35	Exact	15.74	15.29	15.75								
1274	421	Brno	Šipka Cave	C		Exact	27.54	27.54	27.49	Exact	20.83	20.81	20.79								
1275	422	Brno	Šipka Cave	C		Exact	21.66	21.55	21.54	Exact	15.92	15.93	15.92								
1276	423	Brno	Šipka Cave	C		Exact	23.15	23.14	23.13	Exact	17.35	17.46	17.21								
1277	424	Brno	Šipka Cave	C		No measurement				Exact	13.95	14.04	13.98								
1278	425	Brno	Šipka Cave	C		Exact	29.44	29.44	29.44	Close estimate	18.3	18.3	18.3								
1279	426	Brno	Šipka Cave	C		Exact	21.62	21.89	21.54	Exact	15.89	15.99	15.74								
1280	427	Brno	Šipka Cave	C		Close estimate	27.31	27.31	27.31	Close estimate	16.68	16.68	16.68								
1281	428	Brno	Šipka Cave	C		Estimate	25.61	25.61	25.61	Close estimate	18.17	18.17	18.17								
1282	429	Brno	Šipka Cave	C		Close estimate	18.21	18.21	18.21	Close estimate	16.2	16.2	16.2								
1283	430	Brno	Šipka Cave	C		No measurement				Close estimate	19.03	19.03	19.03								
1284	431	Brno	Šipka Cave	C		Exact	27.88	27.51	27.93	Exact	20.92	20.88	20.82								
1285	432	Brno	Šipka Cave	C		Exact	20.7	20.82	20.91	Exact	15.5	15.63	15.71								
1286	433	Brno	Šipka Cave	C		Exact	24.26	24.17	24.28	Exact	16.76	16.8	16.81								
1287	434	Brno	Šipka Cave	C		No measurement				No measurement											
1288	435	Brno	Šipka Cave	C		Exact	29.43	29.45	29.43	Exact	18.03	18.04	18.04								
1289	436	Brno	Šipka Cave	C		No measurement				Close estimate	20.48	20.48	20.48								
1290	437	Brno	Šipka Cave	C		Exact	23.23	23.21	23.11	Exact	15.92	15.95	15.97								
1291	438	Brno	Šipka Cave	C		Exact	21.23	21.27	21.38	Exact	13.89	15.85	13.87								
1292	439	Brno	Šipka Cave	C		Exact	24.46	24.51	24.5	Exact	17.27	17.28	17.23								
1293	440	Brno	Šipka Cave	C		No measurement				Close estimate	18.84	18.84	18.84								
1294	441	Brno	Šipka Cave	C		Exact	18.94	18.94	18.94	Exact	15.37	15.26	15.26								

Table. A1.1. (Continued.)

ID	Tooth ID	Storage ID	Site ID	Tooth Type ID	Left/Right	Measurement Type Length	Length1 (mm)	Length2 (mm)	Length3 (mm)	Measurement Type Breadth	Breadth1 (mm)	Breadth2 (mm)	Breadth3 (mm)	Wear Stage	Articulation ID	Layer	Box Number	Medium Box Number Suffix	Small Box Number Suffix	Log Book Number	Log Book Page Number
1295	442	Brno	Šipka Cave	C		Exact	21.46	21.22	21.35	Exact	15.28	15.31	15.41								
1296	443	Brno	Šipka Cave	C		No measurement				Close estimate	18.08	18.08	18.08								
1297	444	Brno	Šipka Cave	C		Exact	21.44	21.08	21.45	Exact	15.31	15.28	15.32								
1298	445	Brno	Šipka Cave	C		Exact	19.88	19.97	19.82	Exact	15.63	15.52	15.55								
1299	446	Brno	Šipka Cave	C		Exact	19.7	19.54	19.63	Exact	15.12	15.15	15.18								
1300	447	Brno	Šipka Cave	C		No measurement				Exact	15.12	15.15	15.25								
1301	448	Brno	Šipka Cave	C		No measurement				Close estimate	17.22	17.22	17.22								
1302	449	Brno	Šipka Cave	C		No measurement				Close estimate	14.33	14.33	14.33								
1303	450	Brno	Šipka Cave	C		Exact	30.67	30.54	30.53	Exact	22.69	22.7	22.59								
1304	451	Brno	Šipka Cave	C		Close estimate	19.25	19.25	19.25	Close estimate	14.25	14.25	14.25								
1305	452	Brno	Šipka Cave	C		Close estimate	28.45	28.45	28.45	Close estimate	20.05	20.05	20.05								
1306	453	Brno	Šipka Cave	C		Exact	27.62	28.39	27.68	Exact	17.57	17.54	17.55								
1307	454	Brno	Šipka Cave	C		No measurement				Close estimate	21.36	21.36	21.36								
1308	455	Brno	Šipka Cave	C		Close estimate	29.21	29.21	29.21	Exact	20.15	20.07	20.08								
1309	456	Brno	Šipka Cave	C		Close estimate	20.82	20.82	20.82	Exact	13.14	13.12	13.12								
1310	457	Brno	Šipka Cave	C		Exact	24.91	25.26	25.05	Exact	17.46	17.48	17.67								
1311	458	Brno	Šipka Cave	C		Exact	21.62	21.58	21.64	Exact	16.11	15.98	16.38								
1312	459	Brno	Šipka Cave	C		Exact	17.42	17.5	17.56	Exact	13.46	13.57	13.56								
1313	460	Brno	Šipka Cave	C		Exact	17.55	17.56	17.6	Exact	14.3	14.14	14.13								
1314	461	Brno	Šipka Cave	C		Exact	20.66	20.73	20.51	Exact	14.08	13.69	13.78								
1315	462	Brno	Šipka Cave	C		Close estimate	22.93	22.93	22.93	Close estimate	14.18	14.18	14.18								
1316	463	Brno	Šipka Cave	C		Exact	29.24	29.51	29.61	Exact	23.79	23.81	23.84								
1317	464	Brno	Šipka Cave	C		Exact	20.84	20.84	20.84	Exact	14.74	14.81	14.81								
1318	465	Brno	Šipka Cave	C		Exact	18.74	18.83	18.79	Exact	15.01	14.7	15								
1319	466	Brno	Šipka Cave	C		Exact	19.56	20.27	20.04	Exact	15.41	15.42	15.68								
1320	467	Brno	Šipka Cave	C		Exact	18.83	18.73	18.82	Exact	15.07	14.9	15.18								
1321	468	Brno	Šipka Cave	C		Close estimate	24.91	24.91	24.91	No measurement											
1322	469	Brno	Šipka Cave	C		Exact	23.31	23.56	23.65	Exact	16.19	16.19	16.21								
1323	470	Brno	Šipka Cave	C		Exact	19.69	19.75	19.82	Exact	15.42	15.38	15.41								
1324	471	Brno	Šipka Cave	C		Close estimate	22.31	22.31	22.31	Exact	16.66	16.71	16.7								
1325	472	Brno	Šipka Cave	C		Exact	23.78	23.84	23.81	Exact	16.86	16.83	16.86								

Table. A1.1. (Continued.)

ID	Tooth ID	Storage ID	Site ID	Tooth Type ID	Left/Right	Measurement Type Length	Length1 (mm)	Length2 (mm)	Length3 (mm)	Measurement Type Breadth	Breadth1 (mm)	Breadth2 (mm)	Breadth3 (mm)	Wear Stage	Articulation ID	Layer	Box Number	Medium Box Number Suffix	Small Box Number Suffix	Log Book Number	Log Book Page Number
1326	473	Brno	Šipka Cave	C		Exact	22.85	22.86	22.87	Exact	16.28	16.31	16.15								
1327	474	Brno	Šipka Cave	C		Exact	21.59	21.61	21.58	Exact	15.36	15.41	15.55								
1328	475	Brno	Šipka Cave	C		Close estimate	21.6	21.6	21.6	Close estimate	14.92	14.92	14.92								
1329	476	Brno	Šipka Cave	C		No measurement				Close estimate	13.55	13.55	13.55								
1330	477	Brno	Šipka Cave	C		No measurement				Close estimate	25.42	25.42	25.42								
1331	478	Brno	Šipka Cave	C		No measurement				Close estimate	16.51	16.51	16.51								
1332	479	Brno	Šipka Cave	C		Exact	22.77	22.77	22.77	Exact	19.35	19.41	19.58								
1333	480	Brno	Šipka Cave	C		No measurement				Close estimate	14.82	14.82	14.82								
1334	481	Brno	Šipka Cave	C		No measurement				Close estimate	14.61	14.61	14.61								
1335	482	Brno	Šipka Cave	C		No measurement				Close estimate	14.8	14.8	14.8								
1336	483	Brno	Šipka Cave	C		No measurement				Close estimate	18.23	18.23	18.23								
1337	484	Brno	Šipka Cave	C		Exact	17.8	17.56	17.89	Exact	13.38	13.45	13.65								
1338	485	Brno	Šipka Cave	C		No measurement				Exact	18.53	18.69	18.57								
1339	486	Brno	Šipka Cave	C		No measurement				Close estimate	14.51	14.51	14.51								
1340	487	Brno	Šipka Cave	C		No measurement				Exact	14.4	14.76	14.71								
1341	1	Brno	Šipka Cave	C		Close estimate	21.55	21.55	21.55	Close estimate	14.31	14.31	14.31								
1342	2	Brno	Šipka Cave	C		Estimate	24.24	24.24	24.24	Exact	20.65	20.54	20.31								
1343	3	Brno	Šipka Cave	C		Close estimate	27.3	27.3	27.3	Close estimate	19.06	19.06	19.06								
1344	4	Brno	Šipka Cave	C		Exact	27.49	27.53	27.62	Exact	19.61	19.72	19.77								
1345	5	Brno	Šipka Cave	C		No measurement				Close estimate	15.04	15.04	15.04								
1346	6	Brno	Šipka Cave	C		Close estimate	27.58	27.58	27.58	Close estimate	21.59	21.59	21.59								
1347	7	Brno	Šipka Cave	C		No measurement				Close estimate	19.24	19.24	19.24								
1348	8	Brno	Šipka Cave	C		No measurement				No measurement											
1349	9	Brno	Šipka Cave	C		Exact	21.39	21.26	21.34	Exact	16.66	16.05	16.57								
1368	28	Brno	Šipka Cave	C		Close estimate	26.22	26.22	26.22	Exact	19.65	19.65	19.65								
1369	29	Brno	Šipka Cave	C		Close estimate	27.63	27.63	27.63	Close estimate	21.94	21.94	21.94								
1370	30	Brno	Šipka Cave	C		No measurement				No measurement											
1371	31	Brno	Šipka Cave	C		No measurement				No measurement											
1372	32	Brno	Šipka Cave	C		No measurement				Close estimate	15.83	15.83	15.83								
1373	33	Brno	Šipka Cave	C		Close estimate	19.19	19.19	19.19	Close estimate	14.97	14.97	14.97								
1374	34	Brno	Šipka Cave	C		No measurement				Close estimate	13.21	13.21	13.21								

Table. A1.1. (Continued.)

ID	Tooth ID	Storage ID	Site ID	Tooth Type ID	Left/Right	Measurement Type Length	Length1 (mm)	Length2 (mm)	Length3 (mm)	Measurement Type Breadth	Breadth1 (mm)	Breadth2 (mm)	Breadth3 (mm)	Wear Stage	Articulation ID	Layer	Box Number	Medium Box Number Suffix	Small Box Number Suffix	Log Book Number	Log Book Page Number
1385	46	Brno	Šipka Cave	C		Exact	20.77	20.5	20.61	Exact	15.89	15.12	15.18								
1386	47	Brno	Šipka Cave	C		Exact	26.47	26.47	26.47	Close estimate	16.2	16.2	16.2								
1387	48	Brno	Šipka Cave	C		Exact	23.32	23.13	23.15	Exact	15.79	15.79	15.79								
1388	49	Brno	Šipka Cave	C		No measurement				No measurement											
1389	50	Brno	Šipka Cave	C		No measurement				Close estimate	22.12	22.12	22.12								
1390	51	Brno	Šipka Cave	C		Exact	46.95	46.93	47.03	Exact	25.01	25.07	25.05								
1391	52	Brno	Šipka Cave	C		Exact	28.51	28.48	28.53	Exact	17.96	17.89	17.98								
1392	53	Brno	Šipka Cave	C		Exact	30.91	30.95	30.85	Exact	22.61	22.79	22.75								
1394	55	Brno	Šipka Cave	C		Close estimate	25.03	25.03	25.03	Exact	20.69	20.58	20.58								
1395	56	Brno	Šipka Cave	C		Close estimate	28.33	28.33	28.33	Exact	20.78	20.96	21.07								
1396	57	Brno	Šipka Cave	C		Close estimate	29.23	29.23	29.23	Close estimate	21.43	21.43	21.43								
1397	58	Brno	Šipka Cave	C		Exact	22.39	22.31	22.42	Exact	17.09	17.09	17.09								
1398	59	Brno	Šipka Cave	C		Exact	22.14	22.14	22.11	Exact	16.25	16.25	16.25								
1399	60	Brno	Šipka Cave	C		Exact	22.43	22.43	22.51	Exact	16.55	16.55	16.53								
1400	61	Brno	Šipka Cave	C		Exact	27.82	27.15	27.89	Exact	19.37	19.21	19.11								
1401	62	Brno	Šipka Cave	C		Exact	27.86	27.93	27.8	Exact	21.84	22.16	21.93								
1419	80	Brno	Šipka Cave	C		Exact	19.34	19.41	19.4	Exact	14.85	14.65	14.68								
1422	83	Brno	Šipka Cave	C		Close estimate	28.68	28.68	28.68	Exact	20.76	20.65	20.74								
1423	84	Brno	Šipka Cave	C		Close estimate	31.47	31.47	31.47	Exact	21.48	20.62	21.43								
1424	85	Brno	Šipka Cave	C		Exact	29.95	29.98	29.84	Exact	22.09	22.11	22.03								
1425	86	Brno	Šipka Cave	C		Close estimate	31.24	31.24	31.24	Exact	19.18	19.83	19.52								
1426	87	Brno	Šipka Cave	C		Estimate	31.65	31.65	31.65	Estimate	21.85	21.85	21.85								
1427	88	Brno	Šipka Cave	C		Exact	31.65	31.72	31.54	Exact	21.85	21.76	21.82								
1428	89	Brno	Šipka Cave	C		Exact	30.87	30.85	30.92	Exact	21.43	21.05	21.49								
1429	90	Brno	Šipka Cave	C		Exact	27.18	27.52	27.32	Exact	21.49	21.53	21.02								
1430	91	Brno	Šipka Cave	C		Exact	29.05	29.05	29.43	Exact	21.52	21.63	21.54								
1431	92	Brno	Šipka Cave	C		Exact	22	22.04	22.93	Exact	18.24	18.34	18.52								
1432	93	Brno	Šipka Cave	C		Exact	27.46	27.43	27.42	Exact	19.08	19.12	19.53								
1433	94	Brno	Šipka Cave	C		No measurement				No measurement											
1434	95	Brno	Šipka Cave	C		Estimate	23.46	23.46	23.46	Exact	15.8	15.32	15.15								
1435	96	Brno	Šipka Cave	C		Close estimate	24.1	24.1	24.1	Exact	15.41	15.33	15.62								

Table. A1.1. (Continued.)

ID	Tooth ID	Storage ID	Site ID	Tooth Type ID	Left/Right	Measurement Type Length	Length1 (mm)	Length2 (mm)	Length3 (mm)	Measurement Type Breadth	Breadth1 (mm)	Breadth2 (mm)	Breadth3 (mm)	Wear Stage	Articulation ID	Layer	Box Number	Medium Box Number Suffix	Small Box Number Suffix	Log Book Number	Log Book Page Number
1436	97	Brno	Šipka Cave	C		Estimate	21.88	21.88	21.88	Exact	14.87	14.92	14.83								
1437	98	Brno	Šipka Cave	C		Close estimate	23.74	23.74	23.74	Exact	15.64	15.71	15.82								
1438	99	Brno	Šipka Cave	C		Close estimate	24.34	24.34	24.34	Exact	15.33	15.65	15.03								
1439	100	Brno	Šipka Cave	C		Estimate	21.79	21.79	21.79	Exact	16.63	16.61	16.61								
1440	101	Brno	Šipka Cave	C		Exact	19.65	20.02	19.63	Exact	14.91	14.98	15.01								
1441	102	Brno	Šipka Cave	C		Exact	20.79	21.46	20.55	Exact	15.88	15.94	15.94								
1442	103	Brno	Šipka Cave	C		Exact	21.13	21.01	21.1	Exact	15.31	15.41	15.33								
1443	104	Brno	Šipka Cave	C		Exact	21.71	22.57	22.11	Exact	15.56	15.75	15.66								
1444	105	Brno	Šipka Cave	C		Exact	20.04	20.11	19.78	Exact	15.07	14.43	15.01								
1445	106	Brno	Šipka Cave	C		Close estimate	22.29	22.29	22.29	Exact	15.39	15.02	14.97								
1446	107	Brno	Šipka Cave	C		Close estimate	22.97	2.97	22.97	Exact	16.05	16.11	16.32								
1447	108	Brno	Šipka Cave	C		Close estimate	16.86	16.86	16.86	Close estimate	11.28	11.28	11.28								
1448	109	Brno	Šipka Cave	C		Close estimate	21.5	21.5	21.5	Exact	15.59	15.32	15.33								
1449	110	Brno	Šipka Cave	C		Close estimate	21.47	21.47	21.47	Exact	14.49	14.74	14.74								
1450	111	Brno	Šipka Cave	C		Close estimate	21.96	21.96	21.96	Exact	14.32	14.21	14.2								
1451	112	Brno	Šipka Cave	C		Close estimate	19.08	19.08	19.08	Exact	14.04	13.94	14.22								
1452	113	Brno	Šipka Cave	C		Close estimate	21.28	21.28	21.28	Exact	15.73	15.8	15.7								
1453	114	Brno	Šipka Cave	C		Close estimate	20.24	20.24	20.24	Exact	14.04	13.83	13.89								
1454	115	Brno	Šipka Cave	C		No measurement				Estimate	22.54	22.54	22.54								
1455	116	Brno	Šipka Cave	C		Close estimate	20.42	20.42	20.42	Exact	15.76	15.9	15.83								
1456	117	Brno	Šipka Cave	C		Close estimate	18.16	18.16	18.16	Exact	13.23	13.23	13.23								
1457	118	Brno	Šipka Cave	C		Exact	25.82	25.83	25.97	Exact	16.43	16.46	16.46								
1458	119	Brno	Šipka Cave	C		Exact	31.25	31.25	31.25	Exact	21.01	21.01	21.01								
1459	120	Brno	Šipka Cave	C		No measurement				No measurement											
1479	140	Brno	Šipka Cave	C		Exact	21.4	21.33	21.81	Exact	16.02	16.21	16.32								
1485	146	Brno	Šipka Cave	C		Exact	27.76	27.63	27.72	Exact	20.37	20.86	20.12								
1504	165	Brno	Šipka Cave	C		Exact	20.2	20.2	20.13	Exact	14.78	14.59	14.73								
1705	366	Brno	Šipka Cave	C		No measurement				Exact	17.04	17.04	17.04								
1706	367	Brno	Šipka Cave	C		Exact	15.56	15.56	15.56	Exact	14.04	14.04	14.04								
1707	368	Brno	Šipka Cave	C		Exact	22.28	22.28	22.28	Exact	17.42	17.42	17.42								
1708	369	Brno	Šipka Cave	C		No measurement				Exact	18.85	18.85	18.85								

Table. A1.1. (Continued.)

ID	Tooth ID	Storage ID	Site ID	Tooth Type ID	Left/Right	Measurement Type Length	Length1 (mm)	Length2 (mm)	Length3 (mm)	Measurement Type Breadth	Breadth1 (mm)	Breadth2 (mm)	Breadth3 (mm)	Wear Stage	Articulation ID	Layer	Box Number	Medium Box Number Suffix	Small Box Number Suffix	Log Book Number	Log Book Page Number
1350	10	Brno	Šipka Cave	C (J)		Exact	28.7	28.73	28.45	Exact	19.89	19.79	19.84								
1351	11	Brno	Šipka Cave	C (J)		Close estimate	27.4	27.4	27.4	Exact	21.29	21.17	21.85								
1352	12	Brno	Šipka Cave	C (J)		Close estimate	26.29	26.29	26.29	Exact	21.34	21.33	21.32								
1353	13	Brno	Šipka Cave	C (J)		Exact	29.26	29.24	29.28	Exact	19.66	19.59	19.71								
1354	14	Brno	Šipka Cave	C (J)		Exact	22	22.04	22.03	Exact	17.71	17.65	17.71								
1355	15	Brno	Šipka Cave	C (J)		Close estimate	18.98	18.98	18.98	Close estimate	17.39	17.39	17.39								
1356	16	Brno	Šipka Cave	C (J)		No measurement				Close estimate	17.44	17.44	17.44								
1357	17	Brno	Šipka Cave	C (J)		No measurement				Close estimate	16.55	16.55	16.55								
1358	18	Brno	Šipka Cave	C (J)		Close estimate	17.44	17.44	17.44	Close estimate	15.24	15.24	15.24								
1359	19	Brno	Šipka Cave	C (J)		Close estimate	18.94	18.94	18.94	Close estimate	14.91	14.91	14.91								
1360	20	Brno	Šipka Cave	C (J)		Estimate	20.87	20.87	20.87	Close estimate	17.69	17.69	17.69								
1361	21	Brno	Šipka Cave	C (J)		Exact	17.75	17.75	17.75	Close estimate	15.05	15.05	15.05								
1362	22	Brno	Šipka Cave	C (J)		Exact	20.81	20.81	20.81	Close estimate	17.92	17.92	17.92								
1363	23	Brno	Šipka Cave	C (J)		No measurement				No measurement											
1364	24	Brno	Šipka Cave	C (J)		Close estimate	15.5	15.5	15.5	Close estimate	12.61	12.61	12.61								
1365	25	Brno	Šipka Cave	C (J)		No measurement				No measurement											
1366	26	Brno	Šipka Cave	C (J)		Exact	21.42	21.42	21.42	Exact	14.83	14.83	14.83								
1367	27	Brno	Šipka Cave	C (J)		Exact	18.99	18.99	18.99	Exact	13.97	13.97	13.97								
1375	36	Brno	Šipka Cave	C (J)		Close estimate	21.04	21.04	21.04	Close estimate	16.04	16.04	16.04								
1376	37	Brno	Šipka Cave	C (J)		Close estimate	20.89	20.89	20.89	Close estimate	15.85	15.85	15.85								
1377	38	Brno	Šipka Cave	C (J)		Close estimate	18.4	18.4	18.4	Close estimate	17.59	17.59	17.59								
1378	39	Brno	Šipka Cave	C (J)		Close estimate	22.2	22.2	22.2	Close estimate	17.62	17.62	17.62								
1379	40	Brno	Šipka Cave	C (J)		Close estimate	17.84	17.84	17.84	Close estimate	15.81	15.81	15.81								
1380	41	Brno	Šipka Cave	C (J)		Close estimate	22.95	22.95	22.95	Close estimate	18.25	18.25	18.25								
1381	42	Brno	Šipka Cave	C (J)		No measurement				No measurement											
1406	67	Brno	Šipka Cave	LM1	L	Exact	27.3	27.45	27.51	Exact	13.39	13.39	13.39	IV							
1409	70	Brno	Šipka Cave	LM1	R	Exact	28.76	28.76	28.68	Exact	13.71	13.83	13.74	IV							
1414	75	Brno	Šipka Cave	LM1	R	Exact	30.21	30.22	30.01	Exact	13.6	13.54	13.5	VI							
1462	123	Brno	Šipka Cave	LM1	R	Exact	31.22	31.24	31.33	Exact	14.6	14.51	14.62	VII							
1467	128	Brno	Šipka Cave	LM1	L	Exact	30.12	30.12	30.12	Exact	14.36	14.38	14.41	VIII							
1470	131	Brno	Šipka Cave	LM1	L	Exact	27.53	27.58	27.41	Exact	13.06	13.13	13.23	III							

Table. A1.1. (Continued.)

ID	Tooth ID	Storage ID	Site ID	Tooth Type ID	Left/Right	Measurement Type Length	Length1 (mm)	Length2 (mm)	Length3 (mm)	Measurement Type Breadth	Breadth1 (mm)	Breadth2 (mm)	Breadth3 (mm)	Wear Stage	Articulation ID	Layer	Box Number	Medium Box Number Suffix	Small Box Number Suffix	Log Book Number	Log Book Page Number
1472	133	Brno	Šipka Cave	LM1	R	Exact	29.51	29.4	29.33	Exact	13.99	14.01	14.18	III							
1478	139	Brno	Šipka Cave	LM1	R	Exact	29.93	30.03	30.19	Exact	13.99	13.91	14.05	III							
1480	141	Brno	Šipka Cave	LM1	R	Exact	29.37	29.78	29.6	Exact	13.99	14.18	14.47	VI							
1487	148	Brno	Šipka Cave	LM1	R	Exact	27.73	27.64	27.44	Exact	11.85	11.84	11.85	VIII							
1503	164	Brno	Šipka Cave	LM1	L	Exact	26.59	26.63	27.04	Exact	17.06	17.56	17.48	IV							
1507	168	Brno	Šipka Cave	LM1	R	Estimate	31.3	31.3	31.3	Exact	14.55	14.6	14.35	III							
1508	169	Brno	Šipka Cave	LM1	L	No measurement				No measurement				VIII							
1509	170	Brno	Šipka Cave	LM1	L	Estimate	30.73	30.73	30.73	Exact	13.98	14.03	14.01	III							
1510	171	Brno	Šipka Cave	LM1	R	Estimate	29.1	29.1	29.1	Exact	14	14.01	14	IX							
1511	172	Brno	Šipka Cave	LM1	R	Exact	27.9	27.9	27.83	Exact	11.73	11.72	11.85	II							
1512	173	Brno	Šipka Cave	LM1	L	Exact	27.12	27.11	27.14	Exact	12.13	12.14	12.27	III							
1513	174	Brno	Šipka Cave	LM1	R	No measurement				Exact	13.92	13.92	13.92	III							
1514	175	Brno	Šipka Cave	LM1	R	No measurement				Exact	12.49	12.49	12.53	III							
1515	176	Brno	Šipka Cave	LM1	L	Exact	29.35	29.27	29.41	Exact	13.4	13.42	13.45	III							
1516	177	Brno	Šipka Cave	LM1	L	No measurement				Estimate	12.5	12.5	12.5	VIII							
1517	178	Brno	Šipka Cave	LM1	L	Exact	32.59	32.58	32.59	Exact	15.08	15.11	15.07	III							
1518	179	Brno	Šipka Cave	LM1	R	No measurement				Close estimate	13.91	13.91	13.91	III							
1519	180	Brno	Šipka Cave	LM1	R	No measurement				Exact	13.85	13.85	13.86	VI							
1520	181	Brno	Šipka Cave	LM1	R	No measurement				Exact	13.99	14.01	14.01	IX							
1521	182	Brno	Šipka Cave	LM1	L	No measurement				Estimate	12.2	12.2	12.2	III							
1522	183	Brno	Šipka Cave	LM1	L	No measurement				No measurement				V							
1523	184	Brno	Šipka Cave	LM1	R	No measurement				No measurement				V							
1524	185	Brno	Šipka Cave	LM1	R	Close estimate	12.7	12.7	12.7	Exact	30.06	30.04	30.03	VII							
1525	186	Brno	Šipka Cave	LM1	R	No measurement				Exact	14.37	14.32	14.32	IX							
1526	187	Brno	Šipka Cave	LM1	R	No measurement				Exact	17.08	17.12	17.03	IX							
1527	188	Brno	Šipka Cave	LM1	R	Close estimate	31.98	31.98	31.98	Exact	14.59	14.61	14.62	III							
1528	189	Brno	Šipka Cave	LM1	R	Exact	29.78	29.79	29.81	Exact	13.33	13.32	13.32	III							
1529	190	Brno	Šipka Cave	LM1	R	No measurement				No measurement				IV							
1530	191	Brno	Šipka Cave	LM1	R	Exact	31.12	31.21	31.21	Exact	14.63	14.65	14.62	III							
1531	192	Brno	Šipka Cave	LM1	R	No measurement				No measurement				VII							
1532	193	Brno	Šipka Cave	LM1	R	Exact	32.45	32.24	32.31	Exact	14.5	14.5	14.64	III							

Table. A1.1. (Continued.)

ID	Tooth ID	Storage ID	Site ID	Tooth Type ID	Left/Right	Measurement Type Length	Length1 (mm)	Length2 (mm)	Length3 (mm)	Measurement Type Breadth	Breadth1 (mm)	Breadth2 (mm)	Breadth3 (mm)	Wear Stage	Articulation ID	Layer	Box Number	Medium Box Number Suffix	Small Box Number Suffix	Log Book Number	Log Book Page Number
1533	194	Brno	Šipka Cave	LM1	R	No measurement				No measurement				III							
1534	195	Brno	Šipka Cave	LM1	R	No measurement				Exact	14.01	14.01	14.05	III							
1535	196	Brno	Šipka Cave	LM1	L	No measurement				Close estimate	14.49	14.49	14.49	VIII							
1536	197	Brno	Šipka Cave	LM1	R	No measurement				Exact	15	14.99	14.94	III							
1537	198	Brno	Šipka Cave	LM1	L	Exact	27.98	28.01	28.02	Exact	12.44	12.33	12.21	IV							
1538	199	Brno	Šipka Cave	LM1	L	Exact	29.14	29.13	29.14	Exact	12.69	12.72	12.72	III							
1539	200	Brno	Šipka Cave	LM1	L	No measurement				Close estimate	15.22	15.22	15.22	III							
1540	201	Brno	Šipka Cave	LM1	R	No measurement				Close estimate	14.25	14.25	14.25	IX							
1541	202	Brno	Šipka Cave	LM1	R	No measurement				No measurement				III							
1542	203	Brno	Šipka Cave	LM1	R	No measurement				Close estimate	11.59	11.59	11.59	VIII							
1543	204	Brno	Šipka Cave	LM1	L	No measurement				Exact	14.41	14.42	14.43	VIII							
1544	205	Brno	Šipka Cave	LM1	L	No measurement				Close estimate	15.29	15.29	15.29	VII							
1545	206	Brno	Šipka Cave	LM1	L	Close estimate	28.75	28.75	28.75	Close estimate	15.96	15.96	15.96	IX							
1546	207	Brno	Šipka Cave	LM1	R	No measurement				Exact	12.61	12.61	12.61	III							
1547	208	Brno	Šipka Cave	LM1	R	No measurement				Exact	13.18	13.17	13.18	IV							
1548	209	Brno	Šipka Cave	LM1	R	No measurement				Close estimate	13.93	13.93	13.93	VI							
1549	210	Brno	Šipka Cave	LM1	R	No measurement				Close estimate	13.74	13.74	13.74	III							
1550	211	Brno	Šipka Cave	LM1	R	No measurement				Close estimate	12.3	12.3	12.3	III							
1551	212	Brno	Šipka Cave	LM1	L	No measurement				Exact	13.14	13.1	13.06	II							
1552	213	Brno	Šipka Cave	LM1	R	No measurement				No measurement				I							
1383	44	Brno	Šipka Cave	LM2	R	Exact	26.63	26.83	26.81	Exact	15.55	15.54	15.51	II							
1403	64	Brno	Šipka Cave	LM2	R	Exact	30.82	30.84	30.82	Exact	18.42	18.65	18.78	III							
1405	66	Brno	Šipka Cave	LM2	L	Exact	28.02	28.02	28.15	Exact	16.6	16.6	16.6	IV							
1408	69	Brno	Šipka Cave	LM2	R	Exact	30.43	30.42	30.4	Exact	17.9	17.91	17.84	IV							
1410	71	Brno	Šipka Cave	LM2	R	Exact	27.79	27.67	27.71	Exact	17.43	17.48	17.48	IV							
1413	74	Brno	Šipka Cave	LM2	R	Exact	31.22	31.24	31.25	Exact	18.39	18.45	18.63	VI							
1415	76	Brno	Šipka Cave	LM2	R	Exact	29.09	29.12	29.03	Exact	17.3	17.24	17.65	IV							
1417	78	Brno	Šipka Cave	LM2	L	Exact	31.66	31.74	31.6	Exact	19.35	19.34	19.39	III							
1461	122	Brno	Šipka Cave	LM2	R	Exact	29.6	29.58	29.62	Exact	18.29	18.33	18.41	VII							
1464	125	Brno	Šipka Cave	LM2	R	Exact	32.15	31.97	32.05	Exact	18.4	18.46	18.55	III							
1466	127	Brno	Šipka Cave	LM2	L	Exact	30.06	30.08	30.31	Exact	18.09	18.63	18.63	VIII							

Table. A1.1. (Continued.)

ID	Tooth ID	Storage ID	Site ID	Tooth Type ID	Left/Right	Measurement Type Length	Length1 (mm)	Length2 (mm)	Length3 (mm)	Measurement Type Breadth	Breadth1 (mm)	Breadth2 (mm)	Breadth3 (mm)	Wear Stage	Articulation ID	Layer	Box Number	Medium Box Number Suffix	Small Box Number Suffix	Log Book Number	Log Book Page Number
1469	130	Brno	Šipka Cave	LM2	R	Exact	30.93	30.87	30.94	Exact	17.02	17.21	17.34	III							
1471	132	Brno	Šipka Cave	LM2	L	Exact	27.38	27.42	27.33	Exact	16.45	16.76	16.78	III							
1473	134	Brno	Šipka Cave	LM2	R	Exact	28.51	28.18	28.53	Exact	17.84	17.66	17.54	III							
1474	135	Brno	Šipka Cave	LM2	R	Exact	29.91	29.59	29.88	Exact	18	18.01	18	III							
1475	136	Brno	Šipka Cave	LM2	L	Exact	29.18	29.19	29.18	Exact	17.67	17.8	17.64	IV							
1476	137	Brno	Šipka Cave	LM2	L	Exact	30.78	30.74	30.76	Exact	19.1	19.19	19.02	IV							
1481	142	Brno	Šipka Cave	LM2	R	Exact	30.46	30.23	30.26	Exact	17.91	17.84	17.23	VI							
1483	144	Brno	Šipka Cave	LM2	R	Exact	30.2	30.21	30.25	Exact	17.14	17.33	17.41	IV							
1488	149	Brno	Šipka Cave	LM2	R	Exact	28.04	28.42	28.41	Exact	18.56	18.57	18.5	VIII							
1491	152	Brno	Šipka Cave	LM2	R	Exact	30.99	30.51	30.77	Exact	17.52	17.55	17.67	III							
1493	154	Brno	Šipka Cave	LM2	L	Exact	30.8	30.81	30.81	Exact	17.98	18	17.97	VI							
1502	163	Brno	Šipka Cave	LM2	L	Exact	25.69	25.71	25.68	Exact	17.2	17.19	17.18	V							
1608	269	Brno	Šipka Cave	LM2	L	Exact	27.68	27.68	27.68	Exact	15.78	15.78	15.78	III							
1628	289	Brno	Šipka Cave	LM2	R	Exact	30.4	30.4	30.4	Exact	19.19	19.19	19.19	II							
1649	310	Brno	Šipka Cave	LM2	R	Exact	29.51	29.51	29.51	Exact	17.15	17.15	17.15	II							
1654	315	Brno	Šipka Cave	LM2	R	No measurement				Exact	18.9	18.9	18.9	II							
1691	352	Brno	Šipka Cave	LM2	L	No measurement				Exact	21.48	21.48	21.48	II							
1382	43	Brno	Šipka Cave	LM3	L	Exact	27.6	27.92	27.78	Exact	19.36	19.38	19.36	II							
1384	45	Brno	Šipka Cave	LM3	R	Close estimate	25.85	25.85	25.85	Close estimate	17.72	17.72	17.72	II							
1393	54	Brno	Šipka Cave	LM3	L	Exact	27.3	27.22	27.13	Exact	21.01	20.08	20.87	IX							
1402	63	Brno	Šipka Cave	LM3	R	Exact	26.47	26.58	26.64	Exact	18.82	18.85	18.95	III							
1404	65	Brno	Šipka Cave	LM3	L	Exact	29.4	29.35	29.42	Exact	19	19	19	IV							
1407	68	Brno	Šipka Cave	LM3	R	Exact	30.7	30.7	30.64	Exact	21.02	20.95	21.13	IV							
1411	72	Brno	Šipka Cave	LM3	R	Exact	17.18	17.31	17.41	Exact	24.51	24.61	24.41	IV							
1412	73	Brno	Šipka Cave	LM3	R	Exact	28.89	28.91	28.45	Exact	20.31	20.21	20.24	VI							
1416	77	Brno	Šipka Cave	LM3	R	Exact	28.1	28.01	28.15	Exact	18.38	18.21	18.11	IV							
1418	79	Brno	Šipka Cave	LM3	R	Close estimate	28.12	28.12	28.12	Close estimate	22.35	22.35	22.35	II							
1460	121	Brno	Šipka Cave	LM3	R	Exact	27.58	27.44	27.31	Exact	20.47	20.55	20.63	VII							
1463	124	Brno	Šipka Cave	LM3	R	Exact	31.45	31.43	31.47	Exact	20.19	20.14	20.23	III							
1465	126	Brno	Šipka Cave	LM3	L	Exact	25.86	25.32	25.41	Exact	18.71	18.72	18.72	VIII							
1468	129	Brno	Šipka Cave	LM3	L	Exact	25.17	25.32	25.14	Exact	17.89	17.88	17.85	III							

Table. A1.1. (Continued.)

ID	Tooth ID	Storage ID	Site ID	Tooth Type ID	Left/Right	Measurement Type Length	Length1 (mm)	Length2 (mm)	Length3 (mm)	Measurement Type Breadth	Breadth1 (mm)	Breadth2 (mm)	Breadth3 (mm)	Wear Stage	Articulation ID	Layer	Box Number	Medium Box Number Suffix	Small Box Number Suffix	Log Book Number	Log Book Page Number
1477	138	Brno	Šipka Cave	LM3	L	Exact	26.04	25.93	26.02	Exact	19.73	19.73	19.72	IV							
1482	143	Brno	Šipka Cave	LM3	R	Exact	26.33	26.33	26.33	Exact	18.81	18.94	18.81	VI							
1484	145	Brno	Šipka Cave	LM3	R	Exact	25.23	25.18	25.3	Exact	18.56	18.57	18.66	IV							
1486	147	Brno	Šipka Cave	LM3	L	Exact	26.41	26.5	26.49	Exact	18.22	18.01	18.02								
1489	150	Brno	Šipka Cave	LM3	R	Exact	21.73	21.73	21.76	Exact	16.52	16.48	16.51	III							
1490	151	Brno	Šipka Cave	LM3	R	Exact	24.62	24.66	24.63	Exact	17.46	17.45	17.46	III							
1492	153	Brno	Šipka Cave	LM3	R	Exact	24.69	24.54	24.6	Exact	18.24	18.24	18.23	III							
1501	162	Brno	Šipka Cave	LM3	L	Exact	23.1	23.15	23.1	Exact	15.56	15.73	15.73	VIII							
1567	228	Brno	Šipka Cave	LM3	R	Exact	23.33	23.33	23.33	Exact	18.81	18.81	18.81	III							
1609	270	Brno	Šipka Cave	LM3	L	Exact	28.13	28.13	28.13	Exact	18.97	18.97	18.97	III							
1610	271	Brno	Šipka Cave	LM3	L	Exact	26.43	26.43	26.43	Exact	19.78	19.78	19.78	IX							
1611	272	Brno	Šipka Cave	LM3	L	Exact	25.62	25.62	25.62	Exact	18.25	18.25	18.25	III							
1612	273	Brno	Šipka Cave	LM3	R	Exact	23.97	23.97	23.97	Exact	18.22	18.22	18.22	IV							
1613	274	Brno	Šipka Cave	LM3	R	Exact	29.94	29.94	29.94	Exact	21.38	21.38	21.38	VIII							
1614	275	Brno	Šipka Cave	LM3	L	Exact	28.62	28.62	28.62	Exact	18.51	18.51	18.51	IV							
1615	276	Brno	Šipka Cave	LM3	R	Exact	28.52	28.52	28.52	Exact	19.33	19.33	19.33	IV							
1616	277	Brno	Šipka Cave	LM3	R	Exact	28.39	28.39	28.39	Exact	19.1	19.1	19.1	III							
1617	278	Brno	Šipka Cave	LM3	R	Exact	26.4	26.4	26.4	Exact	18.39	18.39	18.39	III							
1618	279	Brno	Šipka Cave	LM3	R	Exact	26.12	26.12	26.12	Exact	18.62	18.62	18.62	V							
1619	280	Brno	Šipka Cave	LM3	R	Exact	31.09	31.09	31.09	Exact	19.6	19.6	19.6	IX							
1620	281	Brno	Šipka Cave	LM3	L	Exact	26.86	26.86	26.86	Exact	18.31	18.31	18.31	III							
1621	282	Brno	Šipka Cave	LM3	R	Exact	25.79	25.79	25.79	Exact	19.41	19.41	19.41	IX							
1622	283	Brno	Šipka Cave	LM3	R	Exact	28.11	28.11	28.11	Exact	18.31	18.31	18.31	IX							
1623	284	Brno	Šipka Cave	LM3	L	Exact	18.23	18.23	18.23	Exact	27.24	27.24	27.24	III							
1624	285	Brno	Šipka Cave	LM3	L	Exact	26.47	26.47	26.47	Exact	19.61	19.61	19.61	III							
1625	286	Brno	Šipka Cave	LM3	L	Exact	27.11	27.11	27.11	Exact	18.13	18.13	18.13	III							
1626	287	Brno	Šipka Cave	LM3	L	Exact	23.92	23.92	23.92	Exact	17.33	17.33	17.33	II							
1627	288	Brno	Šipka Cave	LM3	R	Exact	28.17	28.17	28.17	Exact	16.57	16.57	16.57	VIII							
1629	290	Brno	Šipka Cave	LM3	R	Exact	24.56	24.56	24.56	Exact	17.69	17.69	17.69	IV							
1630	291	Brno	Šipka Cave	LM3	L	Exact	31.21	31.21	31.21	Exact	21.39	21.39	21.39	III							
1631	292	Brno	Šipka Cave	LM3	L	Exact	24.33	24.33	24.33	Exact	16.09	16.09	16.09	III							

Table. A1.1. (Continued.)

ID	Tooth ID	Storage ID	Site ID	Tooth Type ID	Left/Right	Measurement Type Length	Length1 (mm)	Length2 (mm)	Length3 (mm)	Measurement Type Breadth	Breadth1 (mm)	Breadth2 (mm)	Breadth3 (mm)	Wear Stage	Articulation ID	Layer	Box Number	Medium Box Number Suffix	Small Box Number Suffix	Log Book Number	Log Book Page Number
1632	293	Brno	Šipka Cave	LM3	R	Exact	27.17	27.17	27.17	Exact	20.33	20.33	20.33	II							
1633	294	Brno	Šipka Cave	LM3	L	Exact	23.7	23.7	23.7	Exact	17.79	17.79	17.79	IV							
1634	295	Brno	Šipka Cave	LM3	L	Exact	29.46	29.46	29.46	Exact	20.79	20.79	20.79	II							
1635	296	Brno	Šipka Cave	LM3	R	Exact	27.68	27.68	27.68	Exact	20.16	20.16	20.16	III							
1636	297	Brno	Šipka Cave	LM3	L	No measurement				No measurement				VIII							
1637	298	Brno	Šipka Cave	LM3	L	Exact	23.38	23.38	23.38	Exact	17.01	17.01	17.01	IV							
1638	299	Brno	Šipka Cave	LM3	L	Exact	26.17	26.17	26.17	Exact	18.3	18.3	18.3	III							
1639	300	Brno	Šipka Cave	LM3	L	Exact	26.98	26.98	26.98	Exact	15.11	15.11	15.11	VII							
1640	301	Brno	Šipka Cave	LM3	R	Exact	23.34	23.34	23.34	Exact	17.93	17.93	17.93	VI							
1641	302	Brno	Šipka Cave	LM3	R	Exact	26.34	26.34	26.34	Exact	17.93	17.93	17.93	II							
1642	303	Brno	Šipka Cave	LM3	L	Exact	27.95	27.95	27.95	Exact	19.05	19.05	19.05	II							
1643	304	Brno	Šipka Cave	LM3	R	Exact	22.81	22.81	22.81	Exact	18.03	18.03	18.03	II							
1644	305	Brno	Šipka Cave	LM3	L	Exact	28.09	28.09	28.09	Exact	19.79	19.79	19.79	III							
1646	307	Brno	Šipka Cave	LM3	R	Exact	23.96	23.96	23.96	Exact	17.32	17.32	17.32	III							
1647	308	Brno	Šipka Cave	LM3	L	Exact	24.12	24.12	24.12	Exact	17.49	17.49	17.49	II							
1648	309	Brno	Šipka Cave	LM3	L	Exact	27.08	27.08	27.08	Exact	18.09	18.09	18.09	II							
1650	311	Brno	Šipka Cave	LM3	R	No measurement				Exact	17.45	17.45	17.45	VII							
1651	312	Brno	Šipka Cave	LM3	L	Exact	24.08	24.08	24.08	Exact	17.34	17.34	17.34	II							
1652	313	Brno	Šipka Cave	LM3	R	Exact	27.92	27.92	27.92	Exact	19.47	19.47	19.47	II							
1653	314	Brno	Šipka Cave	LM3	L	Exact	22.35	22.35	22.35	Exact	18.77	18.77	18.77	III							
1655	316	Brno	Šipka Cave	LM3	L	Exact	26.93	26.93	26.93	Exact	17.43	17.43	17.43	II							
1656	317	Brno	Šipka Cave	LM3	R	Exact	26.37	26.37	26.37	Exact	16.86	16.86	16.86	V							
1657	318	Brno	Šipka Cave	LM3	L	Exact	28.83	28.83	28.83	Exact	18.91	18.91	18.91	III							
1658	319	Brno	Šipka Cave	LM3	R	Exact	29.47	29.47	29.47	Exact	21.49	21.49	21.49	II							
1659	320	Brno	Šipka Cave	LM3	L	Exact	21.6	21.6	21.6	Exact	15.65	15.65	15.65	VIII							
1660	321	Brno	Šipka Cave	LM3	R	Exact	27.16	27.16	27.16	Exact	17.93	17.93	17.93	III							
1661	322	Brno	Šipka Cave	LM3	L	Exact	26.53	26.53	26.53	Exact	18.98	18.98	18.98	II							
1662	323	Brno	Šipka Cave	LM3	L	Exact	22.95	22.95	22.95	Exact	16.16	16.16	16.16	III							
1663	324	Brno	Šipka Cave	LM3	L	Exact	21.99	21.99	21.99	Exact	16.63	16.63	16.63	III							
1664	325	Brno	Šipka Cave	LM3	L	Exact	24.3	24.3	24.3	Exact	17.79	17.79	17.79	II							
1665	326	Brno	Šipka Cave	LM3	R	Exact	23.04	23.04	23.04	Exact	16.89	16.89	16.89	II							

Table. A1.1. (Continued.)

ID	Tooth ID	Storage ID	Site ID	Tooth Type ID	Left/Right	Measurement Type Length	Length1 (mm)	Length2 (mm)	Length3 (mm)	Measurement Type Breadth	Breadth1 (mm)	Breadth2 (mm)	Breadth3 (mm)	Wear Stage	Articulation ID	Layer	Box Number	Medium Box Number Suffix	Small Box Number Suffix	Log Book Number	Log Book Page Number
1696	357	Brno	Šipka Cave	LM3	L	Exact	27.53	27.53	27.53	Exact	18.59	18.59	18.59	V							
1697	358	Brno	Šipka Cave	LM3	R	Exact	25.6	25.6	25.6	Exact	19.25	19.25	19.25	III							
1698	359	Brno	Šipka Cave	LM3	R	Exact	28.31	28.31	28.31	Exact	19.27	19.27	19.27	III							
1699	360	Brno	Šipka Cave	LM3	R	Exact	21.44	21.44	21.44	Exact	15.05	15.05	15.05	IV							
1700	361	Brno	Šipka Cave	LM3	L	Exact	24.71	24.71	24.71	Exact	17.04	17.04	17.04	IV							
1701	362	Brno	Šipka Cave	LM3	R	Exact	26.08	26.08	26.08	Exact	19.34	19.34	19.34	III							
1702	363	Brno	Šipka Cave	LM3	R	Exact	25.01	25.01	25.01	Exact	18.57	18.57	18.57	III							
1703	364	Brno	Šipka Cave	LM3	R	Exact	22.31	22.31	22.31	Exact	17.5	17.5	17.5	III							
1704	365	Brno	Šipka Cave	LM3	R	Exact	26.57	26.57	26.57	Exact	18.41	18.41	18.41	III							
1421	82	Brno	Šipka Cave	UM1		Exact	28.24	28.17	28.31	Exact	17.31	17.42	17.48	II							
1494	155	Brno	Šipka Cave	UM1	R	Exact	27.51	27.59	27.51	Exact	17.3	17.35	17.2	III							
1496	157	Brno	Šipka Cave	UM1	L	Exact	47.74	47.73	47.74	Exact	24.29	24.23	24.21	VI							
1498	159	Brno	Šipka Cave	UM1	R	Exact	26.46	26.47	26.32	Exact	18.49	18.24	18.59	III							
1500	161	Brno	Šipka Cave	UM1	R	Exact	31.22	30.92	31	Exact	18.93	19.21	19.53	IV							
1506	167	Brno	Šipka Cave	UM1	R	Exact	29.61	29.73	29.42	Exact	18.41	18.52	18.63	IX							
1553	214	Brno	Šipka Cave	UM1	R	No measurement				Close estimate	23.91	23.91	23.91	VII							
1554	215	Brno	Šipka Cave	UM1	L	Exact	39.83	39.83	39.83	Exact	18.73	18.73	18.73	II							
1555	216	Brno	Šipka Cave	UM1	R	Exact	41.45	41.45	41.45	Exact	23.38	23.38	23.38	VII							
1556	217	Brno	Šipka Cave	UM1	L	No measurement				Close estimate	21.37	21.37	21.37	III							
1557	218	Brno	Šipka Cave	UM1	L	No measurement				Close estimate	23.1	23.1	23.1	III							
1558	219	Brno	Šipka Cave	UM1	R	Close estimate	39.45	39.45	39.45	Exact	23.27	23.34	23.21	VIII							
1559	220	Brno	Šipka Cave	UM1	L	No measurement				Exact	20.08	20.08	20.08	III							
1560	221	Brno	Šipka Cave	UM1	L	No measurement				Close estimate	23.74	23.74	23.74	VIII							
1561	222	Brno	Šipka Cave	UM1	L	Exact	42.42	42.42	42.42	Exact	22.32	22.32	22.32	II							
1562	223	Brno	Šipka Cave	UM1	L	No measurement				Exact	22.85	22.85	22.85	II							
1563	224	Brno	Šipka Cave	UM1	R	Close estimate	41	41	41	Exact	22.93	22.93	22.93	II							
1564	225	Brno	Šipka Cave	UM1	R	Close estimate	42.17	42.17	42.17	Exact	23.72	23.72	23.72	III							
1565	226	Brno	Šipka Cave	UM1	L	Estimate	38.38	38.38	38.38	Exact	18.9	18.9	18.9	IX							
1566	227	Brno	Šipka Cave	UM1	R	Estimate	43.77	43.77	43.77	Exact	22.31	22.31	22.31	III							
1568	229	Brno	Šipka Cave	UM1	L	Close estimate	44.62	44.62	44.62	Exact	21.97	21.97	21.97	II							
1569	230	Brno	Šipka Cave	UM1	R	No measurement				No measurement				II							

Table. A1.1. (Continued.)

ID	Tooth ID	Storage ID	Site ID	Tooth Type ID	Left/Right	Measurement Type Length	Length1 (mm)	Length2 (mm)	Length3 (mm)	Measurement Type Breadth	Breadth1 (mm)	Breadth2 (mm)	Breadth3 (mm)	Wear Stage	Articulation ID	Layer	Box Number	Medium Box Number Suffix	Small Box Number Suffix	Log Book Number	Log Book Page Number
1570	231	Brno	Šipka Cave	UM1	L	No measurement				Close estimate	22.14	22.14	22.14	III							
1571	232	Brno	Šipka Cave	UM1	R	No measurement				Close estimate	19.69	19.69	19.69	IV							
1572	233	Brno	Šipka Cave	UM1	L	Exact	41.09	41.09	41.09	Exact	19.92	19.92	19.92	III							
1573	234	Brno	Šipka Cave	UM1	L	No measurement				Close estimate	22.3	22.3	22.3	III							
1574	235	Brno	Šipka Cave	UM1	R	Exact	36.55	36.55	36.55	Exact	18.65	18.65	18.65	III							
1575	236	Brno	Šipka Cave	UM1	R	Exact	42.26	42.26	42.26	Exact	20.37	20.37	20.37	IX							
1576	237	Brno	Šipka Cave	UM1	R	No measurement				Close estimate	21.81	21.81	21.81	III							
1577	238	Brno	Šipka Cave	UM1	L	No measurement				Close estimate	21.15	21.15	21.15	III							
1578	239	Brno	Šipka Cave	UM1	L	No measurement				Close estimate	21.74	21.74	21.74	II							
1579	240	Brno	Šipka Cave	UM1	R	No measurement				Close estimate	19.96	19.96	19.96	II							
1580	241	Brno	Šipka Cave	UM1	L	Estimate	39.9	39.9	39.9	Exact	19.27	19.27	19.27	III							
1581	242	Brno	Šipka Cave	UM1	R	No measurement				Close estimate	20.39	20.39	20.39	IX							
1582	243	Brno	Šipka Cave	UM1	R	Estimate	44.23	44.23	44.23	Exact	23.59	23.59	23.59	II							
1583	244	Brno	Šipka Cave	UM1	R	No measurement				Close estimate	22.08	22.08	22.08	III							
1584	245	Brno	Šipka Cave	UM1	R	No measurement				Close estimate	22.52	22.52	22.52	III							
1585	246	Brno	Šipka Cave	UM1	L	No measurement				Close estimate	18.76	18.76	18.76	II							
1586	247	Brno	Šipka Cave	UM1	R	No measurement				Exact	24.49	24.49	24.49	III							
1587	248	Brno	Šipka Cave	UM1	R	Exact	47.08	47.08	47.08	Close estimate	23.09	23.09	23.09	III							
1588	249	Brno	Šipka Cave	UM1	R	No measurement				Close estimate	23.82	23.82	23.82	III							
1589	250	Brno	Šipka Cave	UM1	L	Exact	47.09	47.09	47.09	Close estimate	21.59	21.59	21.59	VIII							
1590	251	Brno	Šipka Cave	UM1	L	No measurement				Exact	22.29	22.29	22.29	III							
1591	252	Brno	Šipka Cave	UM1	L	Estimate	38.85	38.85	38.85	Exact	19.12	19.12	19.12	III							
1592	253	Brno	Šipka Cave	UM1	R	Close estimate	40.65	40.65	40.65	Exact	20.16	20.16	20.16	III							
1593	254	Brno	Šipka Cave	UM1	R	No measurement				Close estimate	22.8	22.8	22.8	III							
1594	255	Brno	Šipka Cave	UM1	R	Estimate	46.88	46.88	46.88	Exact	23.86	23.86	23.86	III							
1595	256	Brno	Šipka Cave	UM1	R	No measurement				Close estimate	23.4	23.4	23.4	III							
1596	257	Brno	Šipka Cave	UM1	L	No measurement				Exact	22.69	22.69	22.69	II							
1597	258	Brno	Šipka Cave	UM1	L	No measurement				Exact	21.44	21.44	21.44	III							
1598	259	Brno	Šipka Cave	UM1	R	No measurement				Exact	23.2	23.2	23.2	III							
1599	260	Brno	Šipka Cave	UM1	L	Estimate	41.04	41.04	41.04	Exact	21.09	21.09	21.09	III							
1600	261	Brno	Šipka Cave	UM1	R	No measurement				Exact	20.38	20.38	20.38	III							

Table. A1.1. (Continued.)

ID	Tooth ID	Storage ID	Site ID	Tooth Type ID	Left/Right	Measurement Type Length	Length1 (mm)	Length2 (mm)	Length3 (mm)	Measurement Type Breadth	Breadth1 (mm)	Breadth2 (mm)	Breadth3 (mm)	Wear Stage	Articulation ID	Layer	Box Number	Medium Box Number Suffix	Small Box Number Suffix	Log Book Number	Log Book Page Number
1601	262	Brno	Šipka Cave	UM1	L	No measurement				Close estimate	23.59	23.59	23.59	III							
1602	263	Brno	Šipka Cave	UM1	L	Estimate	45.12	45.12	45.12	Exact	22.59	22.59	22.59	II							
1603	264	Brno	Šipka Cave	UM1	R	Estimate	43.88	43.88	43.88	Exact	22.78	22.78	22.78	III							
1604	265	Brno	Šipka Cave	UM1	R	Close estimate	40.41	40.41	40.41	Exact	18.12	18.12	18.12	II							
1605	266	Brno	Šipka Cave	UM1	L	No measurement				Close estimate	21.76	21.76	21.76	III							
1606	267	Brno	Šipka Cave	UM1	L	Exact	46.58	46.58	46.58	Exact	21.61	21.61	21.61	II							
1607	268	Brno	Šipka Cave	UM1	L	No measurement				Exact	22.31	22.31	22.31	VI							
1666	327	Brno	Šipka Cave	UM1	L	Exact	27.72	27.72	27.72	Exact	17.81	17.81	17.81	II							
1667	328	Brno	Šipka Cave	UM1	L	Exact	25.59	25.59	25.59	Exact	18.65	18.65	18.65	IX							
1668	329	Brno	Šipka Cave	UM1	L	Exact	26.58	26.58	26.58	Exact	19.61	19.61	19.61	IX							
1669	330	Brno	Šipka Cave	UM1	R	Exact	30.75	30.75	30.75	Exact	19.15	19.15	19.15	VIII							
1670	331	Brno	Šipka Cave	UM1	L	Exact	26.97	26.97	26.97	Exact	19.38	19.38	19.38	IX							
1671	332	Brno	Šipka Cave	UM1	L	Exact	31.63	31.63	31.63	Exact	19.84	19.84	19.84	II							
1672	333	Brno	Šipka Cave	UM1	L	Exact	30.67	30.67	30.67	Exact	20.56	20.56	20.56	II							
1673	334	Brno	Šipka Cave	UM1	R	Exact	23.74	23.74	23.74	Exact	18.11	18.11	18.11	IX							
1674	335	Brno	Šipka Cave	UM1	R	Exact	29.83	29.83	29.83	Exact	19.13	19.13	19.13								
1675	336	Brno	Šipka Cave	UM1	L	Exact	28	28	28	Exact	18.11	18.11	18.11	III							
1676	337	Brno	Šipka Cave	UM1	R	Exact	29.12	29.12	29.12	Exact	20.72	20.72	20.72	IX							
1677	338	Brno	Šipka Cave	UM1	L	Exact	26.58	26.58	26.58	Exact	18.54	18.54	18.54	VIII							
1678	339	Brno	Šipka Cave	UM1	L	Exact	28.61	28.61	28.61	Exact	19.87	19.87	19.87	VIII							
1679	340	Brno	Šipka Cave	UM1	R	Exact	30.31	30.31	30.31	Exact	17.74	17.74	17.74	III							
1680	341	Brno	Šipka Cave	UM1	L	Exact	29.11	29.11	29.11	Exact	17.62	17.62	17.62	III							
1681	342	Brno	Šipka Cave	UM1	L	Exact	22.01	22.01	22.01	Exact	16.27	16.27	16.27	IX							
1682	343	Brno	Šipka Cave	UM1	R	Exact	26.36	26.36	26.36	Exact	17.1	17.1	17.1	VII							
1683	344	Brno	Šipka Cave	UM1	R	No measurement				Exact	20.33	20.33	20.33	IX							
1684	345	Brno	Šipka Cave	UM1	L	Exact	27.84	27.84	27.84	Exact	18.91	18.91	18.91	IX							
1685	346	Brno	Šipka Cave	UM1	R	Exact	23.41	23.41	23.41	Exact	17.35	17.35	17.35	IX							
1686	347	Brno	Šipka Cave	UM1	R	No measurement				Exact	18.21	18.21	18.21	III							
1687	348	Brno	Šipka Cave	UM1	L	Exact	25.15	25.15	25.15	Exact	17.77	17.77	17.77	IX							
1688	349	Brno	Šipka Cave	UM1	L	Exact	23.28	23.28	23.28	Exact	17.27	17.27	17.27	IX							
1689	350	Brno	Šipka Cave	UM1	L	Exact	25.66	25.66	25.66	Exact	17.34	17.34	17.34	II							

Table. A1.1. (Continued.)

ID	Tooth ID	Storage ID	Site ID	Tooth Type ID	Left/Right	Measurement Type Length	Length1 (mm)	Length2 (mm)	Length3 (mm)	Measurement Type Breadth	Breadth1 (mm)	Breadth2 (mm)	Breadth3 (mm)	Wear Stage	Articulation ID	Layer	Box Number	Medium Box Number Suffix	Small Box Number Suffix	Log Book Number	Log Book Page Number
1690	351	Brno	Šipka Cave	UM1	R	Exact	25.02	25.02	25.02	Exact	17.67	17.67	17.67	II							
1692	353	Brno	Šipka Cave	UM1	R	Exact	29.26	29.26	29.26	Exact	19.31	19.31	19.31	III							
1693	354	Brno	Šipka Cave	UM1	R	Exact	31.14	31.14	31.14	Exact	18	18	18	II							
1694	355	Brno	Šipka Cave	UM1	R	No measurement				Exact	16.4	16.4	16.4	II							
1695	356	Brno	Šipka Cave	UM1	R	No measurement				Exact	18.62	18.62	18.62	III							
1420	81	Brno	Šipka Cave	UM2		Exact	39.35	39.21	39.41	Exact	19.2	19.4	19.3	II							
1495	156	Brno	Šipka Cave	UM2	L	Exact	28.74	28.44	28.63	Exact	18.93	18.01	18.43	VI							
1497	158	Brno	Šipka Cave	UM2	R	Exact	42.22	42.38	42.35	Exact	20.63	21.08	20.84	VIII							
1499	160	Brno	Šipka Cave	UM2	L	Exact	45.89	45.93	45.62	Exact	20.53	20.31	20.21	III							
1505	166	Brno	Šipka Cave	UM2	L	Exact	43.07	43.14	43.19	Exact	21.39	21.6	21.44	IV							
1645	306	Brno	Šipka Cave			No measurement				No measurement				III							

Table. A1.1. (Continued.)

Appendix 2

	Mean δ13C	Std. Dev.	Mean δ15N	Std Dev.
SO1	-21.89	0.10	3.03	0.08
SO2	-21.21	0.31	2.46	0.13
PH01	-21.65	0.03	4.19	0.02
PH02	-21.66	0.05	3.68	0.02
PH04	-21.29	0.01	1.31	0.01

Table A2.1. Cave-bear stable isotope results (δ13C and δ15N). SO1 and SO2 are results associated with cave bears of from Šipka Cave, and PH01, PH02, and PH04 are results associated with cave-bears remains from Pod hradem Cave. Dark shaded areas are void results due to a lack of collagen in the samples, and light shaded areas indicate valid results.

Bibliography

Ábelová, M. 2006. Migration Pattern Inferred from *Ursus Spelaeus* Rosenmüller Tooth from Tmavá Cave (Slovak Republic) Using Strontium Isotope Analyses. In E. Tsoukala and G. Rabeder (eds.), *Scientific Annals of the School of Geology: Proceedings of the 12th International Cave Bear Symposium (I.C.B.S.). (Special Volume)*, 123-126. School of Geology of Aristotle University: Thessaloniki.

Absolon, K. 1938. Die Erforschung Der Diluvialen Mammut-Jüger-Station Von Unter-Wisternitz Anden Pollauer Bergen in Mähren. Verlag Barvič and Novotný: Brünn.

Adams, B. 1998. The Middle to Upper Paleolithic Transition in Central Europe: The Record from the Bükk Mountain Region. B. A. R. 693: Oxford.

Adams, B. 2009. Bükk Mountain Szeletian: Old and New Views on "Transitional" Material from the Eponymous Site of the Szeletian. In M. Camps and P. Chauhan (eds.), *Sourcebook of Paleolithic Transitions: Methods, Theories, and Interpretations*, 427-440. Springer: London.

Adler, D., Bar-Oz, G., Belfer-Cohen, A. and Bar-Yosef, O. 2006. Ahead of the Game: Middle and Upper Palaeolithic Hunting Behaviours in the Southern Caucasus. *Current Anthropology* **47**, 89-118.

Aldenderfer, M. 1996. Introduction. In M. Aldenderfer and H. Maschner (eds.), *Anthropology, Space, and Geographic Information Systems*, 3-18. Oxford University Press: Oxford.

Allen, J. and Huntley, B. 2000. Weichselian Palynological Records from Southern Europe: Correlation and Chronology. *Quaternary International* **73-4**, 111-125.

Allsworth-Jones, P. 1986. The Szeletian and The Transition From Middle to Upper Palaeolithic in Central Europe. Clarendon Press: Oxford.

Anikovich, M., Sinitsyn, A., Hoffecker, J., Holliday, T., Popov, V., Lisitsyn, S., Forman, S., Levkovskaya, G., Pospelova, G., Kuz'mina, I., Burova, N., Goldberg, P., Macphail, R., Giaccio, B. and Praslov, N. 2007. Early Upper Paleolithic in Eastern Europe and Implications for the Dispersal of Modern Humans. *Science* **315**, 223-226.

Argenti, P. and Mazza, P. 2006. Mortality Analysis of the Late Pleistocene Bears From Grotta Lattaia, Central Italy. *Journal of Archaeological Science* **33**, 1552-1558.

Bahn, P. 1982. Inter-Site and Inter-Regional Links During the Upper Paleolithic: The Pyrenean Evidence. *Oxford Journal of Archaeology* **1**, 247-268.

Bahn, P. and Vertut, J. 1988. *Images of the Ice Age*. Winward: Leicester.

Bahn, P. and Vertut, J. 1999. *Journey Through The Ice Age*. Seven Dials: London.

Bailey, G. Carter, P., Gamble, C. and Higgs, H. 1983. Epirus Revisited: Seasonality and Inter-Site Variation in the Upper Palaeolithic of North-West Greece. In G. Bailey (ed.), *Hunter-Gatherer Economy in Prehistory: A European Perspective*, 64-78. Cambridge University Press: Cambridge.

Banks, W., d'Errico, F., Peterson, A.T., Kageyama, M., Sima, A. and Sánchez-Goñi, M-F. 2008. Neanderthal Extinction by Competitive Exclusion. *PLoS ONE* **3**, e3972.

Barbour, M., Burk, J., Pitts, W., Gilliam, F. and Schwartz, M. (3rd edition) 1999. *Terrestrial Plant Ecology*. Benjamin Cummings: Menlo Park, California.

Bar-Yosef, O. 2000. The Middle and Early Upper Paleolithic in Southwest Asia and Neighboring Regions. In O. Bar-Yosef and D. Pilbeam (eds.), *The Geography of Neandertals and Modern Humans in Europe and the Greater Mediterranean*, 107-156. Peabody Museum: Cambridge, Massachusetts.

Bar-Yosef, O. 2002. The Upper Paleolithic Revolution. *Annual Review of Anthropology* **31**, 363-393.

Bar-Yosef, O. and Pilbeam, D. 2000. The Geography of Neandertals and Modern Humans in Europe and the Greater Mediterranean. Peabody Museum: Cambridge, Massachusetts.

Barker, G. 2006. The Agricultural Revolution in Prehistory: Why Did Foragers Become Farmers? Oxford University Press: Oxford.

Barker, W. 1956. *Familiar Animals of America*. Museum Press Limited: London.

Barren, E., van Andel T. and Pollars, D. 2003. Glacial Environments II: Reconstructing the Climate of Europe in the Last Glaciation. In T. van Andel and W. Davies (eds.), *Neanderthals and Modern Humans in the European Landscape During the Last Glaciation*. McDonald Institute for Archaeological Reserch: Cambridge.

Baryshnikov, G., Germonpré, M. and Sablin, M. 2003. Sexual Dimorphism and Morphometric Variability of Cheek Teeth of the Cave Bear (*Ursus spelaeus*). *Belg. J. Zool* **133**, 111-119.

Bell, T. and Lock, G. 2000. Topographic and Cultural Influences on Walking the Ridgeway in Later Prehistoric Times. In G. Lock (ed.), *Beyond the Map: Archaeology and Spatial Technologies*, 85-100. IOS Press: Amsterdam.

Berger, J., Dulamtseren, S., Cain, S., Enkkhbileg, D., Lichtman, P., Namshir, Z., Wingard, G. and Reading, R. 2001. Back-Casting Sociality in Extinct Species: New Perspectives Using Mass Death Assemblages and Sex Ratios. *Proceedings of the Royal Society B* **268**, 131-139.

Biesele, M. 2001. Ju/'Hoan Women's Tracking Knowledge and its Contribution to Their

Husbands' Hunting Success. *African Study Monographs, Supplement* **26**, 67-84.

Binford, L. 1980. Willow Smoke and Dog's Tails: Hunter-Gatherer Settlement Systems and Archaeological Site Formation. *American Antiquity* **45**, 4-20.

Bocherens, H., Billiou, D., Patou-Mathis, M., Bonjean, D., Otte, M. and Mariotti, A. 1997. Paleobiological Implications of the Isotopic Signitures (^{13}C, ^{15}N) of Fossil Mammal Collagen in Scladina Cave (Sclayn, Belgium). *Quaternary Research* **48**, 370-380.

Bocherens, H., Conrad, N., Germonpre, M., Hofreiter, M., Münzel, S., Stephan, E. and Tütken, T. 2007. Isotopic Biogeochemistry and the Evolution of Cave Bear Ecology During Marine Oxygen Isotpic Stage 3 in Western and Central Europe. In R. Musil and V. Vávra (eds.), *Geology* **35**, Proceedings of the 13th International Cave Bear Symposium. Department of Geological Sciences: Brno, 103-106.

Bocherens, H., Drucker, D., Brilliou, D., Geneste, J-M., van der Plicht, J. 2006. Bears and Humans in Chauvet Cave (Vallon-Pont-d'Arc, Ardèche, France): Insights from Stable Isotopes and Radiocarbon Dating of Bone Collagen. *Journal of Human Evolution* **50**, 370-376.

Bocherens, H., Fizet, M. and Maariotti, A. 1994. Diet, Physiology and Ecology of Fossil Mammals as Inferred from Stable Carbon and Nitrogen Isotope Biogeochemistry: Implications for Pleistocene Bears. *Palaeogeography, Palaeoclimatology, Palaeoecology* **107**, 213-225.

Bocquet-Appel, J-P. and Demars, P. 2000. Neanderthal Contraction and Modern Human Colonization of Europe. *Antiquity* **74**, 544-552.

Bond, J. 1996. Burnt Offerings: Animal Bone in Anglo-Saxon Cremations. *World Archaeology* **28**, 76-88.

Boydston, E., Kapheim, K., Szykman, M. and Holekamp, K. 2003. Individual Variation in Space Use by Female Spotted Hyenas. *Journal of Mammalogy* **84**, 1006-1018.

Boyle, K. 1990. Upper Palaeolithic Faunas from South-West France: A Zoogeographical Perspective. B. A. R. 557: Oxford.

Boyle, K. 1998. The Middle Palaeolithic Geography of Southern France. Resources and Site Location. B. A. R. 723. Oxford.

Boyle, K. 2007. Changing Biodiversity and Complexity Across the Middle-Upper Palaeolithic Transition. In P. Mellars, K. Boyle, O. Bar-Yosef and C. Stringer (eds.), *Rethinking the Human Revolution*, 303-316. McDonald Institute for Archaeological Research: Cambridge.

Brantingham, P., Kuhn, S. and Kerry, K. 2004. *The Early Upper Paleolithic Beyond Western Europe.* University of California Press: Berkeley.

Brown, A. 1997. Alluvial Geoarchaeology: Floodplain Archaeology and Environmental Change. Cambridge University Press: Cambridge.

Bruggeman, J., Garrott, R., White, P., Watson, F. and Wallen, R. 2007. Covariates Affecting Spatial Variability in Bison Travel Behaviour in Yellowstone National Park. *Ecological Applications* **17**, 1411-1423.

Bunn, H. 1981.Archaeological Evidence for Meat-Eating By Plio-Pleistocene Hominids From Koobi. *Nature* **291**, 574-577.

Bunnell, F. and McCann, R. 1993. The Brown or Grizzly Bear. In I. Stirling (ed.), *Bears*, 88-95. HarperCollins: London.

Burke, A. 2006. Neanderthal Settlement Patterns in Crimea: A Landscape Approach. *Journal of Anthropological Archaeology* **25**, 510-523.

Burke, A., Ebert, D. and Dauth, D. 2008. Paleoethology As A Tool For The Development of Archaeological Models of Land-Use: The Crimean Middle Palaeolithic. *Journal of Archaeological Science* **35**, 894-904.

Burroughs, W. 2005. *Climate Change in Prehistory: The End of The Reign of Chaos.* Cambridge University Press: Cambridge.

Burt, W. 1943. Territoriality and Home Range Concepts as Applied to Mammals. *Journal of Mammalogy* **24**, 346-352.

Burton, R. 1970. *Animal Senses*. David & Charles: Newton Abbot.

Butzer, K. 1986. Paleolithic Adaptations and Settlement in Cantabrian Spain. *Advances in World Archaeology* **5**, 201-52.

Cansdale, G. 1952. *Animals and Man.* Hutchinson: London.

Cant, J. and Temerin, L. 1984. A Conceptual Approach to Foraging Adaptations in Primates. In P. Rodman and J. Cant (eds.), *Adaptations for Foraging in Nonhuman Primates: Contributions to an Organismal Biology of Prosimians, Monkeys, and Apes*, 304-342. Columbia University Press: New York.

Carbonell, E., Vaquero, M., Maroto, J., Rand, J. and Mallol, C. 2000. A Geographic Perspective on the Middle to Upper Paleolithic Transition in the Iberian Peninsula. In O. Bar-Yosef and D. Pilbeam (eds.), *The Geography of Neandertals and Modern Humans in Europe and the Greater Mediterranean*, 5-34. Peabody Museum: Cambridge, Massachusetts.

Carl, E. 1971. Population Control in Arctic Ground Squirrels. *Ecology* **52**, 395-413.

Champion, T., Gamble, C., Shennan, S. and Whittle, A. 1984. *Prehistoric Europe*. Academic: London.

Chase, P. 1986. The Hunters of Combe Grenal: Approaches to Middle Paleolithic Subsistence in Europe. B. A. R. 286: Oxford.

Chauvet, J-M., Deschamps, E. and Hillaire, C. 1996. *Chauvet Cave: The Discovery of the World's Oldest Paintings*. Thames and Hudson: London.

Cherry, J. 1986. Politics and Palaces: Some Problems in Minoan State Formation. In C. Renfrew and J. Cherry (eds.), *Peer-polity Interaction and Socio-Political Change*, 19-46. Cambridge University Press: Cambridge.

Chilardi, D., Frayer, D., Gioia, P., Macchiarelli, R. and Mussi, M. 1996. Fontana Nuova Di Ragusa (Sicily, Italy): Southernmost Aurignacian Site in Europe. *Antiquity* **70**, 553-563.

Childe, V. 1931. The Forest Cultures of Northern Europe: A Study in Evolution and Diffusion. *Royal Anthropological Institute of Great Britain and Ireland* **LX1**, 325-348.

Christiansen, P. 1999. What Size Were *Arctodus Simus* and *Ursus spelaeus* (Carnivora: Ursidae)? *Annales Zoologici Fennici* **36**, 93-102.

Clark, D. 1954. *Excavations at Star Carr*. Cambridge University Press: Cambridge.

Clark, G. 1983. Boreal Phase Settlement/Subsistence models for Cantabrian Spain. In G. Bailey (ed.), *Hunter-Gatherer Economy in Prehistory: A European perspective*, 96-110. Cambridge University Press: Cambridge.

Cohen, E. 1994. Animals in Medieval Perceptions: The Image of the Ubiquitous Other. In A. Manning and J. Serpell (eds.), *Animals and Human Society: Changing Perspectives*, 59-80. Routledge: London.

Collins, S. and Britten, R. 1924. Variation in Eyesight at Different Ages, As Determined by the Snellen Test. *Public Health Report* **39**, 3189-3195.

Connolly J. and Lake, M. 2006. *Geographical Information Systems in Archaeology*. Cambridge University Press: Cambridge.

Conrad, N. and Bolus, M. 2002. Radiocarbon Dating the Appearance of Modern Humans and Timing of Cultural Innovations in Europe: New Results and New Challenges. *Journal of Human Evolution* **44**, 331-371.

Coward, F. and Gamble, C. 2010. Metaphor and Materiality in Early Prehistory. In L. Malafouris and C. Renfrew (eds.), *The Cognitive Life of Things*, 47-58. McDonald Institute Monographs: Cambridge.

Craighead, J. and Mitchell, J. 1982. Grizzly Bear. In J. Champion and G. Feldhamer (eds.), *Wild Mammals of North America*, 515-556. The Johns Hopkins University Press: Baltimore, Maryland.

Cutler, K., Gray, S., Burr, G., Edwards, R., Taylor, F., Cabioch, G., Beck, J., Cheng, H. and Moore, J. 2004. Radiocarbon Calibration and Comparison to 50 kyr BP with Paired C-14 and Th-230 Dating of Corals from Vanuatu and Papua New Guinea. *Radiocarbon* **46**, 1127-1160.

Dahle, B. and Swensen, J. 2003a. Home Range in Adult Scandinavian Brown Bears (*Ursus arctos*): Effect of Mass, Sex, Reproductive Category, Population Density and Habitat Type. *Journal of Zoology* **260**, 329-335.

Dahle, B. and Swenson, J. 2003b. Seasonal Range Size in Relation to Reproductive Strategies in Brown Bears *Ursus Arctos*. *Journal of Animal Ecology* **72**, 660-667.

Dansgaard, W., Johnsen, S., Clausen, H., Dahl-Jensen, D., Gundestrup, N., Hammer, C., Hvidberg, C., Steffensen, J., Sveinbjörnsdottir, J., Jouzel, J. and Bond, G. 1993. Evidence for General Instability of Past Climate From a 250-kyr Ice-Core Record. *Nature* **364**, 218-220.

Davidson, I. 1983. Site Variability and Prehistoric Economy in Levante. In G. Bailey (ed.), *Hunter-Gatherer Economy in Prehistory: A European Perspective*, 79-85. Cambridge University Press: Cambridge.

Davies, W., Valdes, P., Ross, C. and van Andel, T. 2003. The Human Presence in Europe During the Last Glacial Period III: Site Clusters, Regional Climates and Resource Attractions. In T. van Andel and W. Davies (eds.), *Neanderthals and Modern Humans in the European Landscape During The Last Glaciation*, 191-220. McDonald Institute for Archaeological Research: Cambridge.

Davis, S. 1987. *The Archaeology of Animals*. Batsford: London.

Debeljak, I. 2007. Fossil Population Structure and Mortality of the Cave Bear from the Mokrica Cave (North Slovenia). *Acta Carsologica* **36**, 475-484.

Dell'Arte, G. and Leonardi, G. 2008. Spatial Patterns of Red Fox (*Vulpes vulpes*) Dens in a Semi-Arid Landscape of North Africa. *African Journal of Ecology* **46**, 168-173.

Derocher, A. and Stirling, I. 1993. Bears as Predators. In I. Stirling (ed.), *Bears*, 84-85. HarperCollins: London.

d'Errico, F. and Vanhaeren, M. 2007. Evolution or Revolution? New Evidence for the Origin of Symbolic Behaviour In and Out of Africa. In P. Mellars, K. Boyle, O. Bar-Yosef and Stringer, C. (eds.), *Rethinking the Human Revolution*, 275-286. McDonald Institute for Archaeological Research: Cambridge.

d'Errico, F., Zilhão, J., Julien, M., Baffier, D. and Pelegrin, J. 1998. Neanderthal Acculturation in Western Europe? A Critical Review of the Evidence and Its Interpretation. *Current Anthropology* **39** (Supp.), S1-S44.

Descola, P. 1992. Societies of Nature and the Nature of Society. In A. Kuper (ed.), *Conceptualising Society*, 107-126. Routledge: London and New York.

Descola, P. 1994. *In The Society of Nature: A Native Ecology in Amazonia*. Cambridge University Press: Cambridge.

Dickinson, T. 2005. Symbols of Protection: The Significance of Animal-Ornamented Shields in Early Anglo-Saxon England. *Medieval Archaeology* **XLIX**, 109-164.

Diedrich, C. 2006. Cave Bear Open Air site Remains and Den Caves From Upper Pleistocene of Central Bohemia (Czech Republic). In E. Tsoukala and G. Rabeder (eds.), *Scientific Annals of the School of Geology: Proceedings of the 12th International Cave Bear Symposium (I.C.B.S.). (Special Volume)*, 187-192. School of Geology of Aristotle University: Thessaloniki.

Diedrich, C. 2008. The Holotypes of the Upper Pleistocene *Crocuta crocuta spealaea* (Goldfuss 1823: Hyaenidae) and *Panthera leo spelaea* (Goldfuss 1810: Felidae) of the Zoolithen Cave Hyena Den (South Germany) and their Palaeo-ecology. *Zoological Journal of the Linnean Society* **154**, 822-831.

Diedrich, C. 2009a. Steppe Lion Remains Imported by Ice Age Spotted Hyenas into the Late Pleistocene Perick Caves Hyena Den in Northern Germany. *Quaternary Research* **71**, 361-374.

Diedrich, C. 2009b. Upper Pleistocene *Panthera leo spelaea* (Goldfuss 1810) Remains From the Bilstein Caves (Sauerland Karst) and Contribution to the Steppe Lion Taphonomy, Palaeobiology and Sexual Dimorphism. *Annales de Paléontologie* **95**, 117-138.

Domico, T. and Newman, M. 1988. *Bears of the World.* Facts On File: New York.

Donald, M. 2010. The Exographic Revolution: Neuropsychological Sequelae. In L. Malafouris and C. Renfrew (eds.), *The Cognitive Life of Things*, 71-79. McDonald Institute Monographs: Cambridge.

Doncaster, P. and Woodroffe, R. 1993. Den Site Can Determine Shape and Size of Badger Territories: Implications for Group-Living. *Oikos* **66**, 88-93.

Eerkens, J. and Lipo, C. 2007. Cultural Transmission Theory and the Archaeological Record: Providing Context to Understanding Variation and Temporal Changes in Material Culture. *Journal of Archaeological Research* **15**, 239-274.

Ellen, R. 1972. The Marsupial in Nuaulu Ritual Behaviour. *Man* 7, 223-238.

Fedriani, J., Palomares, F. and Delibes, M. 1999. Niche Relations Among the Sympatric Mediterranean Carnivores. *Oecologia* **121**, 138-148.

Fernández, N., Palomares, F. and Delibes, M. 2002. The Use of Breeding Dens and Kitten Development in the Iberian Lynx (*Lynx pardinus*). *Journal of Zoology* **258**, 1-5.

Fisher, R. 1999. Geographical Information Systems: Today and Tomorrow?. In M. Gillings, D. Mattingly, and J. Van Dalen (eds.), *Geographical Information Systems and Landscape Archaeology*, 5-11. Oxbow: Oxford.

Fletcher, W., Sánchez Goñi, M., Allen, J., Cheddadi, R., Combourieu-Nebout, N., Huntley, B., Lawson, I., Londeix, L., Magri, D., Mardari, V., Müller, U., Naughton, F., Novenko, E., Roucoux, K., Tzedakis, P. (in press). Millennial-Scale Variability During the Last Glacial Vegetation Records from Europe. *Quaternary Science Reviews* (2009), doi:10.1016/j.quascirev.2009.11.015.

Fowler, C. 2004. The Archaeology of Personhood: An Anthropological Approach. Routledge: London.

Gaffney, V. and van Leusen, M. 1995. Postscript – GIS, Environmental Determinism and Archaeology. In G. Lock and Z. Stančič (eds.), *Archaeology and Geographical Information Systems: A European Perspective*, 367-382. Taylor and Francis: London.

Gamble, C. 1986. *The Palaeolithic Settlement of Europe.* Cambridge University Press: Cambridge.

Gamble, C. 1993. Timewalkers: The Prehistory of Global Colonisation. Penguin Books: London.

Gamble, C. 1998. Palaeolithic Society and the Release from Proximity: A Network Approach to Intimate Relations. *World Archaeology* **29**, 426-449.

Gamble, C. 1999. *Palaeolithic Societies of Europe.* Cambridge University Press: Cambridge.

Gamble, C. 2010. Technologies of Seperation and the Evolution of Social Extension. In R. Dunbar, C. Gamble, and J. Gowlett (eds.), Social Brain, Distributed Mind: *Proceedings of the British Academy* **158**, 17-42.

Gamble, C. and Gaudzinski, S. 2005. Bones and Powerful Individuals: Faunal Case Studies from The Arctic and The European Middle Palaeolithic. In C. Gamble and M. Porr (eds.), *The Hominid Individual in Context: Archaeological Investigations of Lower and Middle Palaeolithic Landscapes, Locales and Artefacts*, 154-175. Routledge: London.

Ganskopp, D., Cruz, R. and Johnson, D. 2000. Least-Effort Pathways?: A GIS Analysis of Livestock Trails in Rugged Terrain. *Applied Animal Behaviour Science* **68**, 179-190.

Gargett, R. 1996. Cave Bears and Modern Human Origins: The Spatial Taphonomy of Pod Hradem Cave, Czech Republic. University Press of America: Lanham, Maryland.

Garneau, D., Boudreau, T., Keech, M. and Post, E. 2008. Habitat Use by Black Bears in Relation to Conspecifics and Competitors. *Mammalian Biology* **73**, 48-57.

Garshelis, D. 2000. Delusions in Habitat Evaluation: Measuring Use, Selection and Importance. In L. Boitani and T. Fuller (eds.), *Research Techniques in Animal Ecology, Controversies and Consequences*, 111-164. Columbia University Press: New York.

Gathercole, P. 1995. *Animals in Medieval French Manuscript Illumination.* Edwin Mellen Press: Lewiston, New York.

Gehrt, S. 2003. Racoon. In G. Feldhamer, B. Thompson and J. Chapman (eds.) (2nd edition) 2003. *Wild Mammals of North America: Biology,*

Management, and Conservation, 611-634. John Hopkins University Press: Maryland.

Gell, A. 1995. The Language of the Forest: Landscape and Phonological Iconism in Umeda. In E. Hirsch and M. O'Hanlon (eds.), *The Anthropology of Landscape: Perspectives on Places and Space*, 232-254. Oxford University Press: Oxford.

Gibson, E. 2007. The Archaeology of Movement in a Mediterranean Landscape. *Journal of Mediterranean Archaeology* **20**, 61-87.

Gibson, J. 1982. The Myth of Passive Perception: A Reply to Richards. In E. Reed and R. Jones (eds.), *Reasons for Realism: Selected Essays of James J. Gibson*, 397-400. Lawrence Erlbaum: New Jersey.

Giddens, A. 1984. The Constitution of Society: Outline of the Theory of Structuration. Polity Press: Cambridge.

Gillings, M. 1995. Flood Dynamics and Settlement in the Tisza Valley of North-East Hungary: GIS and the Upper Tisza Project. In G. Lock and Z. Stančič (eds.), *Archaeology and Geographical Information Systems*, 67-84. Taylor Francis: London.

Gillings, M. and Wise, A. 1990. *GIS Guide to Good Practice*. Oxbow: Oxford.

Glenz, C., Massolo, A., Kuonen, D. and Schlaepfer, R. 2001. A Wolf Habitat Suitability Prediction Study in Valais (Switzerland). *Landscape and Urban Planning* **55**, 55-65.

Goebel, T. 2004. The Early Upper Paleolithic of Siberia. In P. Brantingham, S. Kuhn and K. Kerry (eds.), *The Early Upper Paleolithic Beyond Western Europe*, 162-195. University of California Press: Berkeley.

Goffman, E. 1961. Encounters: Two Studies in the Sociology of Interactions. Bobbs-Merrill: Indianapolis.

Goffman, E. 1967. Interaction Ritual: Essays on Face-to-Face Behaviour. Anchor Books: New York.

Goffman, E. 2006. The Presentation of Self. In D. Brissett and C. Edgley (eds.), *Life As Theater: A Dramatical Sourcebook* (2nd Edition), 129-139. Transaction Publishers: New Jersey.

Golledge, R. 2003. Human Wayfinding and Cognitive Maps. In M. Rockman and J. Steele (ed.), *Colonization of Unfamiliar Landscapes: The Archaeology of Adaptation*, 25-43. Routledge: London and New York.

Golonka, J. and Picha, F. 2006. The Carpathians and Their Foreland: Geology and Hydrocarbon Resources. Datapages: Prague.

Goodrich, J. and Berger, J. 1994. Winter Recreation and Hibernating Black Bears *Ursus americanus*. *Biological Conservation* **67**, 105-110.

Gough, M. and Rushton, S. 2000. The Application of GIS-Modelling to Mustelid Landscape Ecology. *Mammal Review* **30**, 197-216.

Grayson, D. and Delpech, F. 2003. Ungulates and the Middle-to-Upper Paleolithic Transition at Grotte XVI (Dordogne, France). *Journal of Archaeological Science* **30**, 1633-1648.

Gremmen, N. 1982. The Vegetation of the Subantarctic Islands Marion and Prince Edward. Dr. W. Junk: The Hague, Boston, and London.

Griffin, D. 1992. *Animal Minds*. University of Chicago Press: Chicago.

Gudeman, S. 1986. Economics as Culture: Models and Metaphors of Livelihood. Routledge: London.

Guthrie, D. 1990. *Frozen Fauna of the Mammoth Steppe: The Story of Blue Babe*. University of Chicago Press: Chicago.

Guthrie, D. 2005. *The Nature of Paleolithic Art*. University of Chicago Press: Chicago.

Hall, E. 1981. *The Mammals of North America* (2nd edition). John Wiley and Sons: New York.

Hallowell, I. 1926. Bear Ceremonialism in the Northern Hemisphere. *American Anthropologist* **28**, 1-175.

Hannerz, U. 1992. The Global Ecumene as a Network of Networks. In A. Kuper (ed.), *Conceptualising Society*, 34-58. Routledge: London and New York.

Hardy, B. 2004. Neanderthal Behaviour and Stone Tool Function at the Middle Palaeolithic Site of La Quina, France. *Antiquity* **301**, 547-565.

Harestad, H. and Bunnell, F. 1979. Home Range and Body Weight – A Re-evaluation. *Ecology* **69**, 389-402.

Harris, M. 1998. The Rhythm of Life on the Amazonian Floodplain: Seasonality and Sociality in a Riverine Village. *The Journal of the Royal Anthropological Institute* **4**, 65-82.

Henshilwood, C. and d'Errico, F. 2005. Being Modern in the Middle Stone Age: Individuals and Innovation. In C. Gamble and M. Porr (eds.), *The Hominid Individual in Context: Archaeological Investigations of Lower and Middle Palaeolithic Landscapes, Locales and Artefacts*, 244-264. Routledge: London.

Hesse, B. and Wapnish, P. 1985. *Animal Bone Archaeology: From Objectives to Analysis*. Taraxacum: Washington.

Heywood, I., Cornelius, S. and Carver, S. 1998. *An Introduction to Geographical Information Systems*. Longman: New York.

Higgs, E. and Jarman, M. 1975. Palaeoeconomy. In E. Higgs (ed.), *Palaeoeconomy*, 1-8. Cambridge University Press: Cambridge.

Higgs, E. and Vita-Finzi, C. 1972. Prehistoric Economies: A Territorial Approach. In E. Higgs (ed.), *Papers in Economic Prehistory*, 22-36. Cambridge University Press: Cambridge.

Hill, W. 1938. *The Agricultural and Hunting Methods of the Navaho Indians*. Yale University Publications in Anthropology, 18. New Haven.

Hillson, S. 2005. *Teeth*. Cambridge University Press: Cambridge.

Hofreiter, M., Capelli, C., Krings, M., Waits, L., Conrad, N., Münzel, S., Rabeder, G., Nagel, D., Paunovic, M., Jambresić, G., Meyer, S., Weiss, G. and Pääbo, S. 2002. Ancient DNA Analyses Reveal High Mitochondrial DNA Sequence Diversity and Parallel Morphological Evolution of Late Pleistocene Cave Bears. *Molecular Biology and Evolution* **19**, 1244-1250.

Holgueras, M. 2009. Mousterian Bone Tools From "D" Level in Axlor (Dima, Vizcaya): Operational Sequence Analysis. *Trabajos de Prehistoria* **66**, 27-46.

Hornborg, A. 2006. Animism, Fetishism, and Objectivism as Strategies for Knowing (or not Knowing) the World. *Ethnos* **71**, 21-32.

Howell, S. 1996. Nature in Culture or Culture in Nature? Chewong Ideas of 'Human' and Other Species. In P. Descola (ed.), *Nature and Society: Anthropological Perspectives*, 127-144. Routledge: London.

Humphrey, C. 1974. Horse Brands of the Mongolians: A System of Signs in a Nomadic Culture. *American Ethnologist* **1**, 471-488.

Humphrey, C. 1996. Shamans and Elders: Experience, Knowledge, and Power Among The Duar Mongols. Clarendon Press: Oxford.

Hunt, E. 1992. Upgrading Site-Catchment Analysis with the Use of GIS: Investigating the Settlement Patterns of Horticulturists. *World Archaeology* **24**, 283-309.

Huntley, B. and Allen, R. 2003. Glacial Environments III: Palaeo-vegetation Patterns in Last Glacial Europe. In T. van Andel and W. Davies (eds.), *Neanderthals and Modern Humans in the European Landscape During the Last Glaciation*, 79-102. McDonald Institute for Archaeological Research: Cambridge.

Ingold, T. 1986a. Hunting, Sacrifice and the Domestication of Animals. In T. Ingold, *The Appropriation of Nature: Essays on Human Ecology and Social Relations*, 243-276. Manchester University Press: Manchester.

Ingold, T. 1986b. Prologue: Concerning the Hunter, and His Spear. In T. Ingold, *The Appropriation of Nature: Essays on Human Ecology and Social Relations*, 1-15. Manchester University Press: Manchester.

Ingold, T. 1986c. The Architect and the Bee: Reflections on the Work of Animals and Men. In T. Ingold, *The Appropriation of Nature: Essays on Human Ecology and Social Relations*, 16-39. Manchester University Press: Manchester.

Ingold, T. 1986d. The Principle of Individual Autonomy and the Collective Appropriation of Nature. In T. Ingold, *The Appropriation of Nature: Essays on Human Ecology and Social Relations*, 222-242. Manchester University Press: Manchester.

Ingold, T. 2000a. Culture, Nature, Environment: Steps to an Ecology of Life. In T. Ingold, *The Perception of the Environment: Essays in Livelihood, Dwelling and Skill*, 13-26. Routledge: London.

Ingold, T. 2000b. Culture, Perception and Cognition. In T. Ingold, *The Perception of the Environment: Essays in Livelihood, Dwelling and Skill*, 157-171. Routledge: London.

Ingold, T. 2000c. Hunting and Gathering as Ways of Perceiving the Environment. In T. Ingold, *The Perception of the Environment: Essays in Livelihood, Dwelling and Skill*, 40-60. Routledge: London.

Ingold, T. 2000d. Stop, Look and Listen: Vision, Hearing and Human Movement. In T. Ingold, *The Perception of the Environment: Essays in Livelihood, Dwelling and Skill*, 243-287. Routledge: London.

Ingold, T. 2000e. The Perception of the Environment: Essays in Livelihood, Dwelling and Skill. Routledge: London.

Ingold, T. 2000f. The Temporality of The Landscape. In T. Ingold, *The Perception of the Environment: Essays in Livelihood, Dwelling and Skill*, 189-208. Routledge: London.

Jankovská, V. 2003. Vegtačnń Poměry Slovenska a Českých Zemí v Posledním Glaciálu Jako Přírodní Prostředí Člověka a Fauny. Vegetation of Slovakia and Czechia During the Last Glacial as an Environment of Human and Fauna. In V. Hašek, R. Nekuda, and J. Unger (eds.), *Ve Službách Archeologie IV. Muzejní a Vlastivědná Společnost v Brně*: Brno.

Jędrzejewski, W., Niedziałkowska, M., Nowak, S. And Jędrzejewski, B. 2004. Habitat Variables and Associated with Wolf (*Canis lupus*) Distribution and Abundance in Northern Poland. *Diversity and Distributions* **10**, 225-233.

Jochim, M. 1976. Hunter-Gatherer Subsistence and Settlement: A Predictive Model. Academic: New York

Jones, P. 2009. Considering Living Beings in the Aceramic Neolithic of Cyprus. *Journal of Mediterranean Archaeology* **22**, 75-99.

Jope, K. 1985. Implications of Grizzly Bear Habituation to Hikers. *Wildlife Society Bulletin* **13**, 32-37.

Joslin, P. 1967. Movements and Home Sites of Timber Wolves in Algonquin Park. *American Zoologist* **7**, 279-288.

Keeling, C. and Nelson, D. 2001. Changes in the Intramolecular Stable Carbon Ratios with Age of the European Cave Bear (*Ursus Spelaeus*). *Oecologia* **127**, 495-500.

Kelly, R. 1995. *The Foraging Spectrum: Diversity in Hunter-Gatherer Lifeways*. Smithsonian Institution Press: Washington, London.

Kelly, R. 2003. Colonization of New Land by Hunter-Gatherers: Expectations and Implications Based on Ethnographic data. In M. Rockman and J. Steele (ed.), *Colonization of Unfamiliar Landscapes: The Archaeology of Adaptation*, 44-58. Routledge: London and New York.

Klein, R. and Cruz-Uribe, K. 1984. *The Analysis of Animal Bones From Archaeological Sites*. University of Chicago Press: Chicago.

Klíma, B. 1963. Dolní Věstonice: Výzkum Tábořiště Lovců Mamutů V Letech 1947-1952. Československé Akademie Věd: Prague.

Knappett, C. 2006. Beyond Skin Layering and Networking in Art and Archaeology. *Cambridge Archaeological Journal* **16**, 239-251.

Knight, R. and Eberhardt, L. 1985. Population Dynamics of Yellowstone Grizzly Bears. *Ecology* **66**, 323-334.

Kohn, E. 2007. How Dogs Dream: Amazonian Natures and The Politics of Trans-Species Engagement. *American Ethnologist* **34**, 3-24.

Kozłowski, J. 2000. The Problem of Cultural Continuioty Between the Middle and Upper Paleolithic in Central and Eastern Europe. In O. Bar-Yosef and D. Pilbeam (eds.), *The Geography of Neandertals and Modern Humans in Europe and the Greater Mediterranean*, 77-105. Peabody Museum: Cambridge, Massachussetts.

Kozłowski, J. 2004. Early Upper Paleolithic Backed Blade Industires in Central Europe. In J. Brantingham, S. Kuhn and K. Kerry (eds.), *The Early Upper Paleolithic Beyond Western Europe*, 14-29. University of California Press: Berkeley.

Kozłowski, J., Mester, Z., Zandler, K., Budek, A., Kalicki, T., Moskal, M. and Ringer, Á. 2009. Le Paléolithique Moyen et Supérieur de la Hongrie du Nord: Nouvelles Investigations Dans la Région d'Eger/Middle and Upper Palaeolithic of Northern Hungary: New Investigations in the Eger Region. *L'Anthropologie* **113**, 399-453.

Kuhn, S., Stiner, M. and Güleç, E. 2004. New Perspectives on the Initial Upper Paleolithic: The View from Üçağizli Cave, Turkey. In J. Brantingham, S. Kuhn and K. Kerry (eds.), *The Early Upper Paleolithic Beyond Western Europe*, 113-128. University of California Press: Berkeley.

Kukla, J. 1968. Comment. In K. Valoch, Evolution of the Paleolithic in Central and Eastern Europe, *Current Anthropology* **9**, 378-380.

Kurtén, B. 1955. Sex Dimorphism and Size Trends in the Cave Bear, *Ursus spelaeus* Rosenmüller and Heinroth. *Acta Zool. Fennica* **90**, 1-48.

Kurtén, B. 1958. Life and Death of the Pleistocene Cave Bears, A Study in Paleoecology. *Acta Zool. Fennica* **95**, 1-59.

Kurtén, B. 1968. *Pleistocene Mammals of Europe*. Weidenfeld and Nicolson: London.

Kurtén, B. 1976. The Cave Bear Story: Life and Death of a Vanished Animal. Columbia University Press: New York.

Kvamme, K. 1995. A View From Across The Water: The North American Experience in Archaeological GIS. In G. Lock and Z. Stančič (eds.), *Archaeology and Geographical Information Systems: A European Perspective*, 1-14. Taylor and Francis: London.

Lau, B., Blackwell, B., Schwarcz, H., Turk, I. and Blickstein, J. 2004. Dating a Flautist? Using ESR (Electron Spin Resonance) in the Mousterian Cave Deposits at Divje Babe I, Slovenia. *Geoarchaeology: An International Journal* **12**, 507-536.

Lee, P. and George, R. 2008. *Body-self-Dualism in Contemporary Ethics and Politics*. Cambridge University Press: Cambridge.

Lee, R. and DeVore, I. 1976. Kalahari Hunter-Gatherers: Studies of the !Kung San and Their Neighbors. Harvard University: Cambridge.

Leiberman, D. and Shea, J. 1994. Behavioural Differences Between Archaic and Modern Humans in the Levantine Mousterian. *American Anthropologist* **96**, 300-332.

Liden, K. and Angerbjörn, A. 1999. Dietary Change and Stable Isotopes: A Model of Growth and Dormancy in Cave Bears. *Proceedings of the Royal Society B* **266**, 1779-1783.

Lindstedt, S., Miller, B. and Buskirk, S. 1986. Home Range, Time, and Body Size in Mammals. *Ecology* **67**, 413-418.

Linklater, W., Cameron, E., Stafford, K. and Veltman, C. 2000. Social and Spatial Structure and Range Use by Kaimanawa Wild Horses (*Equus caballus*: Equidae). *New Zealand Journal of Ecology* **24**, 139-152.

Lippit, A. 1994. Afterthoughts on the Animal World. *Modern Language Notes* **109**, 786-830.

Llobera, M. 2000. Understanding Movement: A Pilot Model Towards the Sociology of Movement. In G. Lock (ed.), *Beyond the Map: Archaeology and Spatial Technologies*, 65-84. IOS Press: Amsterdam.

Macdonald, D. 1985. The Carnivores: Order Carnivora. In R. Brown and D. Macdonald (eds.), *Social Odours in Mammals: Volume 2*, 619-722. Clarendon Press: Oxford.

MacGregor, A. 1989. Bone, Antler and Horn Industries in the Urban Context. In D. Serjeantson and T. Waldron (eds.), *Diet and Crafts in Towns: The Evidence of Animal Remains from the Roman to the Post-Medieval Periods*, 107-128. B. A. R. 199: Oxford.

Magoun, A. and Copeland, J. 1998. Characteristics of Wolverine Reproductive Den Sites. *The Journal of Wildlife Management* **62**, 1313-1320.

Mania, D. and Mania, U. 2005. The Natural and Socio-Cultural Environment of Homo Erectus at Bilzingsleben, Germany. In C. Gamble and M. Porr (eds.), *The Hominid Individual in Context: Archaeological Investigations of Lower and Middle Palaeolithic Landscapes, Locales and Artefacts*, 98-114. Routledge: London.

Manning, A. and Serpell, J. 1994. Introduction. In A. Manning and J. Serpell (eds.), *Animals and*

Human Society, xi-xii. Routledge: London and New York.

Marks, A. and Monigal, K. 2004. Origins of the European Upper Paleolithic, Seen from Crimea: Simple Myth or Complex Reality? In J. Brantingham, S. Kuhn and K. Kerry (eds.), *The Early Upper Paleolithic Beyond Western Europe*, 64-79. University of California Press: Berkeley.

Martin, L. 2000. Gazelle (*Gazella* spp.) Behavioural Ecology: Predicting Animal Behaviour For Prehistoric Environments in South-West Asia. *Journal of Zoology* **250**, 13-30.

Martin, P. and Steadman, D. 1999. Prehistoric Extinctions on Islands and Continents. In R. MacPhee (ed.), *Extinctions in Near Time: Causes, Contexts, and Consequences*, 17-56. Kluwer Academic/Plenum Publishers: New York.

Mason, S., Hather, J. and Hillman, G. 1994. Preliminary Investigation of the Plant Macro-Remains from Dolní Věstonice II, and its Implications for the Role of Plant Foods in Palaeolithic and Mesolithic Europe. *Antiquity* **68**, 48-57.

Massolo, A. and Meriggi, A. 1998. Factors Affecting Habitat Occupancy by Wolves in Northern Apennines (Northern Italy): A Model of Habitat Suitability. *Ecography* **21**, 97-107.

McLoughlin, P., Case, R., Gau, R., Cluff, H., Mulders, R. and Messier, F. 2002. Hierarchical Habitat Selection By Barren-Ground Grizzly Bears in the Central Canadian Arctic. *Oecologia* **132**, 102-108.

McMillion, S. 1998. Mark of the Grizzly: True Stories of Recent Bear Attacks and the Hard Lessons Learned. The Globe Pequot Press: Guilford, Connecticut.

Meignen, L., Bar-Yosef, O., Speth, J. and Stiner, M. 2007. Middle Paleolithic Settlement Patterns in the Levant. In E. Hovers and S. Kuhn (eds.), *Transitions Before the Transition: Evolution and Stability in the Middle Paleolithic and Middle Stone Age*, 149-169. Springer: London.

Meignen, L., Geneste, J-M., Koulakovskaia, L. and Sytnik, A. 2004. Koulichivka and Its Place in the Middle-Upper Paleolithic Transition in Eastern Europe. In J. Brantingham, S. Kuhn, and K. Kerry (eds.), *The Early Upper Paleolithic Beyond Western Europe*, 50-63. University of California Press: Berkeley.

Mellars, P. 1996. The Neanderthal Legacy: An Archaeological Perspective from Western Europe. Princeton University: Princeton.

Mellars, P. 2000. The Archaeological Records of the Neandertal-Modern Human Transition in France. In O. Bar-Yosef and D. Pilbeam (eds.), *The Geography of Neandertals and Modern Humans in Europe and the Greater Mediterranean*, 35-48. Peabody Museum: Cambridge, Massachusetts.

Mellaart, J. 1967. *Çatal Hüyük: A Neolithic Town in Anatolia*. Thames and Hudson: London.

Meyrowitz, J. 1990. Redefining the Situation: Extending Dramaturgy into a Theory of Social Change and Media Effects In S. Riggins, *Beyond Goffman: Studies on Communication, Institution, and Social Interaction*, 66-98. Mouton de Gruyter: New York.

Mills, E. 1919. *The Grizzly Our Greatest Wild Animal*. Houghton Mifflin: Boston.

Miracle, P. 2007. The Krapina Paleolithic Site: Zooarchaeology, Taphonomy, and Catalog of the Faunal Remains. Croatian Natural Museum: Zagreb.

Mitchell, F., Onorato, D., Hellgren, E., Skiles, J. and Harveson, L. 2005. Winter Ecology of American Black Bears in a Desert Montane Island. *Wildlife Society Bulletin* **33**, 164-171.

Mithen, S. 1990. Thoughtful Foragers: A Study of Prehistoric Decision Making. Cambridge University Press: Cambridge.

Morely, I. 2006. Mousterian Musicianship? The Case of the Divje Babe I Bone. Oxford *Journal of Archaeology* **25**, 317-333.

Müller, U., Pross, J. and Bibus, E. 2003. Vegetation Response to Rapid Climate Change in Central Europe During the Past 140,000 yr Based on Evidence from the Furamoos Pollen Record. *Quaternary Research* **59**, 235-245.

Münzel, S. 2001. The Production of Upper Palaeolithic Mammoth Bone Artefacts from Southwestern Germany. *The World of Elephants – International Congress*, 448-454.

Münzel, S. and Conrad, N. 2004. Change and Continuity in Subsistence During the Middle and Upper Palaeolithic in the Ach Valley of Swabia (South-West Germany). *International Journal of Osteoarchaeology* **14**, 225-243.

Musil, R. 1965. Die Bärenhöhle Pod Hradem Die Entwicklung Der Höhlenbären Im Letzten Glazial. *Anthropos* **18**.

Musil, R. 1980a. Ursus Spelaeus-Der Höhlenbär I. Museum für Ur- and Frühgeschichte Thüringens: Weimar.

Musil, R. 1980b. Ursus Spelaeus-Der Höhlenbär II. Museum für Ur- and Frühgeschichte Thüringens: Weimar.

Musil, R. 1980c. Ursus Spelaeus-Der Höhlenbär III. Museum für Ur- and Frühgeschichte Thüringens: Weimar.

Musil, R. 2003. The Middle and Upper Palaeolithic Game Suite in Central and Southeastern Europe. In T. van Andel and W. Davies (eds.), *Neanderthals and Modern Humans in the European Landscape During the Last Glaciation*, 167-190. McDonald Institute for Archaeological Research: Cambridge.

Musil, R. 2007. The Moravian Karst – The Cradle of Karst Research in Central Europe. A Brief Overview and Forward to the Next Contributors.

In R. Musil and V. Vávra (eds.), *Geology* **35**, Proceedings of the 13th International Cave Bear Symposium. Department of Geological Sciences: Brno, 17-30.

Nachtigall, P., Supin, A., Amundin, M., Röken, B., Møller, T., Mooney, T., Taylor, K. and Yuen, M. 2007. Polar Bear *Ursus maritimus* Hearing Measured with Auditory Evoked Potentials. *The Journal of Experimental Biology* **210**, 1116-1122.

Nagel, D. 2005. Mitteilungen Der Kommission Für Quartärforschung. Österreichische. Akademie der Wissenschaften: Wein.

Nelson, D., Angerbjörn, A., Liden, K. and Turk, I. 1998. Stable Isotopes and the Metabolism of the European Cave Bear. *Oecologia* **116**, 177-181.

Nelson, R. 1969. *Hunters of the Northern Ice*. University of Chicago Press: Chicago.

Neruda, P. 2006. Neandertálci Na Kotouči u Štramberka. *Archeologické Památky Střední Moravy*, **11**, 7-65.

Neruda, P. and Valoch, K. 2007. Palaeolithic People and Moravian Caves. In R. Musil and V. Vávra (eds.), *Geology* **35**, Proceedings of the 13th International Cave Bear Symposium. Department of Geological Sciences: Brno, 65-76.

Niven, L. 2007. From Carcass to Cave: Large Mammal Exploitation During the Aurignacian at Vogelherd, Germany. *Journal of Human Evolution* **53**, 362-382.

O'Conner, T. 1989. Bones from Anglo-Scandinavian Levels at 16-22 Coppergate. The Archaeology of York 15/3. Council for British Archaeology: London.

Oliva, M. 1988. A Gravettian Site with Mammoth-Bone Dwelling in Milovice (Southern Moravia). *Anthropologie* **XXVI/2**, 105-112.

Oliva, M. 1991a. The Micoquian Open-Air Site of Ráječko I. *Anthropologie* **29**, 45-61.

Oliva, M. 1991b. The Szeletian in Czechoslovakia. *Antiquity* **65**, 318-325.

Oliva, M. 1993. The Aurignacian in Moravia. In H. Knecht, A. Pike-Tay, and R. White (eds.), *Before Lascaux: Complex Record of the Early Upper Paleolithic*, 37-55. CRC Press: Boca Raton.

Oliva, M. 2000. The Brno II Upper Palaeolithic Burial. In W. Roebroeks, M. Mussi, J. Svoboda and K. Fennema (eds.), *Hunters of the Golden Age: The Mid Upper Palaeolithic of Eurasia 30,000 – 20,000 BP*, 144-153. University of Leiden: Leiden.

Oliva, M. 2005. *Palaeolithic and Mesolithic Moravia*. Moravské Zemské Muzeum: Brno.

Orwell, G. 1946. *Animal farm*. Secker and Warburg: Toronto.

Pacher, M. and Stuart, A. 2008. Extinction Chronology and Palaeobiology of the Cave Bear (*Ursus Spealaeus*). *Boreas* **38**, 189-206.

Palliser, J. 1859. The Solitary Hunter: Sporting Adventures in the Prairies. London.

Pálsson, G. 1996. Human-Environmental Relations: Orientalism, Paternalism and Communalism. In P. Descola and G. Pálsson (eds.), *Nature and Society: Anthropological Perspectives*, 64-81. Routledge: London.

Paquet, P., Strittholt, J., Staus, N., Wilson, P., Grewal, S. and White, B. 2001. Feasibility of Timber Wolf Reintroduction In Adirondack Park. in D. Maehr, R. Noss and L. Larkin (eds.), *Large Mammal Restoration: Ecological and Sociological Challenges for the 21st Century*. Island Press: Washington, D.C.

Parker-Pearson, M. 1999. *The Archaeology of Death and Burial*. Sutton Publishing: Stroud.

Parker-Pearson, M. and Richards, C. 1994. Ordering the World: Perceptions of Architecture, Space, and Time. In M. Parker-Pearson and C. Richards (eds.), *Architecture and Order: Approaches to Social Space*, 1-37. Routledge: London.

Patou-Mathis, M. 2000. Neanderthal Subsistence Behaviours in Europe. *International Journal of Osteoarchaeology* **10**, 379-395.

Patou-Mathis, M. 2009. The Subsistence Behaviours of the Last Crimean Neanderthals. In M. Camps and P. Chauhan (eds.), *Sourcebook of Paleolithic Transitions: Methods, Theories, and Interpretations*, 441-454. Springer: London.

Péan, S. 2001. Mammoth and Subsistence Practices During the Mid Upper Palaeolithic of Central Europe (Moravia, Czech Republic). In G. Cavarretta, P. Gioia, M. Mussi, and M. Palombo (eds.), *The World of Elephants. International Congress*, Rome, 331-336.

Pepper, S. 1942. *World Hypothesis*. University of California Press: Berkeley.

Peterson, B. and Fry, B. 1987. Stable Isotopes in Ecosystem Studies. *Annual Review of Ecology and Systematics* **18**, 293-320.

Petrik, P. and Wild, J. 2006. Environmental Correlates of the Patterns of Plant Distribution at the Meso-scale: A Case Study from Northern Bohemia (Czech Republic). *Preslia* **78**, 211-234.

Pinto, A., Andrews, P. and Etxebarría, F. 2005. *Tafonomía Y Paleoecología De Úrsidos: Cuaternarios Cántabricos*. Fundacion Oso De Asturias.

Pluciennik, M. 2008. The Coastal Mesolithic of the European Mediterranean. In G. Bailey (ed.), *Mesolithic Europe*, 328-356. Cambridge University Press: Cambridge.

Pokorný, P. 2004. Postglacial Vegetation Distribution in the Czech Republic and its Relationship to Settlement Zones: Review From Off-Site Pollen Data. In M. Gojda (ed.), *Ancient Landscape, Settlement Dynamics and Non-Destructive Archaeology*, 395-414. Academia: Prague.

Pope, M. and Roberts, M. 2005. Observations on The Relationship Between Palaeolithic Individuals and Artefact Scatters at the Middle Pleistocene Site of Boxgrove, UK. In C. Gamble and M. Porr (eds.), *The Hominid Individual in Context: Archaeological Investigations of Lower and Middle Palaeolithic Landscapes, Locales and Artefacts*, 81-97. Routledge: London.

Potter, J. 2004. The Creation of Person, The Creation of Place: Hunting Landscapes in the American Southwest. *American Antiquity* **69**: 322-338.

Powell, R., Zimmerman, J. and Seaman, D. 1997. Ecology and Behaviour of North American Black Bears: Home Ranges, Habitat and Social Organisation. Chapman and Hall: London.

Preatoni, D., Mustoni, A., Martinoli, A., Carlini, E., Chiarenzi, B., Chiozzini, S., van Dongen, S., Wauters, L. and Tosi, G. 2005. Conservation of Brown Bear in the Alps: Space Use and Settlement Behaviour of Reintroduced Bears. *Acta Oecologica-International Journal of Ecology* **28**, 189-197.

Přichystal, A., Škrdla, P., Svoboda, J. and Tomášková, S. 1994. Chapter II: The Lithics. In J. Svoboda (ed.), *Pavlov I: Excavations 1952-1953*, 23-94. University of Liège: Liège.

Prus, R. 1996. Symbolic Interactions and Ethnographic Research: Intersubjectivity and the Study of Human Lived Experience. State University of New York Press: Albany.

Putman, W. 1964. *Geology*. Oxford University Press: New York.

Quam, R., Arsuaga, J-L., Maria, J., de Castro, B., Diez, J., Lorenzo, C., Carretero, J., Garcia, N. and Ortega, A. 2001. *Journal of Human Evolution* **41**, 385-435.

Quilés, J., Petrea, C., Moldovan, O., Zilhão, J., Rodrigo, R., Rougier, H., Constantin, S., Milota, Ş., Gherase, M., Sarcină, L. and Trinkaus, E. 2006. Cave Bears (*Ursus spelaues*) From The Peştera Cu Oase (Banat, Romania): Paleobiology and Taphonomy. *Comptes Rendus Palevol* **5**, 927-934.

Rabbett, R. and Barker, G. 2007. Through the Looking Glass: New Evidence on the Presence and Behaviour of Late Pleistocene Humans at Niah Cave, Sarawak, Borneo. In P. Mellars, K. Boyle, O. Bar-Yosef and C. Stringer (eds.), *Rethinking the Human Revolution*, 411-424. McDonald Institute for Archaeological Research: Cambridge.

Ramsay, M. 1993. Cycles of Feasting and Fasting. In I. Stirling (ed.), *Bears*, 62-65. HarperCollins: London.

Reitz, E. and Wing, E. 2008 (2nd edition). *Zooarchaeology*. Cambridge University Press: Cambridge.

Renfrew, C. 1986. Introduction: Peer Polity and Social-Political Change. In C. Renfrew and J. Cherry (eds.), *Peer-polity Interaction and Socio-Political Change*, 1-18. Cambridge University Press: Cambridge.

Reynolds, H., Hansen, R. and Peden, D. 1978. Diets of the Slave River Lowland Bison Herd, Northwest Territories, Canada. *The Journal of Wildlife Management* **42**, 581-590.

Rhodda, H. 2005. The Development and Application of a Flood Risk Model for the Czech Republic. *Natural Hazards* **36**, 207-220.

Richardson, C. 2000. Freshwater Wetlands. In M. Barbour and D. Billings (eds.), *North American Terrestrial Vegetation* (2nd edition), 449-499. University of Cambridge: Cambridge.

Richards, M., Pacher, M., Stiller, M., Quilés J., Hofreiter, M., Constantin, S., Zilhão, J. and Trinkaus, E. 2008. Isotopic Evidence for Omnivory Among European Cave Bears: Late Pleistocene *Ursus Spelaeus* from the Pestera Cu Oase, Romania. *Proceedings of the National Academy of Sciences of the United States of America* **105**, 600-604.

Rink, W., Schwarcz, H., Valoch, K., Seitl, L. and Stringer, C. 1996. ESR Dating of Micoquian Industry and Neanderthal Remains at Kůlna Cave, Czech Republic. *Journal of Archaeological Science* **23**, 889-901.

Ritvo, H. 1994. Animals in Nineteenth-Century Britain: Complicated Attitudes and Competing Categories. In A. Mallard and J. Serpell (eds.), *Animals and Human Society: Changing Perspectives*, 106-126. Routledge: London and New York.

Rival, L. 1996. Blowpipes and Spears: The Social Significance of Huaorani Technological Choices. In P. Descola and G. Pálsson (ed.), *Nature and Society: Anthropological Perspectives*, 145-164. Routledge: London.

Rockman, M. 2003. Knowledge and Learning in the Archaeology of Colonization. In M. Rockman and J. Steele (ed.), *Colonization of Unfamiliar Landscapes: The Archaeology of Adaptation*, 3-24. Routledge: London and New York.

Rockman, M. and Steele J. 2003. Colonization of Unfamiliar Landscapes: The Archaeology of Adaptation. Routledge: London and New York.

Rodseth, L., Wrangham, A., Harrigan, A. and Smuts, B. 1991. The Human Community as a Primate Society. *Current Anthropology* **32**, 221-254.

Roebroeks, W. 2003. Landscape Learning and the Earliest Peopling of Europe. In M. Rockman and J. Steele (ed.), *Colonization of Unfamiliar Landscapes: The Archaeology of Adaptation*, 99-115. Routledge: London and New York.

Rosendahl, W. and Döppes, D. 2006. Trace Fossils From Bears in Caves of Germany and Austria. *Scientific Annals, School of Geology Aristotle University University of Thessaloniki (AUTH)* Special Volume **98**, 241-249.

Rosendahl, W., Kempe, S., and Döppes, D. 2005. The Making of the Cave Bear: Die Wissenschaftliche Entdeckung Des „*Ursus spelaeus*". In D. Nagel (ed.), *Mitt. Komm. Quartärforsch. Österr. Akad. Wiss.* **14**, 89-106. Wien.

Rousseau, D-D., Gerasimenko, N., Matviischina, Z. and Kukla, G. 2001. Late Pleistocene Environments of the Central Ukraine. *Quaternary Research* **56**, 349-356.

Růžíčková, E., Růžíčka, M. Aeman, A., and Kadlec, J. 2001. Quaternary Clastic Sediments of the Czech Republic: Textures and Structures of the Main Genetic Types. Czech Geological Survey: Prague.

Rybníčková, E. and Rybníček, K. 1991. The Environment of the Pavlovian Palaeoecological Results From Bulhary, South Moravia. In Kovar-Eder (ed.), *Palaeovegetational Development in Europe and Regions Relevant to its Palaeofloristic Evolution*, 73-79. Museum of Natural History: Vienna.

Sattler, R. 1997. Large Mammals in Lower Rampart Cave 1, Alaska: Interspecific Utilization of an Eastern Beringian Cave. *Geoarchaeology* **12**, 657-688.

Scheer, A. 2000. The Gravettian in Southwest Germany: Stylistic Features, Raw Material Resources and Settlement Patterns. In W. Roebroeks, M. Mussi, J. Svoboda and K. Fennema (eds.), *Hunters of the Golden Age: The Mid Upper Palaeolithic of Eurasia 30,000 – 20,000 BP*, 257-270. University of Leiden: Leiden.

Senft, R., Coughenour, M., Bailey, D., Rittenhouse, L., Salsa, O. and Swift, D. 1987. Large Herbivore Foraging and Ecological Hierarchies. *BioScience* **37**, 789-799.

Serjeantson , D. 1989. Animal Remains and the Tanning Trade. In D. Serjeantson and T. Waldron (eds.), *Diet and Crafts in Towns: The Evidence of Animal Remains from the Roman to the Post-Medieval Periods*, 129-146. B. A. R. 199: Oxford.

Serpell, J. 1996. In The Company of Animals: A Study of Human-Animal Relationships. Blackwell: Oxford.

Skogland, T. 1984. Wild Reindeer Foraging-Niche Organisation. *Holartic Ecology* **7**, 345-379.

Smith, C., Smits, C. and Slough, B. 1992. Landform Selection and Soil Modification Associated with Arctic Fox (Alopex lagopus) Den Sites in Yukon Territory, Canada. *Arctic and Alpine Research* **24**, 324-325.

Soffer, O. 1985. The Upper Paleolithic of the Central Russian Plain. Academic Press: Orlando.

Steele, J. and Rockman, M. 2003. "Where Do We Go from Here?": Modelling the Decision-Making Process During Exploratory Dispersal. In M. Rockman and J. Steele (ed.), *Colonization of Unfamiliar landscapes: The Archaeology of Adaptation*, 130-143. Routledge: London and New York.

Stewart, J., van Kolfschoten, T., Markova, A. and Musil, R. 2003. The Mammalian Faunas of Europe During Oxygen Isotope Stage Three. In T. van Andel and W. Davies (eds.), *Neanderthals and Modern Humans in the European Landscape During the Last Glaciation*, 103-130. McDonald Institute for Archaeological Research: Cambridge.

Stiner, M. 1999. Cave Bear Ecology and Interactions with Pleistocene Humans. *Ursus* **11**, 41-58.

Stiner, M., Achyuthan, H., Arsebuk, G., Howell, F., Josephson, S., Juell, K., Pigati, J. and Quade, J. 1998. Reconstructing Cave Bear Paleoecology from Skeletons: A Cross-Disciplinary Study of Middle Pleistocene Bears from Yarimburgaz Cave, Turkey. *Paleobiology* **24**, 74-98.

Stiner, M. Arsebük, G. and Howell, C. 1996. Cave Bears and Paleolithic Artifacts in Yarimburgaz Cave, Turkey: Dissecting a Palimpsest. *Geoarchaeology* **11**, 279-327.

Storer, T. and Tevis, L. 1996. *Californian Grizzly*. University of California Press: Berkeley.

Strathern, M. 1992. Parts and Wholes: Refiguring Relationships in a Post-plural World. In A. Kuper (ed.), *Conceptualising Society*, 75-104. Routledge: London.

Sturdy, A. 1972. The Exploitation Patterns of a Modern Reindeer Economy in West Greenland. In E. Higgs (ed.), Papers in Economic Prehistory: Studies By Members and Associates of The British Academy Major Research Project in the Early History of Agriculture, 161-168. Cambridge University Press: Cambridge.

Sturdy, D., Webley, D. and Bailey, G. 1997. The Palaeolithic Geography of Epirus. In G. Bailey (ed.), *Klithi: Palaeolithic Settlement and Quaternary Landscapes in Northwest Greece vol. 2: Klithi in its Local and Regional Setting*, 587-614. McDonald Institute for Archaeological Research: Cambridge.

Suk, M., Bližkovský, M., Buday, T., Chlupáč, I., Cicha, I., Dudek, A., Dvořák, J., Eliáš, M., Holub, V., Ibrmajer, J., Kodym, O., Kukal, Z., Malkovský, M., Menčík, E., Müller, V., Tyráček, J., Vejnar, Z. and Zeman, A. 1984. *Geological History of the territory of the Czech Socialist Republic, Geological Survey of the Czech Socialist Republic and Academia*. Geological Survey, Czech Academy of Sciences: Prague.

Svoboda, J. 1993. The Complex Origin of the Upper Paleolithic in the Czech and Slovak Republics. In H. Knecht, A. Pike-Tay, and R. White (eds.), *Before Lascaux: Complex Record of the Early Upper Paleolithic*, 23-36. CRC Press: Boca Raton.

Svoboda, J. 1995. Environment and Upper Palaeolithic Adaptations in Moravia. In H. Ullrich (ed.), *Man and Environment in the Palaeolithic*, 291-295. Études et Recherches Archéologiques de L'Universite de Liége: Liége.

Svoboda, J. 2004. Continuities, Discontinuities, and Interactions in Early Upper Paleolithic Technologies. In J. Brantingham, S. Kuhn and K.

Kerry (eds.), *The Early Upper Paleolithic Beyond Western Europe*, 30-49. University of California Press: Berkeley.

Svoboda, J. 2005. The Neanderthal Extinction in Eastern Central Europe. *Quaternary International* **137**, 69-75.

Svoboda, J. 2008. The Mesolithic of the Middle Danube and Upper Elbe Rivers. In G. Bailey and P. Sipkins (eds.), *Mesolithic Europe*, 221-238. Cambridge University Press: Cambridge.

Svoboda, J., Klíma, B., Jarošová, L. and Škrdla, P. 2000. The Gravettian in Moravia: Climate, Behaviour and Technological Complexity. In W. Roebroeks, M. Mussi, J. Svoboda and K. Fennema (eds.), *Hunters of the Golden Age: The Mid Upper Palaeolithic of Eurasia 30,000 – 20,000 BP*, 197-217. University of Leiden: Leiden.

Svoboda, J., Ložek, V. and Vlček, E. 1996. *Hunters Between East and West: The Paleolithic of Moravia*. Plenum Press: New York.

Svoboda, J., Péan, S., Wojtal, P. 2005. Mammoth Bone Deposits and Subsistence Practices During Mid-Upper Palaeolithic in Central Europe: Three Cases from Moravia and Poland. *Quaternary International* **126-128**, 209-221.

Svoboda, J. and Simán, K. 1989. The Middle-Upper Paleolithic Transition in Southeastern Central Europe (Czechoslovakia and Hungary). *Journal of World Prehistory* **3**, 283-322.

Svoboda, J., van der Plicht, J. and Kuželka, V. 2002. Upper Palaeolithic and Mesolithic Human Fossils from Moravia and Bohemia (Czech Republic): Some New 14C Dates. *Antiquity* **76**, 957-962.

Svobodová, H. 1991a. Pollen Analysis of the Upper Palaeolithic Triple Burial at Dolní Věstonice. *Archeologicke Rozhledy* **43**, 505-509.

Svobodová, H. 1991b. The Pollen Analysis of Dolní Věstonice II – Western Slope. In J. Svoboda (ed.), *Dolni Věstonice II*, 75–88. Universite de Liege.

Switzer, P. 1993. Site Fidelity in Predictable and Unpredictable Habitats. *Evolutionary Ecology* **7**, 533-555.

Takahashi, K., Shiota, T., Tamatani, H., Koyama, M. and Washitani, I. 2008. Seasonal Variation in Fleshy Fruit Use and Seed Dispersal by the Japanese Black Bear (*Ursus Thibetanus japanicus*). *Ecological Research* **23**, 471-478.

Thackeray, J. 1983. Disguises, Animal Behaviour and Concepts of Control in Relation to Rock Art of Southern Africa. *Goodwin Series* **4**, 38-43.

Tilley, C. 1994. A Phenomenology of Landscape: Places, Paths, and Monuments. Berg: Oxford.

Tilley, C. 1996. An Ethnography of the Neolithic: Early Prehistoric Societies in Southern Scandinavia. Cambridge University Press: Cambridge.

Tilley, C. 2004. *Materiality of Stone*. Berg: Oxford and New York.

Tilson, R. and Henschel, J. 1986. Spatial Arrangement of Spotted Hyaena in a Desert Environment, Namibia. *African Journal of Ecology* **24**, 173-180.

Tindale, N. 1972. The Pitjandjara. In M. Micchieri (ed.), *Hunters and Gatherers Today*, 217-268. Holt, Reinhart and Winston: New York.

Tockner, K., Malard, F., Uehlinger, U. and Ward, J. 2002. Nutrients and Organic Matter in a Glacial River-Floodplain System (Val Roseg, Switzerland). *Limnology and Oceanography* **47**, 266-277.

Tostevin, G. and Škrdla, P. 2006. New Excavations at Bohunice and the Question of the Uniqueness of the Type-Site for the Bohunician Industrial Type. *Anthropologie* **XLIV/1**, 31-48.

Turner, N., Boelscher, M. and Ignace, R. 2000. Traditional Ecological Knowledge and Wisdom of Aboriginals Peoples in British Columbia. *Ecological Application* **10**, 1275-1287.

Ucko, P. and Rosenfeld, A. 1967. *Palaeolithic Cave Art*. Weidenfeld and Nicolson: London.

Valensi, P. and Psathi, E. 2004. Faunal Exploitation During the Middle Palaeolithic in South-Eastern France and North-Western Italy. *International Journal of Osteoarchaeology* **14**, 256-272.

Valoch, K. 1965. *Jeskyně Šipka a Čertova dira u Štramberku.* Moravské museum, Ustav Anthropos: Brno.

Valoch, K. 1987a. Contribution to the Architecture of Upper Palaeolithic Dwellings. *Anthropologie* **XXV/2**, 115-116.

Valoch, K. 1987b. Raw Materials Used in the Moravian Middle and Upper Palaeolithic. *International Conference on Prehistoric Flint Mining.* Budapest.

Valoch, K. 1988. *Die Erforschung Der Kůlna-Höhle, 1961-1976*. Moravské Muzeum, Anthropos Institut: Brno.

Valoch, K. 1993. Vedrivice V, Eiene Siedlung des Szeletian in Südmähren. *Quatär* **43/44**, 7-94.

Van Andel, T. 2003. Glacial Environments I: The Weichselian Climate in Europe Between the End of the OIS-5 Interglacial and the Last Glacial Maximum. In T. van Andel and W. Davies (eds.), *Neanderthals and Modern Humans in the European Landscape During The Last Glaciation*, 9-20. McDonald Institute for Archaeological Research: Cambridge.

Van Andel, T., Davies, W. and Weninger, B. 2003a. The Human Presence in Europe During the Last Glacial Period I: Human Migrations and the Changing Climate. In T. van Andel and W. Davies (eds.), *Neanderthals and Modern Humans in the European Landscape During The Last Glaciation*, 31-56. McDonald Institute for Archaeological Research: Cambridge.

Van Andel, T., Davies, W., Weninger, B. and Jöris, O. 2003b. Archaeological Dates as Proxies for The Spatial and Temporal Human Presence in Europe: A Discourse on the Method. In T. van

Andel and W. Davies (eds.), *Neanderthals and Modern Humans in the European Landscape During The Last Glaciation*, 21-29. McDonald Institute for Archaeological Research: Cambridge.

Van Andel, T. and Tzedakis, P. 1996. Palaeolithic Landscapes of Europe and Environs, 150,000-25,000 Years Ago: An Overview. *Quaternary Science Reviews* **15**, 481-500.

Velichko, A. and Zelikson, E. 2005. Landscape, Climate and Mammoth Food Resources in the East European Plain During the Late Paleolithic Epoch. *Quaternary International* **126**, 137-151.

Verhart, L. 2008. New Developments in the Study of the Mesolithic of the Low Countries. In G. Bailey (ed.), *Mesolithic Europe*, 158-181. Cambridge University Press: Cambridge.

Verpoorte, A. 2005. The Lithic Assemblage of Pavlov I (1954, 1956, 1963, 1964). In J. Svoboda (eds.), *Pavlov I Southeast: A Window into the Gravettian Lifestyles, Dolní Věstonice Studies* **14**, 75-111. Academy of Sciences of the Czech Republic: Brno.

Viveiros de Castro, E. 1998. Cosmological Deixis and Amerindian Perspectivism. *Journal of the Royal Anthropological Institute* **4**, 469-488.

Viveiros de Castro, E. 2004. The Transformation of Objects into Subjects in Amerindian Ontologies. *Common Knowledge* **10**, 463-484.

Vishnyatsky, L. and Nehoroshev, P. 2004. The Beginning of the Upper Paleolithic on the Russian Plain. In J. Brantingham, S. Kuhn and K. Kerry (eds.), *The Early Upper Paleolithic Beyond Western Europe*, 80-96. University of California Press: Berkeley.

Wagner, R. 1991. The Fractal Person. In M. Godelier and M. Strathern (ed.), *Big Men and Great Men: Personifications of Power in Melanesia*, 159-173. Cambridge University Press: Cambridge.

Walton, L., Dean Cluff, H., Paquet, P. and Ramsay, M. 2001. Movement Patterns of Barren-Ground Wolves in the Central Canadian Arctic. *Journal of Mammalogy* **82**, 867-876.

Ward, P. and Kynaston, S. 1995. *Bears of the World*. Blandford: London.

West, D. 1997. Hunting Strategies in Central Europe During the Last Glacial Maximum. B. A. R. 672: Oxford.

Whatmore, S. 2002. Hybrid Geographies: Nature, Cultures, Spaces. Sage: London.

Wheatley, D. and Gillings, M. 2002. Spatial Technology and Archaeology: The Archaeological Applications of GIS. Taylor and Francis: New York.

White, R. 2007. Systems of Personal Ornamentation in the Early Upper Palaeolithic: Methodological Challenges and New Observations. In P. Mellars, K. Boyle, O. Bar-Yosef and C. Stringer (eds.), *Rethinking the Human Revolution*, 287-302. McDonald Institute for Archaeological Research: Cambridge.

Whittle, A. 2000. 'Very Like a Whale': Menhirs, Motifs and Myths in the Mesolithic-Neolithic Transition of Northwest Europe. *Cambridge Archaeological Journal* **10**, 243-259.

Whittle. A. 2003. The Archaeology of People: Dimensions of Neolithic Life. Routledge: London and New York.

Willis, K., Rudner, E. and Sümegi, P. 2000. The Full-Glacial Forests of Central and Southeastern Europe. *Quaternary Research* **53**, 203-213.

Willson, C., Manos, P. and Jackson, R. 2008. Hydraulic Traits are Influenced by Phylogentic History in the Drought-Resistant Invasive Genus *Juniperus* (Cupressaceae). *American Journal of Botany* **95**, 299-314.

Wilson, D. and Ruff, S. 1999. *The Smithsonian Book of North American Mammals*. Smithsonian Institution Press: Washington and London.

Withalm, G. 2004. Pathologies of Cave Bear Bones from Potočka Zijalka (Slovenia). *Mitt. Komm. Quartärforsch. Österr. Akad. Wiss.* **13**, 183-196.

Wojtal, P. 2007. *Zooarchaeological Studies of the Late Pleistocene Sites in Poland*. Institute of Systematics and Evolution of Animals of the Polish Academy of Sciences: Kraków.

Zelený, D. and Chytrý, M. 2007. Environmental Control of the Vegetation Pattern in Deep River Valleys of the Bohemian Massif. *Preslia* **79**, 205-222.

Zilhão, J. 2001. Middle Paleolithic Settlement Patterns in Portugal. In N. Conrad (ed.), *Settlement Dynamics of the Middle Paleolithic and Middle Stone Age*, 597-608. Kerns Verlag: Tübingen.

Ziolkowski, J. 1993. *Talking Animals: Medieval Latin Beast Poetry, 750-1150*. University of Pennsylvania Press: Philadelphia.

Zvelebil, M. 2008. Innovating Hunter-Gatherers: The Mesolithic in the Baltic. In G. Bailey and P. Sipkins (eds.), *Mesolithic Europe*, 18-59. Cambridge University Press: Cambridge.

Internet sources

http://www.clarklabs.org (Visited on 12/07/07.)

http://www.czech.cz (Visited on 7/04/09.)

http://www.esc.cam.ac.uk/research/research-groups/oistage3 (Visited on 8/04/09.)

http://www.esri.com (Visited on 12/07/07.)

http://www.europeetravel.com (Visited on 17/10/08.)

http:// www.fcc.gov/mb/audio/ bickel/DDDMMSS-decimal.html (Visited on 15/05/09.)

http://www.flintsource.net (Visited on 15/05/09.)

http://mapy.geology.cz/website/new_tisk/viewer2.htm (Visited on 12/07/07.)

http://www.grass.itc.it (Visited on 12/07/07.)

http://www.mapinfo.com (Visited on 12/07/07.)

http://www.mapy.cz (Visited on 12/08/09.)

http://www.rook.org (Visited on 03/04/08.)

http://seamless.gov.us (Visited on 12/03/09.)

http://www.southtravels.com (Visited on 7/04/09.)

https://statistics.defra.gov.uk/esg/statnot/jun08_final_eng.pdf) (Visited on 15/06/09.)

http://www.tate.org.uk (Visited on 12/02/09.)

http://www.telegraph.co.uk (Visited on 3/03/09.)
http://www.telegraph.co.uk/earth/earthnews/3520236/Vegetarian-cave-bear-starved-to-death-during-Ice-Age.html
(Visited on 12/05/09.)

Index

www.ingramcontent.com/pod-product-compliance
Ingram Content Group UK Ltd.
Pitfield, Milton Keynes, MK11 3LW, UK
UKHW061848041026
15591UKWH00015B/963

9 781407 309699